# INDIA'S TRYST WITH Bt COTTON
## Learning from the First Decade

## About the Editors

**N.Lalitha** is Professor at the Gujarat Institute of Development Research, Ahmedabad. Her research focuses on industry, globalization, trade and development.
E-mail: lalithanarayanan@gmail.com

**P.K. Viswanathan** is Associate Professor at the Gujarat Institute of Development Research, Ahmedabad. His research interests broadly cover areas, such as Natural Resources Economics; Rural transformation; Sustainable rural livelihoods; Working of technology, institutions, policies and governance in dynamic contexts; Economics of climate change and conservation of natural resources, etc. E-mail: pkviswam@gmail.com

# INDIA'S TRYST WITH Bt COTTON

## Learning from the First Decade

*Edited by*
N. Lalitha
P.K. Viswanathan

**CONCEPT PUBLISHING COMPANY PVT. LTD.**
NEW DELHI-110059

**ISBN-13 : 978-93-5125-108-8**

First Published 2015

*Published and Printed by*

**Concept Publishing Company Pvt. Ltd.**
A/15-16, Commercial Block, Mohan Garden
New Delhi-110059 (India)
*T*: 25351460, 25351794, *F*: 091-11-25357109
E: publishing@conceptpub.com, W: www.conceptpub.com
**Editorial Office:** H-13, Bali Nagar, New Delhi-110 015, India.

Cataloging in Publication Data—*Courtesy:* D.K. Agencies (P) Ltd. <docinfo@dkagencies.com>

**National Seminar on Bt Cotton (2011 : Ahmadābād, India)**
India's tryst with Bt cotton : learning from the first decade / edited by N. Lalitha, P.K. Viswanathan.
pages ; cm
Includes bibliographical references and index.
ISBN 9789351251088

1. Cotton—Genetic engineering—India—Congresses. 2. Cotton growing—Economic aspects—India—Congresses. I. Lalitha, N., editor. II. Viswanathan, P. K., editor. III. Gujarat Institute of Development Research.organizer. IV. Title.

DDC 631.52330954 23

# Foreword

The 11th and 12th Five Year Plans of India have focused on inclusive growth as one of the main development objectives. Agriculture plays a pivotal role in Indian economy and this sector's better performance is vital for inclusive growth. India's agricultural yield growth has been stagnant in the post-reform period except during the period 2005 to 2012. The growth of total factor productivity has been slower in India compared to countries such as China, Brazil and Indonesia. The need for increase in yield and total factor productivity is important for higher and inclusive agricultural growth. It is known that technology is the most important factor for raising agricultural productivity. Even at the global level, new technologies are needed to feed 9 billion plus population by 2050.

India has benefited from green revolution technology which mainly benefited the irrigated areas. Biotechnology has immense potential to raise yields in several parts of the country. The traditional biotechnology like tissue culture and other applications has been followed in several countries of the world. However, application of genetically modified organism (GMOs) created a big debate all over the world. Some of the concerns on GMOs relate to food safety, health safety, environmental safety, regulating corporate control over agriculture, pricing of seeds etc. Despite these concerns, a number of countries have adopted GM crops. In fact, the area under GM crops increased from 1.7 million hectares in 1996 to 170 million hectares in 2012. Out of the total 28 countries which adopted this technology, 18 are developing countries. As mentioned in this volume, the area under biotech crops is the highest in the USA (69.5 million hectares), followed by Brazil (36.6 m.ha), Argentina (23.9 m.ha), Canada (11.6 m.ha) and India (10.8 m.ha). But, the total area under GM is less than 4 per cent of global agricultural land and it is concentrated in four crops such as soybean, maize, cotton and canola.

There are strong proponents and opponents to the application of GM crops. The proponents citing scientific evidences say that there is no problem with safety. The risk of any safety issue is similar to the conventional crop technologies or any new medicines taken by the population for curing diseases in several countries. Moreover, they say that countries like USA, Brazil and China have adopted the GM technologies and one does not hear much about any food safety issues in these countries. The opponents on GM are more vocal than proponents. There is opposition to GM on food and environmental safety issues and the role of multinationals. Europe is in the forefront in leading the opponents to GM. India adopted Bt cotton but opposed Bt Brinjal (egg plant).

One has to take a balanced view on GM technologies. In some cases, genetic modification (GM) has the potential to deliver benefits to farmers and consumers that may not be achieved through conventional breeding methods. This can improve food and nutrition security particularly for developing countries and poor population. However, the research on GM crops must be done responsibly, ethically and keeping in mind health and safety of the consumer and environment. It should also respect national and international laws. Moreover, the research should not be limited to private sector alone. Public sector should also take proactive role in conducting research on GM crops.

In the above context, the research papers in this volume are timely. It is known that Bt cotton was introduced in India in early 2000s. In the words of the editors, this book "is a modest attempt to provide a snapshot of India's tryst with Bt cotton, the learning and experiences of the first decade of its adoption".

Some of the issues addressed in this volume are: (a) issues on India's regulatory framework and institutions to implement the regulatory procedures to facilitate adoption of all types of GM crops; (b) contribution of public sector research institutions regarding cotton development in the country; (c) impact of Bt cotton on livelihoods of the heterogeneous farmers in the major cotton growing states of the country; (d) impact on fertilizers, pesticides, labour use, environment and resource management; (e) impact of Bt seed price policy controls on adoption trends of Bt cotton; (f) Lessons learnt for preparing a road map for the wide scale promotion of agri-biotechnology in India.

The editors give a good overview of the book in Chapter 1 by providing an account of agri-biotechnology scenario in the global and

Indian contexts followed by the adoption of Bt cotton in India. The chapter also highlights the issues relating to Bt cotton experience in India. In the concluding chapter of the volume (Chapter 10), the editors indicate the directions in which policy makers' attention is urgently required. In these two chapters, the editors who have been working on these issues for considerable time made sincere efforts in providing an objective assessment and highlighted both benefits and problems of agri-biotechnology in India.

This volume addresses an extremely important issue faced in global and Indian agriculture today. The papers are highly informative, objective, rich in analysis and also focus on policies. This collection would make a useful reading for students, researchers, civil society and policy makers interested in the vital issue of biotechnology in agriculture.

**S. Mahendra Dev**
Director and Vice Chancellor,
IGIDR, Mumbai

# Preface

The history of cotton cultivation in India has undergone significant changes in the 1900s. Once well known as a producer of some of the very fine varieties of cotton, India's cotton sector had been suffering from the problems of low production and productivity since long. Though India was mostly growing indigenous varieties (*desi*) of cotton, the introduction of hybrid cotton varieties since the 1970's replaced the area under those indigenous varieties significantly, but the problems of low production and productivity still persisted. Thus, while the area sown with hybrid cotton varieties had increased from 45-50 per cent during 1990s to almost 65 per cent by 2000, the production and productivity had remained very low compared to international standards.

At this juncture, the introduction of genetically modified (GM) or *Bacillus thuringiensis* (Bt) cotton during 2002-03 for wider commercial cultivation, has transformed the cotton cultivation scenario in India leading to two major changes having significant bearing on the economy and society. First, the emergence and wider diffusion and adoption of the Bt technology had imparted tremendous dynamism to the country's cotton production sector, as the area, production and productivity of cotton had increased substantially between 2002 and 2012. Second, increased adoption of Bt cotton had resulted in a significant increase in India's relative share in global cotton production from 13 per cent in the early 2000s to 23 per cent by 2012-13.

As the Bt technology has gained wider acceptance in a shorter span of less than a decade, researchers have been highly intrigued by its economic and social impacts, contributing to a number of cross sectional studies across the major cotton growing regions in India and elsewhere. However, a large number of these studies in India and others reportedly suffer from problems of 'selection bias' as well as 'single point analysis'. More importantly, the researchers by and large have been concerned about analyzing the technology impacts in terms of three standard prescriptions, *viz.*, (a) increase in productivity, (b) reduction in the use of chemical fertilizers and pesticides; and (c) the resultant economic and social benefits, including poverty reduction and household welfare impacts. Thus, one could argue that the empirical treat on the subject of adoption of Bt cotton and the entire dynamics revolving around it seriously lack in terms of long-term socio-economic and environmental impacts of the technology involving both cross sectional and longitudinal analytical perspectives. This emerges from the fact that the sustainability of a frontier technology such as Bt cotton which has also been hailed by many as the 'second green revolution' or 'gene revolution', certainly depends on the assessment of long-term economic, social and environmental impacts covering regions and farmers, that are quite diverse in terms of agro-climatic settings and resource endowments.

Against this backdrop, this empirical research based book explores the less researched aspects on the experiences of the Bt cotton farmers in select cotton growing states of India. The different themes as discussed in the book provide a detailed assessment of the technology and its adoption in terms of contribution of public research institutions in the development of cotton research in the country; regulatory regime governing GM crops; decision making process regarding inputs use; impact of price control on technology adoption and profitability of the seed providers; and the influence of Bt on the demand for labour and resource management outcomes of the technology. Notable points of interest that emerge from this volume are: (a) Bt adoption in the Indian context has been scale neutral; and (b) price control by the government and the spread of unapproved cotton has helped in the faster adoption of the technology. Nevertheless, apprehensions prevail over the long-term environmental impact of Bt cotton and sustainability of monoculture cotton farming system. These fears about the legitimacy and the wider social as well as environmental compatibility of the Bt technology also arises from the lack of enforcement of environmental as well as bio-safety regulations across countries, including India. Ideally, the GM technology should have been provided as a package of crop technology and farm management practices rather than merely as a seed input, which would enable the users of the technology realize the optimum benefits while sustaining the natural resources, agri-ecosystems and the environment.

In this regard, one unique feature of the present volume that makes it novel is that it provides a holistic perspective on the state of Bt cotton technology in India, in terms of presenting the farm level adoption scenario; the institutional, R&D, policy regimes and regulatory environments, that are seemingly inadequate to address the veracity of challenges facing the sustainable future of the technology in the country. The volume will also be highly relevant and timely, especially in a context when there are serious apprehensions prevailing in India as regards the promotion of new GM crop varieties.

# Acknowledgements

We, as editors of the volume feel quite privileged for the opportunities provided by individuals and organisations that supported our research on Bt cotton. As we turn back, we recall that a number of agencies and individuals have extended wholehearted support in our long journey of consolidating this work on a theme of topical relevance. This volume bears the testimony and is an outcome of the continued research over the past several years on the dynamics of Bt cotton adoption across the major cotton growing states in India undertaken by several researchers including the contributors to this volume.

For us, the trigger to develop a keen interest in pursuing the research on the trends in Bt cotton adoption in India had come in 2003, when we got an opportunity to look at the data collected by the Jatan Trust, Vadodara, jointly with Prof. Sudarshan Iyengar. This was followed by an opportunity to collaborate with Prof. Bharat Ramaswami and Prof. Carl Prey in a project funded by India Development Foundation during 2004. Subsequent research studies undertaken by us at GIDR, i.e., study supported by the Oxfam USA during 2007-2009; collaborative study with the Centre for Sustainable Agriculture, Jatan Trust and the Gujarat Vidyapith during 2009; and the Government of Gujarat supported study during 2012, all have helped us sharpen our understanding about the farm level dynamics of Bt adoption and its socio-economic and environmental outcomes. In this regard, we specifically thank Prof. Bharat Ramaswami, Prof. Rob Tripp, Prof. Sudarshan Iyengar, Shri. Kapilbhai Shah, Shri. Atul Sharma and Ms. Kavitha Kuruganti for the useful interactions and sharp questions that made us think deeply and probe the issues in greater detail.

That said, the real idea to consolidate the work on India's tryst with Bt cotton over the decade 2002 to 2012 as a volume, has come from the National Seminar on Bt cotton that GIDR organised at Ahmedabad during

3-4 May 2011. This volume contains a few select papers presented at the National Seminar along with couple of additional chapters prepared by the editors. We thank all the Seminar participants, particularly, Dr. Anupam Barik, Dr. V.Kumar and Dr. Keshav Kranthi for their helpful commentson issues discussed there. The occasional interactions and clarifications with Dr. Desai and Shri. J.V. Shah on several aspects of cotton cultivation in the past were very helpful. We also thank the research team at GIDR, particularly, Ms. Ilaben Mehta, and Mr. Ramnikbhai Chavda for their untiring help while undertaking field work and Mr. Bharat Adhyaru for data processing at various time points over the past few years.

This volume would not have been possible without the support of the authors of various chapters and we express our deep sense of gratitude to them for supporting us during the entire process of publication. We thank Prof. Mahendra Dev for positively responding to our request for writing the forward for the volume. We appreciate the excellent editorial support given by Ms. Sugeeta Roy Choudhury and thank Mr. Ashok K. Mittal and the entire team at the Concept Publishing Company for bringing out this volume.

We thank our faculty colleagues at GIDR for the excellent academic environment and quality debates during the presentations. GIDR's library and administrative staff deserve a very special mention. A special word of thanks is due to Prof. Amita Shah, former Director of GIDR for her unstinting support and gently nudging us to consolidate the outcome of the work on Bt cotton, including the national seminar. While placing on record our sincere gratitude to the cotton farmers in Gujarat and Maharashtra, who allowed our intrusion on their time and patiently responded to our queries, we remain solely responsible for any inadequacies that may remain in the volume.

# List of Contributors

**Anchal Arora**, Ph.D. Scholar, Centre for International Trade and Development, School of International Studies, Jawaharlal Nehru University (JNU), New Delhi. E-mail: arora.anchal@yahoo.com

**Ashok, K.R.**, Professor, Department of Agricultural Economics, Tamil Nadu Agricultural University, Coimbatore. E-mail: ashok10tnau@yahoo.com

**A. Indira**, Associate Professor, Chairperson, Prof. C.K. Prahalad Centre for Emerging India, Loyola Institute of Business Administration, Chennai. E-mail: aindira1@rediffmail.com, indira.a@libaedu.in

**Jeyanthi, H.**, Administrative Officer, Agricultural Insurance Company of India Ltd, Bangalore. E-mail: jeyanthi_anand@yahoo.com

**Kamal Vatta**, Director, Center for International Projects Trust, E-mail: kmlvatta@yahoo.com

**Prahadeeswaran, M.**, Assistant Professor of Agricultural Economics, Tamil Nadu Agricultural University, Coimbatore. E-mail: mprahadeeswaran@gmail.com

**Sangeeta Bansal**, Associate Professor, Centre for International Trade and Development, School of International Studies, Jawaharlal Nehru University (JNU), New Delhi. Email: sangeeta.bansal7@gmail.com

**Sidhu R.S.**, Dean, College of Basic Sciences and Humanities, Punjab Agricultural University, Ludhiana. E-mail.deancobsh@pau.edu

**Sarthak Gaurav** is Postdoctoral Researcher, Asia Research Centre, London School of Economics and Political Science (LSE), E-mail: s.gaurav@lse.ac.uk

**Srijit Mishra,** Associate Professor, Indira Gandhi Institute of Development Research (IGIDR), Mumbai. Email: srijit@igidr.ac.in

**Uma, K.,** Professor at Department of Agricultural and Rural Management Tamil Nadu Agricultural University, Coimbatore. E-mail: umaap68@yahoo.co.in

**Vijay Kumar,** Research Scientist & Head (cotton), Main Cotton Research Station, Navsari Agricultural University, Surat. E-mail:drvkumar255@gmail.com

# Contents

# List of Tables

# List of Figures

# List of Charts



# List of Boxes

# 1

# Development of Agri-biotechnology

## An Overview

*N. Lalitha* and *P.K. Viswanathan*

Biotechnology is defined as 'any technological application that uses biological systems, living organisms or derivatives thereof, to make or modify products or processes for specific use' (Article 2, Convention on Biological Diversity1993). While traditional biotechnological tools have been in use for long, modern biotechnology got revolutionised ever since scientists Stanley Cohen and Herbert Boyer published a series of papers during 1973–74 on the process by which pieces of DNA could be transferred between two organisms. These papers were forerunners of solutions for issues in healthcare, agriculture and industry which were hitherto considered impossible. While in the field of healthcare, this technology can act both as a research tool and as a healthcare product, in agriculture, agri-biotech applications range widely from genetic variations, screening and selection of favourable variants to plant management systems including fertilisation and plant protection. Worldwide, biotechnology tools in agriculture have been used to enhance yield, develop resistance to pests, tolerance to drought and cold and salinity, and provide nutritional supplement, value addition and post-harvest quality improvements.

Technological contribution to increasing the productivity of agriculture is essential, particularly for developing countries dependent on agriculture with millions of parcels of landholdings facing problems of low productivity. Research and development in agricultural technologies have wide ranging positive externalities. By improving agricultural production and productivity, these technologies also result in (a) increasing the farm income and food supply, (b) increasing the avenues for agricultural and non-farm employment and wages, and (c) improving the socio-economic status of the farming community. At the same time, it is also true that while possibilities of agri-biotechnology are plenty, there have been some concerns especially surrounding applications in agriculture in both developing and developed countries.

Generally, while medical biotechnology products get universal acceptance without much debate, or, most often, without the knowledge of the consumer, biotechnology applications in agriculture have been subject to controversial debates. Thus, while traditional biotechnology applications such as tissue culture and marker based selections are followed in most countries, genetically modified organisms[1] (GMOs) have not been uniformly accepted in all the countries. The topics of the debate on agri-biotechnology revolve around environmental biosafety concerns, health safety issues, corporate control over agriculture depriving the farmers from their traditional farming practices, and so on. Despite such concerns, the number of countries adopting agri-biotechnology had increased slowly from 6 in 1996 to 9 in 1998 and 17 in 2004 to 28 in 2012, increasing the area under GM crops from 1.7 million hectares in 1996 to 170 million hectares in 2012. Of the 28 countries in 2012, 18 are developing countries, where, for the most part, agriculture is the mainstay (Table 1.1) and they planted more areas with GM crops than the developed countries[2].

With this brief background on agri-biotechnology, this chapter presents the agri-biotechnology scenario in the global and Indian contexts followed by the adoption of Bt cotton in India. Further, this chapter highlights the issues surrounding the Bt cotton experience in India, followed by an overview of the chapters in this volume.

## 1. Global Agri-Biotechnology Business Scenario

Globally, though both the public and private sectors are engaged in the research and development in agri-biotechnology, the public sector R&D focus is on developing varieties that perform well under drought, heat, flood and salinity which is progressing slowly, compared to the research on developing disease and pest resistant varieties by the private sector, particularly dominated by the companies in the US. The private sector's dominance in this field is supported by the fact that: (a) intellectual property rights (IPRs) are actively sought in the agriculture sector, (b) the growing concern for the unprotected use of agro-chemicals, and (c) the

**Table 1.1:** Global Area of Biotech Crops in 2012, by Country

| *Rank* | *Country* | *Crop* | *Area (Million ha)* |
|---|---|---|---|
| 1 | USA | Maize, soybean, cotton, canola, sugarbeet, alfalfa, papaya, squash | 69.5 |
| 2 | Brazil | Soybean, maize, cotton | 36.6 |
| 3 | Argentina | Soybean, maize, cotton | 23.9 |
| 4 | Canada | Canola, maize, soybean, sugarbeet | 11.6 |
| 5 | India | Cotton | 10.8 |
| 6 | China | Cotton, papaya, poplar, tomato, sweet pepper | 4 |
| 7 | Paraguay | Soybean, maize, cotton | 3.4 |
| 8 | South Africa | Soybean, maize, cotton | 2.9 |
| 9 | Pakistan | Cotton | 2.8 |
| 10 | Uruguay | Soybean, maize | 1.4 |
| 11 | Bolivia | Soybean | 1 |
| 12 | Philippines | Maize | 0.8 |
| 13 | Australia | Cotton, canola | 0.7 |
| 14 | Burkina Faso | Cotton | 0.3 |
| 15 | Myanmar | Cotton | 0.3 |
| 16 | Mexico | Cotton, soybean | 0.2 |
| 17 | Spain | Maize | 0.1 |
| 18 | Chile | Maize, soybean, canola | <.01 |
| 19 | Colombia | Cotton | <.01 |
| 20 | Honduras | Maize | <.01 |
| 21 | Sudan | Cotton | <.01 |
| 22 | Portugal | Maize | <.01 |
| 23 | Czech Republic | Maize | <.01 |
| 24 | Cuba | Maize | <.01 |
| 25 | Egypt | Maize | <.01 |
| 26 | Costa Rica | Cotton, soybean | <.01 |
| 27 | Romania | Maize | <.01 |
| 28 | Slovakia | Maize | <.01 |
| | **Total** | | **170.3** |

*Source*: www.isaaa.org, Policy Brief 44, 2012: PPT slide no. 7.

growing possibilities in plant genetic engineering made possible by the rDNA technique. Importantly, on the last point, the proprietary rights over many of the rDNA techniques are with the private sector.

A detailed analysis of the emergence of a few consolidated global agri-biotechnology companies in the US that hold significant influence in agri chemical, seed and plant biotechnology through their R&D, monopoly over products and supply chain (Musselli 2006) indicates that this situation is likely to remain the same for some more years to come due to the mixed response to adoption of GMOs in different parts of the world. One major fear of this corporate control over emerging technologies is that the current research in these important technologies is dictated by the commercial interest of these firms and may be marginally relevant for the problems faced by the developing countries (Alam 1996: 703-704, Rao & Dev 2009: 56–64, Tripp 2009:1–22, Purkayastha and Rath 2010: 42–48). Further, the technologies developed by the private sector in the developed countries are unavailable for the scientists in developing countries and the spillovers of such research has had limited impact on the livelihoods of farmers in the developing countries (Alam 1996, FAO 2011).

Therefore, slowly from the late 1980s onwards, there has been a transformation of the agro chemical companies which entered the arena of plant biotechnology. From a mere strength of about 15 companies in the mid-1990s, they evolved into 6 strong research intensive companies by mid-2000s (UNCTAD, 2006). These big companies also acquired a number of small companies engaged in basic applied research and agri-biotechnology. In 2004, these 6 agro chemical companies, viz. Bayer, Syngenta, BASF, Dow AgroSciences, Monsanto and DuPont accounted for 77 per cent of the agro-chemical market. Such 'acquisitions of seed companies provided a means to better control and market proprietary lines of chemicals, genetic technologies and seeds, often sold in a single bundled package' (Musselli 2006: 7). Further, this study points out that acquisition of seed companies helped the companies surge in plant biotechnology sector fast, as seeds embody the technological development like, improved traits or GMOs. The established distribution network of these biotech companies made it easier for them to launch their products with ease.

The R&D of the consolidated companies focused on introducing input traits that make crops herbicide resistant or insect resistant. The advantage in having such traits is that it will be useful in a number of crops. The consolidation of companies engaged in agro chemicals, seed or crop biotechnology also helped them overcome the hurdles created by the intellectual property owned by each of these companies. In biotechnology, particularly in the inventions which involve genetic engineering, it is feared that strong intellectual protection to innovations would create 'anti commons' and the technology is underutilised due to the protection. This is because patents over genetic material may cover: (1) genes or partial DNA sequences, such as cDNAs, SNPs, ESTs promoters and enhancers; (2) proteins encoded by these genes and their functions in the organism; (3) vectors used in the transfer of genes from one organism to another; (4) genetically modified organisms, cells, plants and animals; (5) process used for making the GM product; (6) uses of genetic sequences or proteins which include specific tests for specific genetic diseases or predisposition to such diseases, drugs developed on the knowledge of proteins and their biological activity, industrial applications of protein functions (OECD 2002: 28).

Hence, companies will have to look for ways of patent pooling, cross licensing or voluntary licensing of the intellectual property to take advantage of the intellectual assets each one possesses and avoid engaging in law suits. Thus again in the seed segment/crop biotechnology area, Monsanto, Bayer, Syngenta and Dow Agro Sciences have strengthened their position by cross licensing each other the relevant technologies and were holding majority of the approvals from the US Department of Agriculture for release of GM crops in 2005 and Monsanto was leading the rest of the groups by holding 14 per cent of the agricultural biotechnology patents (Musselli 2006). This stronghold is evident in 2012 also, with Monsanto and its associates being the developer of important GM traits.

The latest available figures on GM approval database maintained by the International Service for the Acquisition of Agri-biotech Applications (ISAAA)[3], include approvals for: (a) 25 crops that include food and commercial crops; (b) 319 GM crop events of which maize and cotton top the list with 121 and 48 events respectively; (c) 28 GM traits that include, drought tolerance, delayed fruit softening, anti-allergy and antibiotic resistance etc.; (d) 6 commercial traits that include abiotic stress tolerance, disease resistance, herbicide tolerance, insect resistance, modified product quality and pollination control system; (e) 46 developers and (f) 33 countries. Importantly, of the 46 developers, 20 are from the USA and four from India. Of the four from India, three are private sector companies and the other belongs to the public sector (jointly by the Central Institute of Cotton Research and University of Agricultural Sciences, Dharwad) and all are focusing on cotton. All the three private sector companies from India (JK Agri Genetics, Nath Seeds and MetahelixLife Sciences) are focusing on two GM traits that would provide lepidopteran insect resistance and antibiotic resistance and thus involve two genes. The public sector is focusing on single gene, i.e. lepidopteran insect resistance.

## 2. Status of Agri Biotechnology in India

The Approach Paper to the 12th Five Year Plan (2012–2017) by the Government of India (GOI hereafter), titled 'Faster, Sustainable and More

Inclusive Growth', talks of various measures to reduce the disparities across states and sectors. It is suggested that if India is to achieve the target of 9 per cent growth in gross domestic product (GDP) during the 12th Five Year Plan, the agriculture sector has to grow at 4 per cent per annum. However, it has fallen short of this target, as growth hovered around 3.2 per cent per annum during the 11th Five Year Plan and at 2.3 per cent during the 10th Five Year Plan, widening the income disparities in the country. This is a concern for India, since, more than half of the 1.2 billion population is dependent on agriculture and related activities. In this context, the Approach Paper observes that 'global development experience, especially from the BRIC countries, reveals that one percentage point growth in agriculture is at least two or three times more effective in reducing poverty than the same magnitude of growth emanating from non-agriculture sector' (GOI 2011: 7) and 'one-third of the future growth in productivity should come from crop technologies' (GOI 2011: 71). Moreover, a major chunk of Indian agriculture (56% of the total cropped area) is rainfed which constitutes 48 and 68 per cent under food and non-food crops respectively. Recognising that more areas could come under the rainfed category due to climatic factors, GOI 2011 emphasises that 'it is necessary to remain abreast with latest advances in bio-technology and make full use of marker assisted breeding methods. An acceptable set of protocols and a clearly defined regulatory mechanism are required so that transgenic food crops can be cleared if advantages of doing so vastly outweigh any precautionary misgivings' (GOI 2011: 71).

While this is the broad philosophy of the government, the actual emphasis on the growth of biotechnology in India began with the setting up of the National Biotechnology Board (NBTB) in 1982 in the Central Ministry of Science and Technology to 'identify priorities, co-ordinate, oversees and plan for required manpower, integrated industrial development and large scale use of biotech products and processes' (Rao 2004: 127). Subsequently, in 1986, the NBTB became a separate department under the Ministry of Science and Technology and henceforth named as the Department of Biotechnology (DBT). Since then, concerted efforts were taken by the Government of India towards setting up research institutions, biotech infrastructure, improving the human resource and in terms of policy formulations. The biotechnology industry in India comprises bio-pharma, bio-services, bio-agri, bio-industrial and bio-informatics. Bio-agri ranks third in terms of revenue as compared to the other sectors. Importantly, bio-agri's share in the total biotech sector which was 4.6 per cent in 2002–03, has increased to 14.9 per cent in 2011–12 (Table 1.2). The spurt in the growth of modern biotechnology in India came in the mid-1990s with companies announcing their new health products particularly in vaccines and later in diagnostic kits segment and the early 2000s saw the adoption of genetically modified crops in the agricultural sector.

The Central Government plays a larger role in the regulation of research, import, production and the open release of GMOs in India. The state governments while following these guidelines, play a role of the facilitator and formulate their own policies of incentives, infrastructure and to an extent venture capital funds also to support biotechnology ventures.

A few important measures adopted in India for the promotion of biotech deserve mention here. Biotechnology research in India is mainly funded by the Department of Biotechnology (DBT), the Council of Scientific and Industrial Research (CSIR), the Indian Council of Agricultural Research (ICAR) and the Indian Council of Medical Research (ICMR).

In November 2007, the GOI announced the National Biotech Development Strategy which was done after a two-year long nationwide consultation process involving different stakeholders. This strategy proposed the setting up of a national biotech regulatory authority (NBRA), which provides single window clearance of genetically modified products and processes. As part of the biotech communication strategy with the different stakeholders, the draft of the NBRA formation was placed in public domain for suggestions. The formation of the NBRA of India bill is yet to be approved by the Parliament. As proposed in the strategy, a Biotechnology Industry Research Assistance Council (BIRAC) has been set up mainly to stimulate, foster and enhance the innovation capabilities of small and medium enterprises and foster linkages between academic and private sectors. The GOI has also set up the Biotech Consortium Limited (BCL) as a public limited company in 1990 to facilitate linkages among research institutions, industry, government and funding institutions and to facilitate commercialisation of biotechnology.

Also, on 23 August, 2012, the GOI approved the proposal of the Ministry of Agriculture for the establishment of the Indian Institute of Agricultural Biotechnology. The institute is planned to be set up at Ranchi in Jharkhand during the 12th Five Year Plan under the Department of Agricultural Research (DAR) at a cost of Rs. 287.50 crore. The mandate of the institute would be to undertake multi-disciplinary basic and strategic

**Table 1.2:** Growth of Biotech Industry 2002-12

*(Rs. crores)*

| | *2002-03* | *2003-04* | *2004-05* | *2005-06* | *2006-07* | *2007-08* | *2008-09* | *2009-10* | *2010-11* | *20011-12* |
|---|---|---|---|---|---|---|---|---|---|---|
| Bio-pharma | 1790<br>(76.30) | 2752<br>(79.20) | 3570<br>(75.20) | 4708<br>(72.20) | 5973<br>(69.90) | 6899<br>(67.20) | 7833<br>(65.20) | 8829<br>(62.20) | 10655<br>(61.80) | 12679<br>(62.00) |
| Bio-services | 135<br>(5.80) | 275<br>(7.90) | 425<br>(9.00) | 720<br>(11.00) | 1102<br>(12.90) | 1572<br>(15.30) | 2062<br>(17.20) | 2639<br>(18.60) | 3245.97<br>(18.80) | 3749<br>(18.30) |
| Bio-agri | 110<br>(4.70) | 130<br>(3.70) | 330<br>(7.00) | 598<br>(9.20) | 926<br>(10.80) | 1201<br>(11.70) | 1413<br>(11.80) | 1936<br>(13.60) | 2480<br>(14.40) | 3050<br>(14.90) |
| Bio-industrial | 235<br>(10.00) | 238<br>(6.80) | 320<br>(6.70) | 375<br>(5.80) | 395<br>(4.60) | 410<br>(4.00) | 478<br>(4.00) | 564<br>(4.00) | 625.94<br>(3.60) | 696<br>(3.40) |
| Bio-informatics | 75<br>(3.20) | 80<br>(2.30) | 100<br>(2.10) | 120<br>(1.80) | 145<br>(1.70) | 190<br>(1.80) | 220<br>(1.80) | 231<br>(1.60) | 242.43<br>(1.40) | 266<br>(1.30) |
| Total | 2345 | 3475 | 4745 | 6521 | 8541 | 10272 | 12006 | 14199 | 17249.3 | 20440 |

*Note:* Figures in parentheses indicate the percentages to column total.
*Source*: Various issues of *Bio-spectrum*.

research with a view to developing future crops for traits such as increased yield, or increased tolerance to biotic and abiotic stresses. It will also provide both curricula and course material to India's agricultural universities currently running or trying to establish successful agricultural biotechnology graduate and post-graduate programmes.[4]

It may be argued that the birth of agri-biotech companies in India was essentially due to the liberalisation of seed policies. The GOI had followed a very restrictive seed policy till the 1990s. For instance, in the 1970s, under the import substitution policy, the GOI banned commercial imports of all agricultural inputs that were produced in India. So while the seed crops could be produced only by the Government of India, vegetable crops could be imported by private firms after obtaining the required permission from the government. The seed sector was one of the beneficiaries of the partial liberalisation measures introduced in the early 1980s, where GOI gave seed and biotech industries the 'core industry' status as per the seeds policy of 1988. Further, this policy allowed seed firms to import commercial vegetable seeds with no quotas and also to import commercial seeds of foreign varieties of coarse grains and oilseeds for a period of two years after which the firm was expected to produce the seeds in India. It also allowed import of germplasm for research purposes. Transgenic seeds developed locally or imported were to be deposited with the gene bank.

According to the Indian GMO Research Information System (IGMORIS), field trials are in different stages for cotton, maize, rice, rubber, sorghum, groundnut, watermelon, tomato, papaya, cauliflower, cabbage, sugarcane, brinjal and okra. About 64 agri-based corporates have been listed as those working with transgenic crops. Interestingly, a sizeable percentage of the companies are located in Andhra Pradesh and Maharashtra. Besides cotton, genetically modified brinjal was also reported to have successfully completed the field trials and was about to be commercially released, when the moratorium on Bt brinjal was imposed by the GOI[5]. Thus, cotton remains as the only GM crop in India which is the focus of this book. The two events that have been introduced in India are Mon 531 (*Cry 1 Ac*) and 15985 (*Cry1 Ac+Cry2Ab*). It is important at this point to mention about the gene that has been introduced in cotton. A series of soil organisms, *Bacillus thuringiensis* (Bt) that produced specific toxins specific to particular types of insect were the early target for plant genetic engineering (Tripp 2009: 23–48). Tripp points out that though the early research demonstrated the presence of the Bt gene in the transformed plants, the insecticidal performance was very modest. A major breakthrough came in 1988, when Monsanto discovered that the bacterial gene needed to be 'codon modified' so that its genetic code was more compatible with that of plant systems. Once this was done, the insecticidal activity was significant in the transgenic plants.

'Crystalline (Cry) toxins from the soil bacterium Bt have been the most extensively studied and used in transgenics. After activation by insect proteases in the insect midgut, Bt proteins bind to receptors in the midgut. Such binding leads to the formation of pores in the midgut membrane that ultimately cause cell lysis and insect death. Five *cry* genes (*cry* 1A, *cry* 1Ac, *cry* 2Ab, *cry* 1F and *cry* 1EC) are commercially available or nearly commercialised in cotton' (Showalter *et al.* 2009:53).

It should be noted that while Mon 531 event also known as Bollgard I offers protection against bollworm alone, event 15985 (Bollgard II) offers protection against both bollworm and spodoptera. Bollgard II is also known as the pyramid strategy because here the transgenic plant will produce two distinct toxins to delay pest resistance (Showalter 2009).

Bollworms have been a major pest in cotton against which farmers had to spray insecticides a number of times. The loss due to bollworm is estimated at around 50 to 60 per cent of the yield. The introduction of these traits in the seed is expected to reduce the pesticide use on the plants and, thus, reduce cost for the farmers, and by reducing the yield loss due to bollworm attack, the technology is expected to improve the yield performance of the crop.

## 3. Cotton in India

India is the accredited home of the cotton plant. 'Four hundred fifty years before Christ, Herodotus testified that India had wild trees that bore fleeces as their fruit... of these the Indians made their clothes' (Dantwala 1947: 1). Two important events that changed the course of India's political scenario, viz. industrial revolution and the consolidation of the political power by the British in India, have been closely associated with cotton.

The post-World War II economy of India needed more food supply and the 'Government of Bombay enacted the Growth of Food Crops Act, which drastically restricted the cultivation of non-food crops. A control was instituted over prices of cotton, one of the objects being to weaken the incentive to grow cotton. The net result was the reduction in the acreage under cotton from 23.5 million acres in 1938–39 to 14.4 million in 1945–

46, the production falling from 5 million to 3.6 million bales' (Dantwala 1947: 10). India had a crucial advantage over other countries that grew cotton such as USA, China, Egypt, Brazil and Russia, as India was home to Asiatic or old world cottons (*Gossypium arboreum* and *G. herbaceum*) and American cottons (*G. hirsutum* and *G. barbadense*) which are the four cultivable cotton species. Presently in India, *G. arboreum* and *G. hirsutum* are the principal species that are being cultivated (Table 1.3). Desi varieties and in particular, *G. arboreum,* are known for their drought tolerance and resistance to bollworms and sucking pests. On the other hand, American cottons usually have long and extra long staple and better spinning potential (higher counts) than desi cottons. They were introduced in India by the colonial administrators to meet the demands of English textile manufacturers anxious to secure an alternative and cheaper source than the United States (Guha 2007).

**Table 1.3:** Area under Different Cotton Species

*(% of Total Cotton Area)*

| *Species* | *1947* | *1970* | *1980* | *1990* | *2000* | *2007* |
|---|---|---|---|---|---|---|
| Gossypium Arboreum | 65 | 30 | 20 | 30 | 17 | 4 |
| G. Herbaceum | 32 | 17 | 14 | 12 | 11 | 5 |
| G. hirsutum | 3 | 53 | 54 | 48 | 69 | 90 |
| G. barbadence | | | 11 | 10 | 3 | 1 |

*Note:* G. arboreum and herbaceum are known as Asiatic cotton, G. Hirsutum as American cotton and the G. barbadence is known as Egyptian cotton. While G. arboreum and G. hirsutum are grown in all the states, G. herbaceum is limited to Gujarat and Karnataka.

*Source*: Barik and Gautam (2009).

Cotton is vulnerable to a large number of insects/pests and natural enemies and most of these occur at different stages of crop growth. A few of them are: (1) Wilt where the leaves turn brown and drop off; (2) Root- and complete wilting of the plant; (3) Anthracnose—reddish brown depression spots on leaves and bolls; (4) Bacterial blight at all stages which causes secondary infection; (5) Alternaria which causes leaf spots and affects the plant at all stages; (6) Areolate or grey mildew, and (7) Carcospora or leafsop—both occur at maturity stage where the leaves become yellowish and finally fall. Further, American bollworm, pink bollworm, spotted bollworm, tobacco caterpillar, jassids and spider mite are active throughout the year on the cotton plant, necessitating pesticide sprays.

### *3.1. Institutions for Cotton Development*

Lives of more than 60 million people are associated with cotton farming, processing, ginning and in textile industry, etc. In view of the direct and indirect employment opportunities and livelihoods dependent on cotton-processing, etc., the GOI has set up a number of agencies for the promotion and development of cotton, viz. Ministry of Agriculture, Directorate of Cotton Development, Central Institute for Cotton Research, and the ones at the state level, Department of Agriculture and state Agricultural Universities (Chart 1.1)[6]. Till the late 1990s, the public sector played a

**Chart 1.1:** Institutions for Cotton Research and Development in India

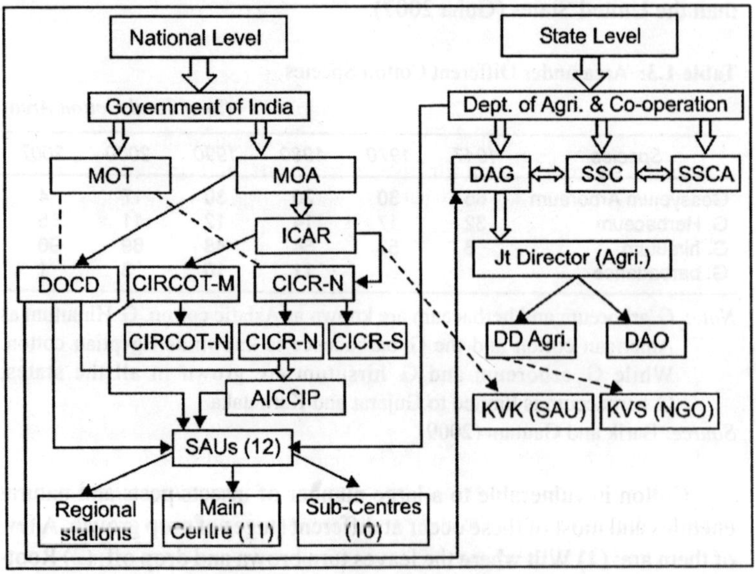


*Note:* MoT: Ministry of Textiles; MOA: Ministry of Agriculture; ICAR: Indian Council of Agricultural Research; DoCD: Directorate of Cotton Development; CIRCOT: Central Institute of Research on Cotton Technology, Mumbai and Nagpur; CICR: Central Institute of Cotton Research, Nagpur, Coimbatore and Sirsa; AICCIP: All India Coordinated Cotton Improvement Project; SAU: State Agricultural Universities; DAG: Directorate of Agriculture; SSC: State Seed Corporation; SSCA: State Seed Certification Agency; DD Agri: Deputy Director Agriculture, DAO: District Agricultural Officer; KVK SAU KrishiVigyan Kendra at State Agricultural Universities; KVS: Krushi Vikas Sansthan; NGO: Non Governmental Organisation.

critical role in the provision of a very fundamental input in agriculture, viz. seeds. Particularly in the case of cotton, the public sector institutions in India have many feathers in their cap, as they hold the credit in introducing the world's first hybrid variety in 1970 known as H4.

The Directorate of Cotton Development is associated with monitoring the centrally sponsored Intensive Cotton Development Programme (ICDP) and Technology Mission in Cotton (TMC). A few of the important terms of the mandate of the Central Institute for Cotton Research are (a) to conduct basic research to improve yield, fibre quality and by-products, (b) create genetic variability for location specific adoption, and (c) to collect, conserve, evaluate and cataloguing of cotton germplasm, development of appropriate farming/cropping system for different cotton growing zones. Hence, the CICR is linked to the different state agricultural departments and agricultural universities.

The National Agriculture Innovation Project (NAIP), which is under the Central Institute for Research on Cotton Technology, focuses on (a) value chain for cotton fibre, seed and stalks to ensure higher economic returns to farmers and allied stakeholders, (b) research into development of decision support system for insects and pests, and (c) geo-referenced soil information system for land use planning and monitoring soil and land quality for agriculture[7].

### *3.2. Role of Private Sector in Cotton Development in India*

Discussing the institutional mix in India's seed system, Jaffe and Srivastava (1992) list the various crops with clear roles played by public and private sector in: (a) variety sourcing (import)/development, (b) seed production, processing, and (c) seed marketing and distribution. In the case of cotton, they observed that: 'while, no direct seed import was done, there was both public and private sector involvement with separate roles for each in all other stages. The private sector seed sector comprises both domestic private sector, multinationals, and collaborative companies of multinationals and private sector. There are about 76 companies in the private sector that produce foundation and certified seed (http://seednet.gov.in/Material/Seed_Companies_Pvt_FS_CS.pdf, accessed on 19 March, 2013), an increase by around four times since the early 1990s when there were only 17 firms in the private sector producing the breeder and foundation seeds (Jaffe and Srivastava 1992). Cotton in India is one of the major field crops where the share of private sector hybrids in total hybrids is 88.5 per cent. While the private sector developed 150 hybrids till 2001–02, the public sector has developed only 15 hybrids. Between 2002–03 and 2009–10, while the private sector contributed 43 hybrids, those of the public sector were only 10. The number of notified varieties and hybrids has been slowly increasing (72, 78 and 95 in 1980–89, 1990–99 and 2000–10 respectively) (Dadlani 2012).

Investment for crop improvement in cotton which accounted for a meagre 9.59 per cent of the total R&D investment in crop improvement in 1987, increased to a sizeable 50 per cent of the total R&D investment in 2009–10. The number of companies engaged in cotton research also increased from 9 to more than 50, of which 37 are engaged in research on GM cotton. The increase in the R&D expense is also due to the fact that the companies were focusing on introducing hybrid varieties (Pray and Nagarajan 2010). This change has been brought in by the liberalisation policy measures of the 1990s which allowed the import of genetic material.

**Table 1.4:** Private Sector R&D Investment in Cotton-crop improvement in India over years (Real rupee terms)

| *R&D Investments* | *1987*[A] | *1995*[B] | *2003*[C] | *2009/10*[D] |
|---|---|---|---|---|
| Total R&D Investment by seed firms in crop improvement (Rs. Millions) | 417 | 1549 | 2000 | 6000 |
| R&D investment in cotton-crop improvement (Rs. Millions) | 40 | 270 | 500-750 | 2500-3000 |
| Number of firms with cotton R&D | 9 | 27 | 30-32 | >50* (37 Bt cotton) |

*Note:** More than 50 companies are engaged in cotton-crop improvement and breeding aspects out of which 37 companies have directly engaged in Bt transformation.

A - Pray, Ribeiro, Mueller and Rao (1991); B - Pray and Kelley (1998), Pray and Ramaswami, (2001); C - Biospectrum (2009), James (2008); D - Pray and Nagarajan (2010) and National Seed Association of India/ Industry estimates (2009–10).

*Source*: Table 1.1 in Pray and Nagarajan (2010).

Randhawa and Chhabra (2009: 115–130) note that of the 12 different transgenic lines imported since 1997 through the National Bureau of Plant Genetic Resources (NBPGR), New Delhi, transgenic cotton lines topped the list followed by transgenic maize and rice. Of the total 79 transgenic lines imported then, 84.6 per cent was by the private sector in eight different crops. The predominant trait in these imported crops being the resistance to lepidopteran insects followed by herbicide tolerance. Of the six events released for commercialisation, four involve single gene and two involve double genes. Cry1Ac Bt gene was first introduced by Mahyco in India which was licensed from Monsanto, US, in 2002 (event Mon-531). The same gene was sourced from IIT Kharagpur in 2006 and developed by JK Nath seeds, known as Event-I. The third event known as BNLA-601, involving Cry1Ac, was developed by the Central Institute of Cotton Research, Nagpur and University of Agricultural Sciences, Dharwad in 2008. A different single gene event known as event MLS-9124 was developed by MetahelixLife Sciences in 2009[8].

Mon 15985 (Bollgard II) licensed from Monsanto has Cry 1Ac and Cry 2Ab2 in it and was developed by Mahyco in 2006. The other multiple gene is the event-GFM sourced from China in 2006 and includes Cry 1Ab and Cry 1Ac, fused and developed by Nath Seeds. The latest in the pipeline that has completed the stipulated field trials and awaits the green signal for commercialisation is Mahyco's Bollgard II Roundup Ready Flex Cotton II (RRF). RRF provides herbicide tolerance in the form of weed management and inbuilt tolerance for bollworm and Spodoptera caterpillar. It should be noted that the background genotype is still the product of non-transgenic biotechnology.

**Table 1.5:** R&D Expenses Incurred by Technology Providers and Sub-licensees of Bt gene

| | *4 firms with own Bt events (Rs. Millions)* | *5 major sub-licensees of Bt events (Rs. Millions)* |
|---|---|---|
| 2003 | 289.5 | 26.8 |
| 2004 | 437.7 | 34.9 |
| 2005 | 54.03 | 33.5 |
| 2006 | 669.9 | 59.5 |
| 2007 | 724.8 | 82.6 |
| 2008 | 912.7 | 87.1 |
| 2009 | 1164.5 | 159.6 |

*Source*: Pray and Nagarajan (2010).

As agri R&D involves developing the technology and adapting it to the local context, it requires large resources. The R&D expenditure of companies with own Bt events is several times higher than the companies which have sub-licensed the Bt events from somewhere (Table 1.5). As cotton is the only GM crop technology that has been approved, a number of companies have backcrossed the GM cotton events to their hybrids and have released these hybrids for sale. There has been a substantial jump in the number of varieties that were released through the years as evident from Table 1.6. Figure 1.1 shows the cumulative number of hybrids released by GM events. It is also evident from the table that after Bollgard II has become popular, there has been a drastic decline in the number of hybrids released with single gene.

**Table 1.6:** Number of Hybrids Released by GM Event in India

| *Year* | *Event 531 (Bollgard 1)* | *Event 15985 (Bollgard II)* |
|---|---|---|
| 2002 | 3 | – |
| 2004 | 1 | – |
| 2005 | 16 | – |
| 2006 | 32 | 7 |
| 2007 | 59 | 13 |
| 2008 | 53 | 74 |
| 2009 | 55 | 171 |
| 2010 | 27 | 186 |
| 2011 | 5 | 95 |
| 2012 | – | 206 |

*Source*: Compiled from igmoris.nic.in

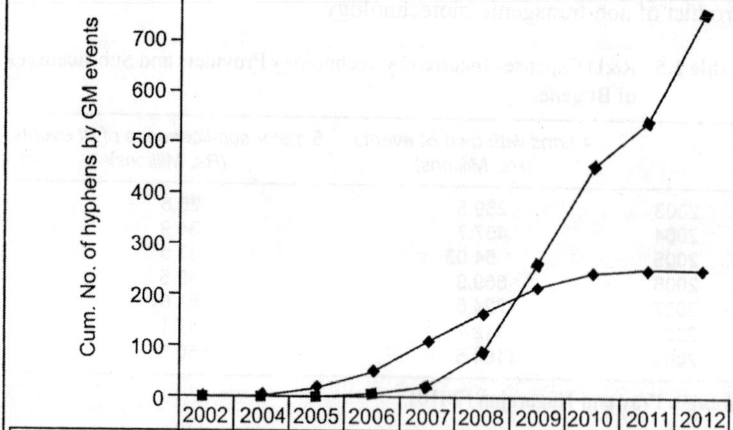


| | 2002 | 2004 | 2005 | 2006 | 2007 | 2008 | 2009 | 2010 | 2011 | 2012 |
|---|---|---|---|---|---|---|---|---|---|---|
| Event 531 | 3 | 4 | 20 | 52 | 111 | 164 | 219 | 246 | 251 | 251 |
| Event 15985 | 0 | 0 | 0 | 7 | 20 | 94 | 265 | 451 | 546 | 752 |

*Source*: Compiled from http://www.igmoris.nic.in/Files2/YearWise_List2002_May2012.pdf, accessed on 23 January 2013.

**Figure 1.1**: Cumulative Number of Hybrids Released by GM Events

In India, Bt has been introduced only in hybrids. The percentage of hybrids which was 24 per cent of the cotton seed sales in 1996–97 increased to 95 per cent of the total cotton seeds in 2009–10 (Pray and Nagarajan 2010). Introduction of Bt gene in the popular hybrids has resulted in pushing the revenues upwards for the companies concerned, which are lead by Nuziveedu and followed by Rasi seeds. Interestingly, the share of revenue of Mahyco (which introduced the Bt cotton in India) declined till 2007 and then started increasing, and ranked 3rd in 2010–11 (Table 1.7).

**Table 1.7:** Total Sales Revenue of Top Agri-bio Companies

*(Rs. Crores)*

| Companies | 2004-05 | 2005-06 | 2006-07 | 2007-08 | 2008-09 | 2009-10 | 2010-11 |
|---|---|---|---|---|---|---|---|
| Rasi | 86.9 | 309.5 | 333.3 | 293.3 | 376.84 | 358.78 | 378.88 |
| Nuziveedu | – | 62.5 | 226.4 | 291.9 | 369.96 | 476.86 | 610.0 |
| Mahyco | 166 | 117.8 | 110.7 | 135.9 | 211.1 | 312.0 | 364.9 |
| Ankur | – | – | 69.5 | 55.7 | 80.35 | 109.5 | 325.0 |
| Krushidhan seeds | – | – | 18.0 | 47.7 | 63.2 | 133.2 | 276.13 |
| Nath seeds | – | | 11.9 | 62.0 | 48.5 | 85.0 | 27.49 |
| JK Agri. | – | | 18.7 | 46.0 | 26.0 | 35.0 | 50.6 |
| Metahelix | – | | 6 | 6.6 | 9.7 | – | – |
| Bayer crop | – | – | – | – | – | 22.5 | 31.5 |

*Source*: Compiled from Pray and Nagarajan (2010) and *Biospectrum* June 2011: 48.

In what follows, we provide a brief overview of the trends and status of adoption of Bt cotton across states in India.

## 4. Growth in Cotton Area, Production and Productivity

Officially, Bt technology in three cotton hybrids of Mahyco was introduced in the year 2002, in the central and southern zones in India. But much before that, in 2001, farmers had started cultivating an unapproved variety of Bt cotton, which suited Gujarat particularly well. Though the company responsible for the release of unapproved variety was penalised, the network that was producing the unapproved variety was not broken and this resulted in the unapproved varieties penetrating into other cotton growing areas as well, thus leading to the rapid adoption of the technology. It is quite important to note that since the introduction of Bt technology in cotton, India's cotton area had increased from 7.63 million ha during 2003–04 to 11.1 million ha during 2010–11. While cotton production has more than doubled from 137 lakh bales to 335 lakh bales, productivity increased from 321 kg/ha to 518 kg/ha during the same period (Table 1.8). If expressed in annual terms, the post-Bt cotton period witnessed an annual increase in cotton area by 3.71 per cent, 14.6 per cent increase in production and 8.3 per cent rise in productivity during the reporting period.

**Table 1.8:** Trends in Area, Production and Productivity of Cotton in India

| *Year* | *Area (Million Ha)* | *Production (Lakh bales of 170 kgs)* | *Yield (Kg/ha)* |
|---|---|---|---|
| 2000–03* | 8.32 | 145 | 296 |
| 2003–04 | 7.63 | 137 | 306 |
| 2004–05 | 8.79 | 164 | 321 |
| 2005–06 | 8.68 | 185 | 354 |
| 2006–07 | 9.14 | 226 | 421 |
| 2007–08 | 9.41 | 259 | 470 |
| 2008–09 | 9.41 | 223 | 419 |
| 2009–10 | 10.31 | 240 | 395 |
| 2010–11 | 11.10 | 335 | 518 |

*Note:* * Figures indicate the three year average for the period 2000–2003.
*Source*: Directorate of Economics and Statistics, Ministry of Agriculture.

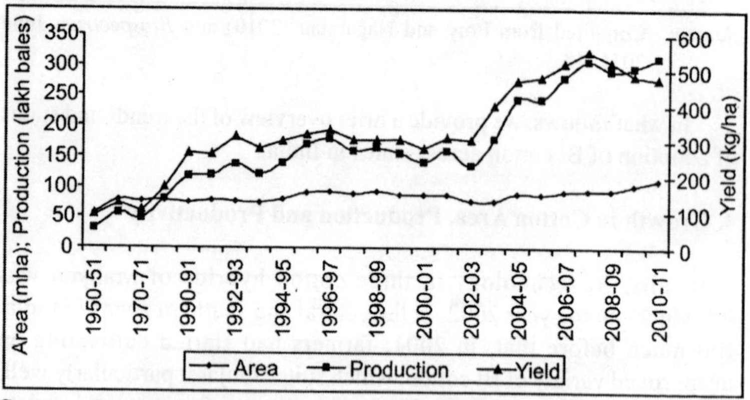


*Source*: Compiled from Directorate of Economics and Statistics, Ministry of Agriculture, Government of India.

**Figure 1.2:** Trends in Area, Production and Productivity of Cotton in India

The long-term trends in area, production and productivity of cotton are shown in Figure 1.2, which clearly brings out the substantial increase in production and productivity of cotton following the introduction of Bt cotton (2002–03), while increase in cotton area was only marginal. Among the states, growth in area, production and productivity of cotton has been quite significant in the states of Maharashtra, Gujarat, Andhra Pradesh and Punjab, though other states, viz. Tamil Nadu, Karnataka, Madhya Pradesh, Rajasthan and Haryana had also shown a perceptible shift towards the new technology. This is further evident from Table 1.9, which presents the comparative performance of the three major indicators of cotton in the pre and post-Bt adoption scenarios.

**Table 1.9:** Indicators of Performance of Cotton across States in India, Pre and Post-Bt Scenarios

| State | 2002–03 (Pre-Bt scenario) | | | 2010–11 (Post-Bt scenario) | | |
|---|---|---|---|---|---|---|
| | Area/ % | Production/ % | Yield/ Rank | Area/ % | Production/ % | Yield/ Rank |
| Andhra Pradesh | 8.03 (10.5) | 19.8 (14.5) | 418 (3) | 17.76 (15.9) | 53.0 (17.0) | 507 (4) |
| Gujarat | 16.34 (21.3) | 30.5 (22.4) | 317 (4) | 26.33 (23.6) | 102.0 (32.7) | 659 (1) |
| Haryana | 5.19 (6.8) | 8.8 (6.4) | 287 (5) | 4.95 (4.4) | 14.0 (4.5) | 481 (5) |
| Karnataka | 3.93 (5.1) | 5.0 (3.7) | 216 (8) | 5.34 (4.8) | 10.0 (3.2) | 318 (9) |
| Madhya Pradesh | 5.45 (7.1) | 18.0 (13.2) | 561 (2) | 6.51 (5.8) | 17.0 (5.4) | 444 (7) |
| Maharashtra | 28.00 (36.5) | 26.0 (19.1) | 158 (9) | 39.73 (35.6) | 82.0 (26.3) | 351 (8) |
| Punjab | 4.49 (5.9) | 7.5 (5.5) | 284 (6) | 5.30 (4.7) | 16.0 (5.1) | 513 (3) |
| Rajasthan | 3.86 (5.0) | 5.0 (3.7) | 220 (7) | 3.34 (3.0 | 9.0 (2.9) | 458 (6) |
| Tamil Nadu | 0.85 (1.1) | 3.0 (2.2) | 600 (1) | 1.30 (1.2) | 5.0 (1.6) | 654 (2) |
| All India | 76.67 (100.0) | 136.0 (100.0) | 302 | 111.61 (100.0) | 312.0 (100.0) | 475 |

*Note:* Area in Lakh ha; Production in Lakh bales of 170 kgs each; Yield in kg/ha.
*Source*: Cotton Advisory Board.

There has been a significant increase in production and productivity of cotton in all the nine major cotton growing states, though there was a marginal decline in cotton area in Haryana and Rajasthan between the pre and post-Bt cotton scenarios. Among the major states, Maharashtra continued to occupy the largest share in area during both the periods (36.5% and 35.6% respectively), while Gujarat has emerged as the largest producer in the post-Bt scenario with a relative share of 33 per cent in the national cotton output, followed by Maharashtra (26.3%) and Andhra Pradesh (17%). Gujarat also significantly improved its status as number one in terms of cotton productivity (659 kg/ha) in the post-Bt scenario. Gujarat, with an average productivity of 317 kg/ha, was ranked fourth in the pre-Bt scenario and the state has pushed backwards the other leading producers, viz. Andhra Pradesh and Madhya Pradesh as well as Tamil Nadu, which had the highest productivity (600 kg/ha) in the pre-Bt scenario.

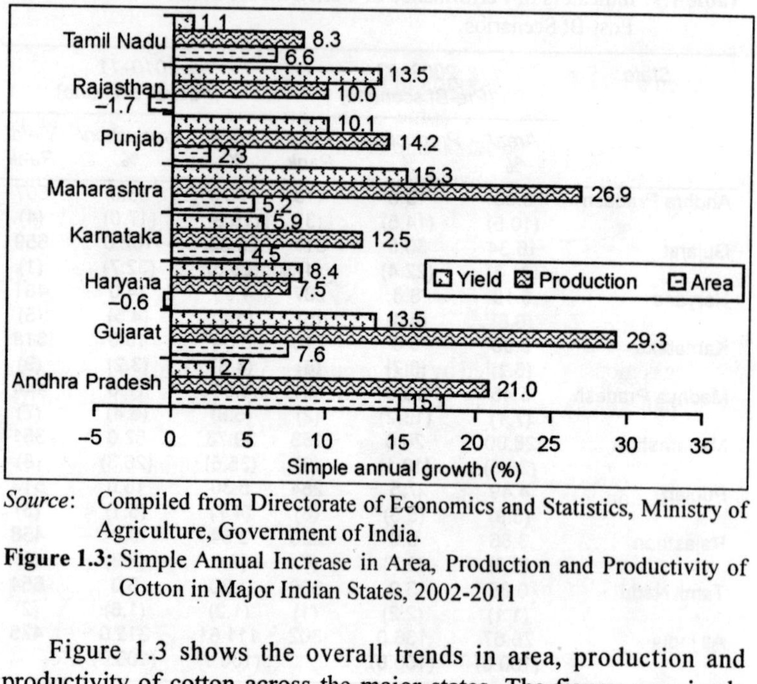


*Source*: Compiled from Directorate of Economics and Statistics, Ministry of Agriculture, Government of India.

**Figure 1.3:** Simple Annual Increase in Area, Production and Productivity of Cotton in Major Indian States, 2002-2011

Figure 1.3 shows the overall trends in area, production and productivity of cotton across the major states. The figures are simple annual average growth rates in area, production and productivity of cotton in the major states during the period between 2002–03 and 2010–11.

### *4.1 Research Issues in Bt Cotton*

By and large, most of the empirical studies on the impact of Bt cotton come to a consensus on three major positive outcomes of the technology. First of all, with the widespread adoption of the technology, India's cotton sector could overcome the historic crisis of low production and productivity. In this context, the emergence and wider diffusion of Bt technology imparted tremendous amount of dynamism to the country's cotton production sector, as the area under cultivation, production and productivity of cotton had increased substantially between 2002 and 2012.

The second major outcome of the technology adoption has been that India's relative contribution to the global cotton production significantly increased from hardly 13 per cent in the early 2000s to more than 20 per cent by the close of the decade (Baffes 2011). In fact, India's cotton production has surpassed that of the US, relegating the latter to the third position in the global cotton production. As a result, India turned as a net exporter of cotton with a consistent increase in India's cotton exports and even a reduction in imports in recent years.

The third important outcome of the diffusion of Bt technology in India was the almost universal and complete adoption of the technology across the major cotton growing regions without any significant difference in the scale of adoption among the different size classes of farmers. To that extent, one could even argue that the technology has shown the attributes of 'inclusiveness', though the economic performance as well as income gains due to the technology appear to be significantly determined by the resource endowment status of the farmers and the cotton growing regions, as revealed by a large number of empirical studies within the Indian context (Lalitha *et al.* 2009: 135-167, Rao and Dev 2010, Herring and Rao 2012: 45-53).

The income gains realising from the improved economic performance of the Bt cotton are considered to be substantial (Pray and Nagarajan 2010). It also clearly emerges from a large number of empirical studies in the Indian context that have analysed the resource use and net returns in cotton cultivation like Pushpavalli (2004), Gandhi and Namboodiri (2007: 56-76 and 2010: 39-62), Rao and Dev (2010), Herring and Rao (2012), and in particular, that the income gains are more than the costs incurred, increasing the net returns for the farmers. It is also essential to know whether the economic gains have also been uniform in all farm size classes or has it resulted in increased income disparities among the adopters and non-adopters. All these positive outcomes provide encouraging signals of continued economic dynamism of India's cotton production sector in the near future, unless major crisis happen in the global market. India's cotton sector also seems to be enjoying a comparative advantage in terms of cost of production in the international scenario as revealed by the relatively lower cost of production *vis-à-vis* other major cotton producing countries, including USA, China, Mexico

**Table 1.10:** Cost of Cotton Production across Major Countries (US$/ha)

| *Country* | *Pre-sowing* | *Sowing* | *Growing* | *Harvesting* | *Ginning* | *Economic* | *Fixed* | *Total* |
|---|---|---|---|---|---|---|---|---|
| India-North zone (Irrigated) | 212.96 | 71.32 | 196.91 | 64.24 | — | 73.58 | — | 619.01 |
| India-Central zone (Irrigated) | 108.59 | 119.35 | 190.98 | 62.39 | — | 66.32 | — | 547.63 |
| India-Central zone (Rainfed) | 76.65 | 80.98 | 156.02 | 45.73 | — | 48.48 | — | 407.86 |
| India-South zone (Rainfed) | 121.74 | 109.78 | 203.00 | 63.75 | — | 67.27 | — | 565.54 |
| Pakistan (Punjab) | 235.95 | 165.62 | 429.07 | 82.62 | 63.36 | 73.72 | — | 1050.34 |
| Australia, Irrigated upland | 44.76 | 48.96 | 670.72 | 122.4 | 367.88 | 986.85 | 13.29 | 2254.86 |
| Israel, Upland/Pima | 169.00 | 544.00 | 1220.00 | 747.00 | 380.00 | 320.00 | 570.00 | 3950.00 |
| USA (national average) | 117.67 | 116.31 | 554.71 | — | 147.30 | 158.02 | 253.30 | 1347.31 |
| China (Mainland) | 601.90 | 385.08 | 410.41 | 34.50 | 68.72 | 181.60 | 181.60 | 1863.81 |
| Brazil (Cerrado) | 195.57 | 278.62 | 676.50 | 109.84 | 100.96 | 142.53 | 85.74 | 1589.76 |
| Mexico (Sonara) | 142.36 | 373.19 | 773.54 | 289.52 | 97.82 | 327.95 | — | 2004.38 |
| Peru (Central coast) | 682.86 | 97.89 | 647.71 | 304.97 | 167.96 | 199.2 | — | 2100.59 |

*Source*: Cotton Statistics at a Glance, (GOI 2010).

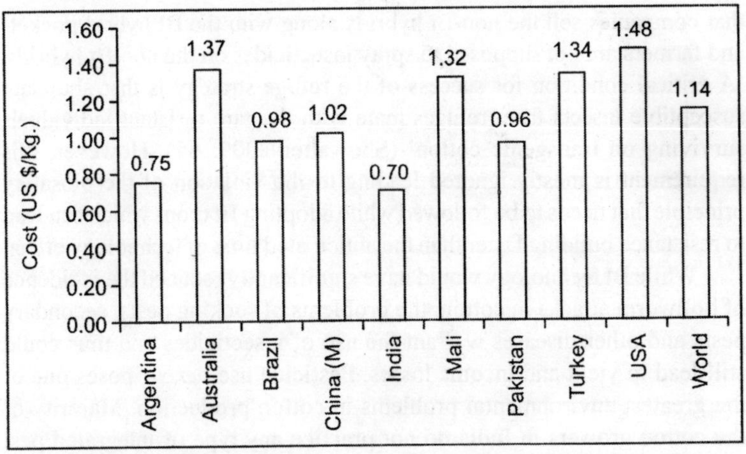


*Source*: ICAC, 2003–04.

**Figure 1.4:** Net Cost of Lint Cotton Production across Major Countries (US$/kg).

and Pakistan as evident from Table 1.10. India's comparative advantage in cotton production is also evident from Figure 1.4, which presents the net cost of lint cotton production across countries as reported by the ICAC (International Cotton Advisory Committee) for the year 2003–04.

As Bt technology has been introduced in hybrids, it would warrant the use of critical inputs, such as fertilisers, pesticides and irrigation which are demanded not individually but as a package, because use of one input determines the need for other inputs. It is quite likely that the input-intensive agro-management regime as being widely followed by the farmers in the post-Bt scenario might affect the comparative advantage of cotton producers in India in the near future.

Despite cotton being a valuable cash crop for small and marginal farmers in particular, Bt cotton cultivation in India is beset with several challenges, including environmental and social problems. The environmental implication of cultivating GM cotton is still unknown. In the context of India, in particular, it is not very clear how the environmental externalities (positive or otherwise) of Bt technology that emerge from the field are continuously monitored and assessed in the country to provide feedbacks to the farmer. Scientifically, Bt cotton is recommended to be grown with the refuge plants that consist of non-Bt hybrids of the same Bt hybrids. The refuge strategy is commonly used to delay the evolution of pest resistance to Bt toxins in transgenic cotton. Hence, it is recommended that companies sell the non-Bt hybrids along with the Bt hybrid pockets and farmers are not supposed to spray insecticides on the non-Bt hybrids. 'A critical condition for success of the refuge strategy is that abundant susceptible insects from refuges mate with the rare resistant individuals surviving on transgenic cotton' (Showalter 2009: 65). However, this requirement is mostly ignored leading to the violation of the biosafety principle that needs to be followed while adopting Bt crops which can lead to resistance building faster than the anticipated time of technology effect.

While Bt technology would have significantly reduced the incidence of bollworm attacks in cotton, the problems of sucking pests, secondary pests and other diseases warrant the use of insecticides and thus could still lead to yield and income losses. Pesticide use *per se*, poses one of the greatest environmental problems in cotton production. Majority of the cotton growers in India do not practice any type of integrated pest management (IPM) or insecticide resistance management (IRM) practices. Hence, pesticide use among Bt cotton growers is widely prevalent. The widespread use of highly toxic pesticides is a major concern both for the natural environment and for the workers who are exposed to them while applying these pesticides and fertilisers. Intensive and inefficient irrigation techniques, poor cultivation practices, excessive and improper use of pesticides and misuse of fertilisers threaten the availability of clean water, soil fertility, biodiversity and human health. Fertilisers and pesticides enter the water supply through run-offs and leaching into the groundwater table and pose a hazard to human, animal and plant population. Some of these chemicals include several substances considered extremely hazardous by WHO (World Health Organisation) and are banned or under strict control in developed countries. Studies on the River Ganga indicate the presence of chemicals such as HCH, DDT, endosulfan, methyl malathion, malathion, dimethoate, and ethion in levels greater than those recommended by the international standards (cited in GOI 2009).

With any new technology, the implication on labour use is important for a surplus labour economy like India. Hence, when the Bt technology was introduced, there were questions by anti-Bt activists on Bt cotton's impact on the labour use in different operations, like whether it would reduce the need for manual labour in weeding (Sahai and Rahman 2003). As Bt technology is logically intended to reduce the pesticide use in cotton, it is expected that the labour used for the same would decline after technology adoption. Similarly, by increasing the yield (by avoiding boll damage and restricting the insects, it is expected that the labour use for harvesting would increase. However, if the technology results in reduced use of labour, but increased income for the farmers, it would not benefit the landless labourers and other casual labourers dependent on cotton cultivation, and would result in loss of livelihood opportunities and severe income disparities.

Besides these issues surrounding Bt cotton, agri biotech research in India suffered a double whammy when, in August 2012, the Parliamentary Standing Committee on Agriculture (PSCA) (GOI 2012) in its report had recommended among other things, an immediate ban on any field trials on transgenic crops. The Chairman of the Committee Basudeb Acharia said that 'the committee has come to the conclusion that since concerns on the potential and actual impacts of GM crops to our food, farming, health and environment are valid, GM crops are just not the right solution for our country'.

The next was the interim report of the Technical Expert Committee (TEC) appointed by the Supreme Court of India, which was set up in response to the public interest litigations filed by anti-GM groups challenging the current regulatory practices in India. The TEC has recommended a ten year moratorium on all field trials in India to revamp the regulatory process followed by India (Chaturvedi and Srinivas 2013: 19–23). The final report from the TEC which came out in July 2013 has also recommended a ten year moratorium and industry sources point out that already states have become wary of providing green signals for the field trials of GM crops. Thus a blanket ban on GM research would adversely affect the current research that is pursued both in the public and private sector and divert the research attention in the future.

There are also some concerns raised about the adverse socio-economic impacts in the cotton seed production in terms of arduous working conditions (particularly for women workers), child labour and forms of forced or bonded labour, as well as indebtedness due to high input prices, crop failure, delayed payment and/or unaffordable rates of interest. The incidence of pesticide poisoning has been one of the major social problems widely reported from cotton growing areas in recent years. In fact, these issues are yet to receive adequate attention.

It is in the context of these issues that this volume, if not to provide an all-encompassing analysis of the whole gamut of the issues surrounding diffusion of Bt technology, is a modest attempt to provide a snapshot of India's tryst with Bt cotton, the learnings and experiences of the first decade of its adoption.

Some of the critical issues that have been addressed in this book are:

1. Has India laid out appropriate regulatory framework and

institutions to implement the regulatory procedures to facilitate adoption of all types of GM crops? What are the governing principles of the Indian regulatory framework? As agriculture is a state subject, what sorts of consultation processes exist in involving the various states and the different stakeholders?

2. What has been the contribution of the public sector research institutions with reference to cotton development in the country, so far? What is the focus of the public sector research particularly with reference to genetically modified organisms? Does it overlap with the private sector's focus?
3. Who are the adopters of the technology and how do the farmers decide on a particular technology to be adopted? Have there been differences in the intensity of adoption among different size classes of farmers? What have been the information gaps identified in the context of the diffusion of Bt technology?
4. How has Bt cotton adoption impacted the livelihoods of the heterogeneous farmers in the major cotton growing states? Is there uniformity across empirical studies in the results obtained or is there any divergence in the conclusions arrived at by these studies?
5. How has the price control policy on the Bt seeds impacted the adoption trends among the farmers and how has it impacted the profitability of the technology providers?
6. Has Bt technology been able to reduce the use of pesticides and fertilisers in cotton cultivation? What is the role and importance of integrated pest management practices in Bt cotton cultivation across states?
7. How has the Bt technology impacted the labour use practices in cotton cultivation? Have there been any changes in the demand for labour due to Bt technology in cotton? Did the use of Bt hybrid seed have any impact on the prevailing agricultural wages?
8. Does the development of Bt technology result in better environmental as well as resource management outcomes in the face of growing crisis in natural resources and climate change induced uncertainties?
9. From the experience of Bt cotton, what lessons could be learned so as to draw a roadmap for the widescale promotion of the agri biotechnology in India in the future without compromising on

the generic concerns arising from food, livelihoods and the environmental securities?

**Organisation of the Book**

The present volume tries to shed light on some of the critical issues arising from the development and diffusion of Bt technology in the Indian context in particular. Chapter 2 by V. Kumar presents the historical development of cotton research in the country since the pre-independence period to the present times and highlights the achievements of the public sector in cotton research and developments in plant biotechnology. This chapter provides the fundamental knowledge on the foundation seed production process and contribution of the various public sector research institutions. Some of the varieties introduced by the public sector have been grown for fifty years in a row indicating their versatility and sturdiness. Besides the varietal development, the public sector has contributed in developing production technology which included nutrient requirement, crop geometry, sowing time, irrigation management, weed control, intercropping, micronutrients, use of plant genetic resources, integrated pest management for sucking pest and bollworms, pest and disease monitoring and forewarning and their management, evaluation of new chemistries, insecticide resistance management and many more, both for conventional cotton as well as *Bt* cotton.

Chapter 3 by A. Indira analyses the regulatory regime that consists of a three-tier structure—the RCGM under the Department of Biotechnology (DBT), the GEAC under the Ministry of Environment and Forestry (MoEF), and the MEC under the DBT. The DBT provides the secretariat for the RCGM and the MEC, and the MoEF for the GEAC. With regards to the functioning of the regulatory regime, Indira identifies two distinctive phases: the first up to 2002 and the second phase that began in 2005. Following a public interest litigation highlighting the limitations in the regulatory regime and the subsequent intervention of the Supreme Court, there was a serious relook at the functioning of the regime. At a broad level, the regulatory structure has remained the same, though many changes have been adopted, shaped by the experience gained while adopting Bt-cotton in various regions of India. This Chapter sets the context of the biosafety regulatory regime in India and looks at the institutional framework for biosafety that has evolved over time.

Chapter 4 by K.R. Ashok, Uma, Prahadeeswaran and Jeyanthi examines the technology impact of Bt adoption in Maharashtra, Gujarat,

Andhra Pradesh and Tamil Nadu and focuses on pesticide use, yield and income, profitability, distributional impact and environmental consequences. Information was gathered from 480 cotton farmers from the four states using a multi-stage sampling procedure. The data were analysed through the Analysis-of-Variance model, Decomposition model, Economic Surplus model and Environmental Impact Quotient method. The study revealed that the impact of Bt technology on increase in yield of cotton was statistically significant in all states. The expenditure on pesticides in Bt was less than non-Bt and statistically significant in all states. Decomposition analysis revealed that 17 per cent of the increased yield was due to Bt technology and 15 per cent of the increased yield was due to higher input use. Under relatively inelastic supply assumption, producers benefit more from the technology than the consumers.

Chapter 5 by Sarthak Gaurav and Srijit Mishra points out that the significant acreage increase under Bt cotton lead to discussions that focused on increments in productivity and income, without much analysis on risk. The authors argue that claims on productivity gains seem to be misplaced, as appropriate counterfactuals do not exist for the same hybrid varieties. This chapter analyses production costs and crop incomes in drought years to test a simplistic theory of risk based on first principles. Employing a mixed-methods framework to draw inferences by combining data from two cross-sectional surveys in Gujarat (Saurashtra and Southern-Plains) and Maharashtra (Western Vidarbha) for the period 2009-10 and compare it with unit-level data for the corresponding regions from a nationally representative sample for the period 2002-03. The analysis brings out the problem of how a high cost technology could be associated with higher risks and may be dominated by traditional alternatives under certain conditions. The authors also use ethnographic accounts from the field to provide qualitative support to their analysis.

Chapter 6 by Anchal Arora and Sangeeta Bansal analyses the effects of price control on the Bt seeds. They examine the impact of such price controls on the diffusion of Bt cotton in India and also the revenue and profitability of the seed providers. Using a panel data for nine cotton growing states in India over 2002-2008, the authors have developed a dynamic logistic model to predict percentage area under Bt cotton. It is shown that the price controls have positively impacted the diffusion of technology in India and were also successful in increasing the revenue of the seed providers in the short run. They also examine the impact of seed price control on profitability of the seed providers by postulating alternative cost conditions and find that the impact of price controls on profitability would depend on the cost of providing Bt cotton seeds.

Chapter 7 by N. Lalitha and P.K. Viswanathan focuses on the seeds and pesticides selection process by the cotton growers in Gujarat. Using information from two farmer surveys (conducted in 2007-08 and 2010-11) and one dealer survey (2007-08), this chapter analyses the selection and use of seeds and pesticides. Both the seed and the pesticide markets provide a variety of choice to the farmers. Farmers adopt a variety after observing it in their field which reflects the experimentation process with a variety. While dis-adoption and re-adoption of varieties is common, it is evident that farmers experiment with a variety for two or three years. Longer adoption cases are very few. Intensity of Bt adoption is higher for small and large farmers than medium size land holders. Regarding plant protection, it was evident that though farmers do not go for blanket spraying, the intensity of sprays by all size groups of farmers was higher in the initial two–three months of plant life than the later period. Overall, the reduction in the pesticide use targeted for bollworm is offset by the requirement to spray for sucking pests.

In Chapter 8, Kamal Vatta and R.S. Sidhu examine the current labour use scenario in cotton in the light of rising wages and decline in the supply of labour over time, using the primary data that was collected from four districts of Punjab. The chapter reports that the cost of labour use accounted for 48 per cent of the cost of cultivation of cotton in 2009–10 and registered an increase by about 65 per cent over 2004–05 owing largely to a sharp increase in the wage rates during this period. Cotton picking and manual weeding emerged as two major components of labour use in crop with the respective shares of 55 per cent and 27 per cent. The extent of labour shortages varied between 20 per cent and 35 per cent for manual weeding and cotton picking and the medium and large farmers were more vulnerable to such shortages. The current situation of labour use and supply in cotton points towards the possibility of further rise in wage rates in the future according to the author. The net returns from cotton cultivation will decline if there is a little rise in cotton prices but a sharp rise in the wages. It may result into a significant long-term change in the pattern of labour use in the crop and may also encourage the mechanisation of important labour-intensive operations such as cotton picking and weeding.

Chapter 9 by P.K. Viswanathan and N. Lalitha critically reviews the development and diffusion of Bt technology in India in the broader theoretical perspective of technology induced sustainable environmental management outcomes. In this regard, the chapter takes stock of the status of implementation and adoption of the state-of-the-art Integrated Pest Management (IPM) as well as Insecticide Resistance Management (IRM) among the major cotton growing states in the post-Bt scenario. The chapter brings out the apparent contradictions in the adoption of Bt technology in India *vis-à-vis* other major cotton growing countries, such as the US, where adoption of Bt technology has been strictly regulated with respect to compliance of IPM/IRM as well as refuge management practices. Even in the states of Maharashtra and Punjab where IPM/IRM practices are being adopted by Bt cotton farmers, it was observed that participation in the programmes is quite voluntary with negligible adoption by small and marginal farmers. The lack of extension support systems along with knowledge intensity of the IPM/IRM interventions make them less popular among the farmers who are constrained by lack of education and basic skills to adopt the scientific ways of farm management prescriptions as laid out in the programme. Based on a panel data of 66 farmers over three time periods (2003-04, 2007-08 and 2010-11), the chapter also examines the interface between Bt technology adoption and the agro-ecological management practices of the farmers in Gujarat. Based on the emerging scenario of resource intensive way of Bt adoption which will have potential implications on the natural resources and environment, the Chapter brings out the imperative of streamlining policies and programmes (including broad-basing of IRM/IPM) as well as evolving/ revamping regulatory/ institutional interventions for effecting sustainable environmental management outcomes in the post-Bt context.

The concluding chapter by the editors discusses the future scenario of agri-biotechnology in the country as emerging from the different studies and indicates the directions in which policy makers' attention is urgently required.

## Endnotes

1. In this chapter and in the book in general, genetically modified organisms and transgenic are used interchangeably.
2. www.isaaa.org
3. www.ISAAA.org, accessed on 13 March, 2013.
4. *Biospectrum*.2012b. 'Cabinet Approves Agri-Biotech Institute'. September, Vol. 10, No. 9: 14.

5. In view of the moratorium recommended by the Technical Expert Committee, it is not very clear whether the approvals already given would be allowed to continue their field trials.
6. The specific roles of these institutions are discussed in Chapter 3 of this book. We acknowledge Dr. Kumar's help in constructing the organogram.
7. Compiled from www.CICR.org.in on 22 March, 2013.
8. Compiled from 'Events Approved in India for Commercialisation' in 'Beyond the Boom—10 Years of Bt Cotton', *Biospectrum*, April, Vol. 10, No. 4: 15–27.

## References

Alam, Ghayur. 1996. 'Biotechnology, Agriculture and Developing Countries'. *Economic and Political Weekly*, Vol. 31, No. 12: 703-704.

Baffes, John. 2011. 'Cotton, Biotechnology, and Economic Development'. Policy Research Working Paper 5896, The World Bank, December. Available at: http:/econ.worldbank.org.

Barik, Anupam and H.C. Gautam. 2009. *Revolution in Indian Cotton*. Directorate of Cotton Development, Government of India, Mumbai and National Centre of Integrated Pest Management, Indian Council of Agriculture Research, Pusa Campus, New Delhi.

Biospectrum. 2012a. 'Beyond the Boom—10 Years of Bt Cotton'. April, Vol. 10, No. 4: 15-27.

Biospectrum.2012b. 'Cabinet Approves Agri-Biotech Institute'. September, Vol. 10, No. 9: 14

Chaturvedi, S. and Krishna Ravi Srinivas.2013. 'Genetically Modified Crops: Policy Logjam'. *Economic and Political Weekly*, XLVIII, No. 16: 19-23.

Dadlani, N.K. 2012. 'Seed Scenario in India'. *Indian Journal of Fertilisers*, 8 (11): 14-21, November.

Dantwala, M.L. 1947. *A Hundred Years of Indian Cotton*. The East India Cotton Association Ltd., Bombay.

FAO. 2011. *Biotechnologies for Agricultural Development*. Food and Agriculture Organisation (FAO) of the United Nations, Rome.

Gandhi, Vasant and N.V. Namboodiri. 2007. 'Cotton Revolution in Gujarat: The Impact of Bt Technology'. In Ravindra Dholakia (ed.), *Frontiers of Agricultural Development in Gujarat*, Centre for Management in Agriculture, Indian Institute of Management, Ahmedabad, pp. 56-76.

Gandhi, Vasant and N.V. Namboodiri. 2010. 'The Economics and Contribution of Cotton Biotechnology in the Agricultural Growth in Gujarat'. In Ravindra Dholakia and Datta K. Samar (eds), *High Growth Trajectory and Structural Changes in Gujarat's Agriculture*, Centre for Monitoring Agriculture, Publication No. 234, Indian Institute of Management, pp. 39-62.

GOI. 2009. *State of Environment Report, India*. Ministry of Environment and Forests, Government of India.

GOI. 2011. *Faster, Sustainable and More Inclusive Growth*. Planning Commission, New Delhi.

GOI. 2012. 'Cultivation of Genetically Modified Food Crops: Prospects and Effects'. Report No. 37, Fifteenth Lok Sabha, Ministry of Agriculture (Department of Agriculture and Co-operation).

Guha, Ramachandra. 2007. 'Genetic Change and Colonial Cotton Improvement in 19th and 20th Century India'. In Ranjan Chakrabarti (ed.), *Situating Environmental History*. Delhi: Manohar Publications.

Herring, R.J. and Chandrasekhara N. Rao. 2012. 'On the Failure of Bt Cotton—Analysing a Decade of Experience'. *Economic and Political Weekly*, Vol. XLVII (18): 45-53.

Jaffe, S. and J. Srivastava. 1992. 'Seed System Development: The Appropriate Roles for Private and Public Sectors'. World Bank Discussion Paper No. 167, Washington DC.

Lalitha, N., Bharat Ramaswami and P.K. Viswanathan. 2009. 'India's Experience with Bt Cotton: Case Studies from Gujarat and Maharashtra'. In Robert Tripp (ed.), *Biotechnology and Agricultural Development: Transgenic Cotton, Rural Institutions and Resource-poor Farmers*, UK: Routledge, pp. 135-167.

OECD. 2002. 'Genetic Inventions, Intellectual Property Rights and Licensing Practices—Evidence and Policies, Paris.

Musselli, Moretti. 2006. 'Tracking the Trend towards Market Concentration: The Case of the Agricultural Input Industry'. United Nations Conference on Trade and Development (UNCTAD), Geneva.

Pray, C. and L. Nagarajan. 2010. 'Price Controls and Biotechnology Innovation: Are State Government Policies Reducing Research and Innovation by the Ag-Biotech Industry in India'. *AgBioForum*, 13 (4).

Purkayastha, P. and S. Rath. 2010. 'Bt Brinjal Need to Refocus the Debate'. *Economic and Political Weekly*, Vol. XLV, No. 20: 42-48.

Pushpavalli, A. 2004. *Returns of Bt. Cotton vis-à-vis Traditional Cotton in Tamil Nadu*, Agro Economic Research Centre, University of Madras, Chennai.

Randhawa, G. and R. Chhabra. 2009. 'Import and Commercialisation of Transgenic Crops: An Indian Perspective'. *Asian Biotechnology and Development Review*, March, Vol.11, No. 2: 115–130.

Rao, Chandrasekhara N. and Mahendra S. Dev. 2009.'Biotechnology and Pro-Poor Agricultural Development'. *Economic and Political Weekly*, Vol. XLIV, No. 52: 56-64.

Rao, Chandrasekhara N. and Mahendra S. Dev. 2010. *Biotechnology in Indian Agriculture-Potential, Performance and Concerns*. New Delhi: Academic Foundation.

Rao, S.R. 2004. 'Investment, Capacity and Incentives in Biotechnology, Indian Experience' in Sachin Chaturvedi and S.R. Rao (ed.), *Biotechnology and Development: Challenges and Opportunities for Asia*. Published by

Institute of Southeast Asian Studies, Singapore and Academic Foundation, New Delhi.

Showalter, Ann M., Shannon Heuberger, Bruce E. Tabashnic, Yves Carriere. 2009. 'Transgenic Cotton for Insect Control'. In R. Tripp (ed.), *Biotechnology and Agricultural Development: Transgenic Cotton, Rural Institutions and Resource-poor Farmers*. Oxon/New York: Routledge, pp. 49-71.

Tripp, Robert. 2009. 'Biotechnology and Agricultural Development'. In R. Tripp (ed.), *Biotechnology and Agricultural Development: Transgenic Cotton, Rural Institutions and Resource-poor Farmers*. Oxon/New York: Routledge, pp. 1-22.

# 2

# Status of Cotton Development in India

## Role of Public Research Institutions

*V. Kumar*[1]

### 1. Introduction

India, the largest democracy of the world, is an agrarian society. Despite dramatic changes in the last 66 years since independence, Indian agriculture stands alone as the single largest private enterprise with more than 100 million farm holdings that sustain livelihood of about two-thirds of the population. Factors like changes in cropped area, cropping pattern, crop inputs (both in quality and quantity), irrigation facilities, knowledge base of cultivators, widening network of research institutes, high yielding varieties and hybrids and associated agro-technologies, development of post-harvest processing and food industry, future trading of agricultural commodities, market demand, liberal export and import of goods and commodities, conducive government policies, quota free regime and spread of mass media have transformed Indian agriculture into a vibrant private enterprise although still an unorganised sector.

Public sector institutions have played a significant role in each of these factors and have contributed significantly in the introduction and adoption of modern technology based inputs in agriculture. This chapter focuses on the role of the public sector in the development of cotton in the country. In the second section following this introduction, it provides an overview of cotton development in India from a historical perspective. It also details the characteristics of the different cotton zones of the country. The third section presents the network of institutions set up for cotton development and their role. The fourth section discusses the role played by the public sector in providing the most important input, namely seed and the process of producing seeds meant for release to farmers. The fifth section delineates important contributions made by different institutions in developing popular cotton varieties. The sixth section highlights the public sector's status in plant biotechnology research. The concluding section looks at the future thrust areas of public sector research.

### *1.1 Cotton in India: A Historical Overview*

India has a long and glorious history of cotton, right from the days of prehistoric civilisation to modern day India. History, recorded or unrecorded, bears testimony to the fact that cotton, particularly *G. arboreum* and *G. herbaceum*, are indigenous to this country. The British attempted to introduce *G. hirsutum* cotton in the mid-eighteenth century but were not very successful. Therefore, simultaneous attempts were made to improve Asiatic cotton. India was the main producer of cotton textile with substantial exports during the early modern period. However, due to modernisation and mechanisation, the British industry was able to produce textile at a cheaper rate and posed a threat to handlooms produced by weavers and artisans in India. The great famine of the early nineteenth century led to the closure of several mills in Mumbai and Ahmedabad and the textile industry was badly crippled. However, the Swadeshi movement and World War II provided the much needed impetus to the Indian textile industry and the industry was not only able to get back on track but also grew handsomely.

In pre-independent India, a lot of developmental research took place as far as *herbaceum* and *arboreum* cotton are concerned but these were not very suitable for mills in undivided India and Britain; therefore, the focus was mainly on *hirsutum* cotton. Some American varieties found suitable in Punjab, Karnataka, Uttar Pradesh and Tamil Nadu were either introduction or selections from material imported from Malta, Mauritius, New Orleans (USA) and Cambodia as early as 1790 and afterwards. Egyptian cotton or Sea Island *Gossypium barbadence* was also introduced in 1790 but it took 150 years to adapt to Indian conditions.

When India gained independence, 2.3 million bales (each bale is 170 kg lint) comprising 67 per cent medium staple and 33 per cent short staple cotton was produced from *desi* varieties cultivated in 97 per cent of the area (65% *G. arboreum* and 32% *G. herbaceum*) occupying 4.3 million hectares. The cotton mills remained in India and the regions that were suitable for long staple cotton went to Pakistan. Therefore, efforts were intensified to bring more area under quality cotton. By 1965, 40 per cent of the area was under *G. hirsutum*, 36 per cent under *G. arboreum* and 24 per cent under *G. herbaceum*. Despite intense and concerted efforts to acclimatise and promote American cotton in India, the two *desi* species continued to occupy at least 25 per cent of the area (18 per cent of *G. arboreum* and 7 per cent of *G. herbaceum*) until 2000-01 (Kranthi *et al.* 2011). However, after the introduction of Bt cotton in 2002, *G. hirsutum* cotton currently occupies nearly 95 per cent cotton area in the country.

## 2. History of Cotton Development in India

Cotton *(Gossypium)*, belonging to the family Malvaceae, is reported to have 50 different species, of which four are cultivated. India has the distinction of being the only country in the world to grow all the four cultivated species, viz. *G. hirsutum*, *G. barabadence*, *G. herbaceum* and *G. arboreum*. The former two are tetraploid species (2n = 52) known as upland/new world cotton. The latter two species are diploid cotton (2n = 26) and also known as old world or Asiatic/desi cotton. Each of the species has its own characteristics.

The *hirsutum* cotton presently account for about 90 per cent of total global production. They have relatively high productivity, wide adaptability and wide variations in yield, quality and pest susceptibility and need lot of inputs for cultivation.

The *G. barbadence* is cotton generally known as Egyptian cotton and also as Sea island/Peruvian cotton. The Egyptian cotton is usually extra long, fine and strong fibred. These cottons are more susceptible to biotic and abiotic stresses than upland cottons, low yielders but high quality in produce.

The *G. arboreum* is grown in tropical and subtropical regions of Asia. The species is a native of India and is cultivated here since time immemorial. *Arboreum* cotton is usually short duration, suitable even for shallow marginal soils, coarse and short fibred and low yielder. It has a high degree of resistance to pest, diseases and drought stress.

*G. herbaceum*, also known as levant cotton, is grown in India since time immemorial. This cotton has wide adaptability and high degree of resistance to abiotic and biotic stresses. It has a deep root system, long duration and relatively poor quality fibre and low yield.

In addition to these species, inter-specific (between two different species) and intra-specific (within the species) hybrids are also cultivated in India for advantage of yield and/or quality or other specific traits developed through exploitation of heterosis by hybridisation.

The history of different groups of cotton cultivated in India till around 1950 is briefly narrated in the following paragraphs.

### 2.1 Asiatic Cotton

Improvement of *herbaceum* cotton started at Surat (1896), Dharwad (1904), Bharuch (1913) and Viramgam (1922). Dharwad-I (1918) was the first improved variety developed by selection from Kumpta. Later, Dharwad-2 was also released. In Gujarat, a selection from Surat local 1-ALB[2] (1919) was released which was high yielding with high GOT (ginning out turn). It was followed by 1027 ALF (1923) a cross between Kumpta and Goghari. In Karnataka, *var.* Jaywant (Dharwad-1 × Dharwad-2) was released in 1930. Later, a high yielding and drought tolerant resistant variety Jayadhar (1950) was released which is still grown in some parts of Karnataka. For North Gujarat, a variety Wagad-8 (1933) was released but had low spinning value and was followed by Wagotar (1942) and Kalyan (1947). For central Gujarat BD-8 (1938), a selection from Bharuch local was released followed by Vijay (1943) and Digvijay (1951). The later is grown in parts of central Gujarat even now. In South Gujarat *var.* Suyog (1945) a cross between 1027 ALF × 1-ALB was released followed by Vijalpa (1951).

The *arboreum* cotton grown in south Saurashtra known as Mathio cotton was in fact Khandesh jadi mixture which along with Wagad (*G. herbaceum*) used to be called Dholleras by the trade after the name of the port from where it was exported. Several *arboreum* strains like *N. Roseum*, *Jarilla*, *Benilla*, *Verum* and *Cawnpore* 520 were tried in Saurashtra. Of these, *Cawnpore* 520 (C-520) was distributed for cultivation in 1942. Further selection led to the release of an improved variety Pratap in 1945. A cross between C-520 and Jarilla lead to the development of C-J73 known as Sanjay. Similar developments took place in other parts of the country. The development of Asiatic cotton went on the backfoot once *hirsutum* varieties became successful, especially the hybrids.

### 2.2 American Cotton

The American cotton *var.* Bourbon *G. hirsutum punctatum* race was introduced from Malta and Mauritius into Bombay State and Madras Presidencies of India in 1790. Subsequently, efforts were intensified to improve and promote the cultivation of American cotton in India. The British started putting efforts in improving American cotton in India to suit their mills. *G. hirsutum* (*latifolium* race) *var.* New Orleans was introduced in Gujarat, Deccan and Konkan in 1840. By 1862, *G. hirsutum* (Bourbon) was grown successfully in Hubli which was not so in Punjab. In 1905, *G. hirsutum var.* Cambodia was introduced successfully in Madras Presidency. By this time, 'Punjabi Narma' (Bourbon) was found to have established itself in parts of Punjab. Experiments in Uttar Pradesh starting from 1826 established the American cotton variety in 1909. In 1912, Milne selected a variety 4F from Punjabi Narma which was resistant to jassids. Sardar Labh Singh developed a jassid resistant late variety LSS from 4F in 1933. In 1920–29, Hilson and Ramanatha Iyer released Co.1 and Co.2 cotton from Cambodia in south India. Simultaneous attempts to cultivate *hirsutum* were not successful in Gujarat. Interspecific hybridisation of *G. hirsutum* and red *arboreum* lead to the first successful *G. hirsutum* variety Deviraj (1951) having the background of Asiatic cotton.

### 2.3 Hybrid Cotton

The history of hybrid cotton in India is almost a century old. In the early nineteenth century, efforts were basically made to improve Asiatic cotton which yielded some improved varieties. With the success of the introduction of *hirsutum* cotton in some parts of the country, efforts began in the early thirties of the last century to improve *hirsutum* varieties through hybridisation as also to exploit heterosis. Surat (a part of the then Bombay state) used to be main centre of cotton improvement. The first such hybrid (Hybrid-1), a cross between Co.2 (*G. hirsutum*) and Sea Island (*G. barbadense*), was made in 1951. However, it could not be successful. The second Hybrid-2 was developed in 1960 (BC-68 x SIV 135- *G. hirsutum*). However, this too could not reach the farmers. Hybrid-3, a cross between Gujarat-67 and Moco (a perennial cotton), released in 1963, also could not find a footing amongst farmers. The efforts of the Late Dr C.T. Patel finally succeeded in the release of the world's first commercially successful hybrid cotton, Hybrid-4, in 1971. The hybrid was a cross between Gujarat-67 and American Necteriless-*G. hirsutum* variety. The first parent, the female of the Hybrid-4 is a *G. hirsutum* variety developed by interspecific hybridisation of red *arboreum* (*G. arboreum*) and Co.2 (*G. hirsutum*) which has high photosynthetic efficiency, luxuriant plant growth with good fibre quality and big boll with limitations of long duration and low yield (because of poor harvest index). The other parent, the male is a short duration compact plant type with very good translocation efficiency (harvest index). The product Hybrid-4 combined all the desirable traits of the two parents and that is why it became successful. This hybrid opened new vistas in agriculture. It helped in generating employment among village folks engaged in seed production. It also boosted the fertiliser industry, as it responded to fertilisers, the pesticides industry, as it needed pesticides to control pest, the ginning and processing industry, as it got a fillip due to ample supply. In fact, the success of Hybrid-4 changed the entire scenario of cotton research, cotton cultivation related industry and textile sector in India. Almost simultaneously, Dharwad also succeeded in releasing first extra long staple (ELS)—more than 32.5 mm 2.5 per cent span length *hirsutum* × *barbadense* hybrid Varlaxmi in 1972. The world's first genic male sterility (GMS) based hybrid Suguna was released by CICR (1978). The first cytoplasmic genic male sterility (CGMS) based hybrid PKV Hy-3 was released by PDKV, Akola. Surat was again the first in the world to have released the inter-specific desi hybrid (h × A)$^3$ G. Cot. DH 7 in 1984. These opened a plethora of hybrids in public institutes as also in the private sector, which changed the entire cotton scenario in the country.

## 3. Contribution of the Public Sector in Cotton Research

Three research accomplishments are considered as milestones of Indian agriculture. The first was the introduction of Norin-10 Brevor-14 dwarfing gene in wheat, which led to the green revolution in the mid-sixties. The second was the development of the world's first successful hybrid of cotton, Hybrid-4 which led to the white revolution in the early seventies and the third was the introduction of Cry gene (Bt) in cotton which is considered as the beginning of the gene revolution in the early years of the current century (Kumar 2011).

The cotton research programme in India had been systematically pursued since the early years of nineteenth century, first with the establishment of the Department of Agriculture in the states in 1904. Further strengthening was done through the Project for Intensification of Research on Cotton, Oilseeds and Millets (PIRCOM) and subsequently, through the Indian Central Cotton Committee (ICCC) in 1923 (Narayanan *et al.* 2011). In post-independent India, many projects were awarded specific to cotton development that were to be implemented through central and state government agencies. The real fillip to the programme came after the inception of the All India Coordinated Cotton Improvement Project (AICCIP) in 1967. This was the first unified effort in a networking mode with the objectives to develop high yielding varieties and hybrids, crop husbandry for the same, seed production and extension activities. Subsequently, the research programmes got strengthened under the Intensive Cotton Development Project (ICDP), and various research scheme/projects by the Indian Council of Agriculture Research (ICAR) through the Hybrid Research Project, National Agricultural Research Project (NARP), National Agricultural Technology Project (NATP), Technology Mission on Cotton Mini Mission-I (TMC-MM-I), National Agriculture Innovation Projects (NAIP) and Rashtriya Krishi Vikas Yojana (RKVY) of the Ministry of Agriculture. Several other projects/schemes funded by various other government agencies/departments supported cotton research in the country.

Cotton in India is grown under varied agro-climatic environments. It is well known that cotton is sown and picked around the year in some or the other part of the country. The cultivation in the country is divided into three zones, viz. North Zone, Central Zone and South Zone. The major characters of each zone are described in Table 2.1.

Table 2.2 presents the trends in area, production and productivity of cotton during the 10th and 11th Five Year Plans, which may be regarded as the golden period of India's cotton sector as the Bt technology received wide-scale adoption across states, irrespective of the notable differences in agro-climatic characteristics as described above.

### 3.1 Network of Cotton Research Institutions in India

Cotton research in India is carried out through a well-knit system of research institutes/centres spread across the country. The entire cotton research programme at the national level is carried out under the aegis of the Central Institute for Cotton Research (CICR) located at Nagpur. The Institute has two regional stations, one at Coimbatore (Tamil Nadu) in the South and other at Sirsa (Haryana) in the North. The Institute is the hub of cotton research in the country which undertakes basic and fundamental research in genetics, breeding, biotechnology, physiology, biochemistry, entomology, pathology, bio control and IPM (Integrated Pest Management), bio pesticides, new gene discovery for biotic and abiotic stresses, agronomy, soil science, climate change, fertiliser and water use efficiency, extension and economics, and transfers the findings to the SAUs for application at the field level.

**Table 2.1:** Profile of Cotton Growing Zones in India (By Zones)

| *Characteristics* | *North Zone* | *Central Zone* | *South Zone* |
|---|---|---|---|
| *States covered* | *Punjab, Haryana, Rajasthan and Western Uttar Pradesh* | *Madhya Pradesh, Gujarat, Maharashtra, Orissa and South East Rajasthan* | *Andhra Pradesh, Karnataka and Tamil Nadu* |
| Soil type | Alluvial | Mostly black to medium black, also sandy loam | Black cotton to red laterite |
| Topography | Plain | Plain and undulating | Plain and undulating |
| Rainfall pattern | 300-600 mm, precipitation in 30 to 35 days | 300 to 1200 mm, precipitation in 35 to 55 rainy days | 300-1500 mm, precipitation in 40 to 60 days |
| Irrigation status of cotton grown area | 100 % | Nearly 75%, area is rainfed | 40%, area is irrigated. In rainfed areas, by and large, the distribution of rainfall is quite suitable for cotton. |
| Varieties grown | G. hirsutum, G. arboreum and intra-hirsutum, mostly Bt hybrids. | G. herbaceum, G. arboreum varieties and mostly intra-hirsutum Bt hybrids. | G. herbaceum, G. hirsutum, G. barbadense varieties and intra- and inter-specific hybrids, mostly Bt. |
| Staple characteristics/ | Mostly short and medium staple varieties are grown; lately, long staple Bt cottons. | Short, medium and long staple category and occasionally extra long staple. | Short to extra long superior cotton varieties and hybrids. |
| Sowing time | Around mid-May | In irrigated areas - First fortnight of May; Unirrigated areas: after onset of monsoon, usually third week of June to first week of July. In North West coastal areas of Gujarat, herbaceum cotton is sown in August end to September. | In Irrigated tracts: Mid-April to mid-August; Rainfed areas: from mid-July to mid-October |
| Crop rotation | Cotton-Wheat rotation is common | Intercropping with short duration pulse crops is common. Cotton-wheat rotation is gaining popularity in irrigated areas. | Intercropping with chillies, garlic, onion and coriander is popular in many areas of the zone |
| Contribution of the zone to the country | 14.8% of cotton output from 14.2% area | 61.8% cotton output from 65.1% area | 22.4% cotton output from 20.1% area |
| Risk of pests in the zone | CLCuV and Jassids are serious threat to hirsutum cotton. | Jassids and mealy bugs, of late, have appeared as damaging pests. | Jassids and mealy bugs are major pests. |
| Climatic character | Crop encounters high temperature and wind during early stage of growth and low temperature at maturity. | The crop encounters high temperature and water stress in October during peak flowering/bolling. | Crop often faces water stress during terminal stage. |
| Average yield range | 15-40 qtl/ha | 10-30 qtl/ha | 10-30 qtl/ha |

*Note:* Compiled from different sources.

**Table 2.2:** Trends in Cotton Area, Production and Productivity by States and Zones

| *Major cotton growing states/ Zone* | *Average Area (Million ha)* | | *Average Production (Million bales)* | | *Average Productivity (kg/ha)* | |
|---|---|---|---|---|---|---|
| | *X FYP (2002–07)* | *XI FYP (2007–12)* | *X FYP (2002–07)* | *XI FYP (2007–12)* | *X FYP (2002–07)* | *XI FYP (2007–12)* |
| 1. Punjab | 0.514 | 0.546 | 1.567 | 1.750 | 501.3 | 542.9 |
| 2. Haryana | 0.555 | 0.516 | 1.275 | 1.745 | 388.2 | 567.7 |
| 3. Rajasthan | 0.397 | 0.384 | 0.843 | 1.122 | 364.4 | 488.5 |
| **NORTH ZONE** | 1.468 | 1.446 | 3.685 | 4.617 | 421.9 | 540.0 |
| 1. Gujarat | 1.889 | 2.599 | 6.910 | 10.484 | 602.1 | 686.2 |
| 2. Maharashtra | 2.877 | 3.579 | 3.880 | 7.030 | 228.0 | 333.5 |
| 3. Madhya Pradesh | 0.594 | 0.644 | 1.833 | 1.779 | 524.9 | 469.9 |
| **CENTRAL ZONE** | 5.361 | 6.822 | 12.623 | 19.293 | 412.8 | 481.3 |
| 1. Andhra Pradesh | 0.964 | 1.534 | 2.983 | 5.380 | 524.6 | 601.4 |
| 2. Karnataka | 0.403 | 0.473 | 0.584 | 1.087 | 244.3 | 389.2 |
| 3. Tamil Nadu | 0.111 | 0.113 | 0.435 | 0.554 | 666.9 | 823.6 |
| **SOUTH ZONE** | 1.479 | 2.120 | 4.002 | 7.021 | 468.2 | 564.7 |
| 1. Orissa | – | 0.068 | – | 0.176 | – | 385.7 |
| 2. Other | 0.644 | 0.328 | 0.100 | 0.135 | – | – |
| **INDIA** | 8.380 | 10.490 | 21.58 | 32.97 | 432.8 | 518.1 |

*Source*: Annual report AICCIP 2012–13: A2–A3.

The Central Institute for Research on Cotton Technology (CIRCOT) at Mumbai acts as a link between the textile industry and research institutes, and undertakes basic research on fibre development, ginning and pressing, processing, industrial use of cotton by-products and large scale testing for breeding material developed by the SAUs, CICR and also by the private sector.

The Directorate of Cotton Development, Mumbai is the nodal agency for extension and dissemination of technology developed by the CICR/SAUs, through organising Front Line Demonstrations, IPM, technology transfer and Breeder Seed Production programmes for the states as well as research institutes.

The All India Coordinated Cotton Improvement Project (AICCIP) located at Regional Stations CICR, Coimbatore operates through a network of 12 Agricultural Universities at 11 main centres and 10 sub-centres in the states. Some of the landmark achievements in cotton research by public sector institutes since independence are listed in Table 2.3.

**Table 2.3:** Landmark Achievements in India's Cotton Sector under the Public Sector

| *Year* | *Event Reported* |
|---|---|
| 1950 | *Gossypium herbaceum* variety 'Jayadhar' released (cultivated even today) by Dharwad. |
| 1951 | *Gossypium hirsutum* variety Deviraj with *G. arboreum* background released by Surat. |
| 1963 | First long staple *hirsutum* variety Gujarat 67 released from Surat. |
| 1968 | *Gossypium hirsutum* variety 'MCU' 60 counts released (finest quality *G. hirsutum variety*). |
| 1970 | World's first Hybrid 'H-4' released (highest yield with superior fibre traits) from Surat. |
| 1972 | World's first inter-specific tetraploid hybrid 'Varalakshmi' 80 counts released by Dharwad. |
| 1974 | *Gossypium barbadense* variety 'Suvin' 120 counts released (finest Indian variety ever). |
| 1976 | Bikaneri Narma and its selections F 414 and H 777 identified for adaptability and high yields in North India. |
| 1978 | World's first GMS based hybrid 'Suguna' released from CICR. |
| 1981 | Inter specific tetraploid hybrid DCH 32 with 80 counts released. |
| 1982 | Highly adaptable *G. hirsutum* variety LRA 5166 released. |
| 1983 | World's first inter specific diploid hybrid G. Cot DH 7 released by Surat. |
| 1983 | Highly adaptable hybrid NHH 44 released for rainfed conditions by Nanded. |
| 1988 | First long staple desi hybrid G. Cot DH 9 released from Surat. |
| 1993 | First cytoplasmic genic male sterility based hybrid PKV Hy 3 released from Akola. |
| 1999 | Gossypium arboreum variety AKA 8401 of 40 counts released from Akola. |
| 2008 | Highly adaptable superior fibre variety Suraj released from CICR. |
| 2008 | First Bt-cotton variety, Bikaneri Narma (Cry1AC) released by CICR Nagpur and UAS Dharwad. |
| 2012 | First two BG II Bt-cotton hybrids G.Cot.Hy-6 and G.Cot.Hy-8 from public sector released from Surat. |

*Source*: Kranthi *et al.*, 2011.

## 4. Seed Production, Maintenance and Distribution

The seed is the most vital and critical input in agriculture on which the efficacy of other agriculture inputs depends. Therefore, the quality of seed should meet all the required standards. Its purity and vigour should be maintained at all times; at the same time, enough quantity should be produced through multiplication for distribution to the end users.

### *4.1 Release of a Variety*

Cotton or for that matter ail crop varieties are released after a thorough assessment of their value for cultivation and use under a unique and a most successful Indian model of multi-location testing system in diverse agro-climatic conditions under the All India Co-ordinate Cotton Improvement Project, across the country. For release of a variety, the concerned research station/institute generates data for performance for at least two years at one or more locations and contributes to the national trial in the AICCIP in irrigated and/or rainfed condition. Based on its performance in yield and quality, the entry is promoted to the higher level trial at the zonal level. Again the yield and quality is considered at the zonal level bigger evaluation trial.

The different levels/stages of trials for evaluation of *hirsutum*, *barbadence*, *arboreum* and *herbaceum* varieties and hybrid in rainfed/ irrigated condition are represented in Table 2.4.

**Table 2.4:** Details of Trials for Evaluation of Cotton Varieties and Hybrids

| *Cotton* | *National Trial I year* | *Zonal Trial II year* | *Zonal Trial III year* |
|---|---|---|---|
| *Hirsutum* varieties | Br 02a IET | Br 03a PVT | Br 04a CVT |
| | Br 02b IET | Br 03b PVT | Br 04bCVT |
| H x H hybrids | Br 05a PHT | Br 05a CHT | Br 05a CHT |
| | Br 05b PHT | Br 05b CHT | Br 05b CHT |
| Compact genotypes | Br 06a IET | Br 06a PVT | Br 06a CVT |
| | Br 06b IET | Br 06b PVT | Br 06b CVT |
| *Barbadense* varieties | Br 12a IET | Br 13a PVT | Br 14a CVT |
| H x B/ B x B hybrids | Br 15a PHT (IS) | Br 15a CHT | Br 15a CHT |
| *Arboreum* varieties | Br 22a/b IET | Br 24a PVT | Br 24a CVT |
| | | Br 24b PVT | Br 24b CVT |
| A x A / h x A hybrids | Br 25a/b PHT (DH) | Br 25a CHT | BR 25a CHT |
| | | Br 25b CHT | BR 25b CHT |
| *Herbaceum* varieties | Br 34b IET | Br 34b PVT | Br 34b CVT |

*Note:* * IET- *Initial evaluation trial;* PHT- *Preliminary hybrid trial; PVT- Preliminary varietal trial, CHT – Coordinated hybrid trial; CVT- Coordinated varietal trial ; a-Irrigated; b-Rainfed*
*H-Hirsutum; B-Barbadence; A-Arboreum; h-Herbaceum ;*
*IS-interspecific hybrid;*
*DH-Desihybrid*

*Source*: Annual report, AICCIP, 2012: C1.

Simultaneously, data on agronomic requirement for fertilisers, spacing and date of sowing etc. is generated. Disease and pest reaction of the entry is evaluated at all levels for its suitability for release. Fibre quality analyses for 2.5 per cent span length, strength (g/tex)[4], fineness and maturity is also done. Spinning performance and mill test is conducted. In the meantime, the research stations generate additional data on yield and quality in multi-location trials in the concerned state. The proposal in a prescribed proforma is put before the variety evaluation committee headed by the Assistant Director General ADG (CC[5]), ICAR, Director CICR, Project Coordinator (AICCIP), Director CIRCOT, Director DOCD, Principal Investigators of Plant Breeding, Agronomy, Plant Protection during the Annual Group Meeting of the AICCIP every year. Entries fulfilling required norms are cleared for release.

The research stations then put the performance data of an entry along with/without endorsement of the AICCIP in the Agricultural Research Council Meeting (AGRESCO) of the SAU for consideration for release. Simultaneously, the material along with passport data is submitted to the National Bureau of Plant Genetic Resources (NBPGR), New Delhi for registration as per the requirement of the procedure. Once the proposal is cleared by the AGRESCO, it is sent to the Director of Agriculture of the state for consideration and recommendation by the State Seed Sub-Committee (SSSC) which, if satisfied, sends it to the Central Sub-Committee on Crop Standard, Notification and Release of Varieties for Agricultural Crops (CCCSN & RVAC), New Delhi for consideration and release. Finally, the CVRC issues notification for release of the variety.

For a hybrid to be eligible for release, it should either be significantly superior or yield at least 10 per cent more than the check hybrid. The normal criteria for release of cotton variety is significant superiority or at least 10 per cent more yield than the existing check variety, provided it fulfils required norms of fibre quality. If the proposed entry has any specific trait value or characteristic, then also it is considered.

At times, a variety is released exclusively for a region of/or the state. In such situation, the proposal, after approval of the AGRESCO, is sent for endorsement of the project coordinator of the AICCIP and subsequently, to the SVRC and the CVRC for release and notification.

For making the new variety/hybrid popular, front line demonstrations are conducted by the research station throughout the region/state at farmers' field and performance/evaluation report is submitted to the AICCIP and the state as the final step of the entire process.

### *4.2 System for Seed Production and Supply*

The public sector is entrusted with the supply of quality seed of HYV/ Hybrid released by it. The public sector comprises the Indian Council of Agriculture Research, the State Agricultural Universities (SAUs), the National Seed Corporation (NSC), the States' Farm Corporation of India Ltd. and state-level Seed Corporations. In cotton, it is mostly the CICR, SAUs and some seed corporations, which are involved in seed production and distribution. The indents for breeder seed are placed with the Department of Agriculture and Cooperation (DAC), Government of India, which passes it on to the ICAR. Breeder seed production as per the DAC indent is undertaken by the ICAR Institutes/SAUs who, in turn, distribute to the indenters for production of the foundation and eventually certified seed. At times, the Directorate of Agriculture of the states also accepts the indents directly and passes it on to the concerned SAUs for production of breeder seed. The rates for breeder seed are decided by the DAC. The indenters are required to deposit 50 per cent advance of the cost of seed with the indent one season before, to ensure seed availability in the next season. For hybrids, the parental lines are produced as breeder seed by the ICAR/SAUs which are made available to the indenters for production of the foundation and certified seed.

### *4.3 Maintenance of Seed Purity*

The breeders usually maintain one or more progenies from the individual plant selection by selfing. The yield and quality is assessed for individual progenies. If the variation is within acceptable limits, the progenies are bulked to make the breeder seed. If any progeny is found errant in any traits, it is rejected. Maintenance of the breeder seed was until recently supported by the Nucleus and Foundation Scheme (NFS) of the Ministry of Agriculture, Government of India; now the funding is done by the ICAR through the Technology Mission on Cotton (TMC) Mini Mission-II.

After the release and notification of a cotton variety, its seed is multiplied in sufficient quantity, before it reaches the farmers for commercial use. At the same time, its genetic purity is maintained for future use. There are different categories of seed which a new variety has to go through before certification and distribution.

**Nucleus seed:** The nucleus seed is the initial seed obtained from the selected individual plants of a particular variety or parental line for the purpose of purifying and maintaining that variety or parental lines by the originating breeder and its further multiplication is done under his own supervision or the supervision of a qualified plant breeder. It is used to produce the breeder seed and forms the basis of the total seed production chain.

**Breeder Seed:** It is the seed directly controlled by the originating or sponsored plant breeder or institution. Its production is personally supervised by a qualified plant breeder. It serves as the source for production of the foundation seed. The breeder seed is genetically pure so as to ensure that the subsequent foundation and certified seed conform to the prescribed standards of genetic purity.

**Foundation Seed:** The foundation seed is the progeny of the breeder seed. The production of the foundation seed is supervised and approved by the Central/State Seed Certification Agency and so handled as to maintain specific genetic identity and genetic purity required to conform to the certification standards for the crop that is being certified.

**Certified Seed:** It is the progeny of the foundation seed and so its production is done to maintain specific genetic identity and purity according to the standards prescribed for cotton crop certification.

**Seed Certification:** Seed certification is a legal system which ensures production of high quality seeds in terms of purity and germination; the task of seed certification is undertaken by the State Seed Certification Agency. The requirement for seed certification, in terms of seed purity, genetic purity and germination for different category of cotton seed is shown in Table 2.5.

**Table 2.5:** Seed Standard for Foundation and Certified Seed Classes of Cotton

| *Particulars* | *Foundation Seed* | *Certified Seed* | *Hybrid Seed (Certified)* |
|---|---|---|---|
| Pure Seed % (Minimum) | 98 | 98 | 98 |
| Inert matter % (Maximum) | 2.0 | 2.0 | 2.0 |
| Other Crop Seeds (Maximum) | 5/kg | 10/kg | 10/kg |
| Weed Seeds (Maximum) | 5/kg | 10/kg | 10/kg |
| Genetic Purity % (Minimum) | 100 | 100 | 90 |
| Germination % (Minimum) | 65 | 65 | 75 |
| Moisture Ordinary Container | 10 | 10 | 10 |
| Moisture Vapour Proof Container | 6 | 6 | 6 |
| Off type (%) | 0.10 | 0.20 | 1.5 |
| Self plant (%) | – | – | 8.5 |

*Source*: Handbook of Agriculture (2006).

### *4.4 Seed Production Methodology*

Seed production requires the availability of proper location and season, maintenance of isolation distance, rouging, proper harvesting, threshing and processing, and all other related operations necessary to meet the seed quality standards.

**Cultivation:** The seed production is carried out in well drained good fertile soil, under irrigated conditions to raise a healthy crop. Recommenced agronomic and plant protection measures are taken. For hybrid seed production, wider spacing (plant to plant) is followed as compared to that followed for the stable variety.

**Crop Rotation:** Proper crop rotation is advised to prevent contamination from the previous year's variety.

**Isolation**: Proper isolation distance is maintained to avoid contamination through natural outcrossing. In hybrid cotton, the minimum isolation distance requirement is 50 m for foundation seed and 30 m for certified seed. Isolation will be 5 m from the varieties of other species (different ploidy level) and 5 m isolation in certified classes between blocks of parental lines of the same hybrid.

**Rouging:** The process of removal of off type (phenotypically different) plants from the field of a particular variety is known as rouging. In cotton, rouging is done at three different stages at least, viz. before flowering, twice during flowering to avoid any contamination by off type plant and at peak bolling/boll bursting. Use of marker characters makes the rouging work simple.

**Picking:** The seed quality of the first two pickings is always better than later pickings.

**Ginning:** The bad quality cotton is separated and removed from the lot before ginning. The machine is thoroughly cleaned and adjusted to obtain the good seed.

### *4.5 Processing, Packing and Labelling*

After ginning, the seed is cleaned properly to remove broken/damaged/hollow seeds and other contaminants/impurities. Then a sample from each lot is taken for certification. If conforming to the norms of certification of the concerned seed stage, the seed is packed, appropriately labelled and sealed before marketing. This procedure is normally followed for stable varieties of cotton. For hybrid cotton seed and parent seed, usually the fuzzy seeds after ginning are delinted (usually acid delinted and neutralised) and dried to bring the moisture to desired level (depends on packing). The seed is then subjected to thorough cleaning by machine as also by human labour. Each lot of seed is then sampled by the certification agency. If conforming to the standard, the seed is labelled and packed for marketing. The certification agency takes up grow out test to ensure that seed possesses all morphological marker characters as described by the originating breeder while releasing a variety.

### 4.6 Male Sterility (MS) Based Hybrid

Besides the conventionally produced hybrid seed (Doak's method), male sterility based hybrid seed production technology is also used in cotton (Singh 2012). The conventional hybrid seed production method is costly, labour intensive and low yielding. The crossing involves emasculation (removal of male floral parts by thumb nail) on the first day and pollination on the second day. Further in this process, the chances of injury to the ovarian wall are high which results in shedding and poor setting. In the MS based hybrid seed production, the emasculation step is totally eliminated, therefore, setting percentage is high, less labour is involved and it is preferred by the seed producers. There are mainly two types of MS based seed production system used in cotton.

In the genic male sterility (GMS) based hybrid seed production, the pollen sterility is caused by nuclear genes. The genic male sterile line is maintained by sib-mating with isogenic fertile plant. In the following generation, the plant segregates into 1:1 ratio for sterile and fertile plant. The male sterile plants are used as female for pollinating and male fertile plants are rouged out. Presence of marker character facilitates identification of male sterility. The GMS hybrid 'Suguna' (CICR, Coimbatore) was the first such product from the public sector. This system has its own drawbacks.

The second such system is cytoplasmic genic male sterility (CGMS) based hybrid. Pollen sterility in controlled by both cytoplasmic and nuclear genes. In this seed production system, 'A' line is male sterile and 'B' line is isogenic fertile. The male sterile line is maintained by crossing with maintainer 'B' line. Hybrid seed is produced by crossing with a pollinator 'R' line. The 'R' line restores fertility in the F1 hybrid and is hence called the restorer line. The first CGMS based hybrid PKV Hy-3 was released by PKV, Akola in 1993.

## 5. Contribution of Public Sector in Development of Cotton Varieties

The role of public sector institutes began as far back as 1904 when the Departments of Agriculture were established in the states and continued unabated until ICAR/SAUs were set up and research component was transferred to SAUs/Institutes. In fact, it is only after the release of the hybrid cotton H-4 in 1970 from Surat that the private sector stepped into the cotton seed business. The public sector played a major role until the late eighties when the private sector captured a good market share in hybrid cotton seed. But, after 2002, when Bt cotton was introduced, the public sector share in hybrid cotton seed dwindled to a bare minimum. However, as far as stable varieties of *arboreum*, *herbaceum* and to some extent *hirsutum* and *desi* hybrids are concerned, the public sector is the major player. For organic cotton cultivation, the public sector is the sole source of seed supply. All in all, the public sector is continuing to play a major role in the overall cotton development in the country. The contributions made by the major public sector institutes are briefly given hereunder.

**Central Institute for Cotton Research:** The Central Institute of Cotton Research (CICR), Nagpur under the ICAR is a premier institute on cotton research that conducts and also coordinates basic strategic and applied research and development activities on cotton to enhance productivity and profitability of cotton farmers and the value chain associated with cotton. The institute has been recognised all over the world for its outstanding work on plant breeding and crop improvement, immunological diagnostic kits and insecticide resistance management. The CICR has developed and released 26 improved genotypes including 12 varieties of *Gossypium hirsutum*, two varieties of *G. arboreum*, one variety of *G. barbadense*, eight intra-*hirsutum* hybrids and one intra-*arboreum* hybrid. It also has the world's largest germplasm collection and a large collection of wild species. The institute developed the desi Bt-cotton incorporating *cry1Ac*, *cry1F* and *cry1Aa3* into *G. arboreum* and *G. hirsutum* cotton. Three lectins have been identified by the institute for control of sap-sucking insect pests (like aphids, jassids and whiteflies). The institute has developed many implements and devices including a planter, battery operated sprayer, which have been commercialised. The Ministry of Agriculture declared the CICR as the 'National Bt-referral Labouratory' to provide expertise for regulatory purposes to maintain seed purity and seed quality of Bt-cotton. The institute has done outstanding work in the development and dissemination of IRM strategies in about 200,000 hectares in 30 districts of nine cotton growing states in the fields of about 90,000 farmers, resulting in a net financial benefit of Rs. 5000 lakh per year due to 50–60 per cent reduction in pesticide use and enhanced yields.

**Punjab:** Cotton improvement work is carried out in the Punjab Agricultural University at the main centre Faridkot and sub-centres at Bhatinda and Ludhiana. The university has developed many high yielding varieties of *G. hirsutum*, *G. arboreum*, intra-*arboreum* and intra-*hirsutum* hybrids, in addition to several useful recommendations for farmers.

**Haryana:** Cotton research in the state is carried out in CCS[6] Haryana Agricultural University, Hisar with the main centre at Hisar and sub-centre at Sirsa. The university has released several high yielding varieties of *G. hirsutum*, *G. arboreum*, H × H[7] hybrids and A × A[8] hybrids, besides, many agro technique recommendations for farmers.

**Rajasthan:** Cotton research is conducted under the Rajasthan Agricultural University, Bikaner at its main centre Sriganganagar and sub-centre at Banswara (Maharana Pratap University of Agriculture and Technology, Udaipur). High yielding varieties of *G. hirsutum*, *G. arboreum* and two hybrids, one each of H×H and A×A have been released for farmers, in addition to several agro technique recommendations for farmers.

**Uttar Pradesh:** There is a small centre of the northern zone at CSA[9] University of Agriculture and Technology, Kanpur. Cotton is grown in only a few thousand hectares in the state. So far, two varieties have been released by the centre.

**Madhya Pradesh:** Cotton research is carried out at the main centre at Khandwa and sub-centre at Indore now under MVRS[10] University of Agriculture and Technology, Gwalior. Many high yielding varieties/hybrids of *G. hirsutum*, *G. arboreum*, H × B[11] and H × H hybrids have been released by the university, besides many useful agronomic and plant protection recommendations for farmers.

**Maharashtra:** Cotton research in the state is carried out at three main centres at Dr. Punjabrao Deshmukh Krishi Vidyapith, Akola, Mahatma Phule Krishi Vidyapith, Rahuri and Marathwada Agricultural University, Nanded, and two sub-centres at Jalgaon and Pune. The SAUs in the state have released many high yielding varieties of *hirsutum*, *arboreum* and several H × B hybrids, A × A hybrids, inter-specific desi hybrid and H × H hybrids for cultivation, in addition to several agro technique recommendations for farmers.

**Gujarat:** Cotton research in the state is carried out at the Main Cotton Research Station, Surat of the Navsari Agricultural University with a sub-centre at the Junagadh Agricultural University (JAU) besides regional research stations of the state at Talod, Viramgam, Anand, Amreli and Bharuch. After the formation of the SAU, several *hirsutum, arboreum, herbaceum* varieties and H × H hybrids, desi intraspecific hybrid, GMS A × A hybrid, ELS H × B hybrid, one budded cotton hybrid and recently, two Bt cotton hybrids (BG II) have been released in the state, besides over hundred agro technique recommendations for farmers.

**Andhra Pradesh:** The research on cotton is carried out at the main centre Lam, ANGR[12] Agricultural University and the sub-centre at Nandyal. Several high yielding varieties of *hirsutum, arboreum, herbaceum* and hybrid (H × B, H × H) have been released by the university. Several recommendations for farmers on agronomic requirement and plant protection have also been made for the benefit of farmers.

**Karnataka**: The University of Agricultural Sciences (UAS), Dharwad is the main centre for cotton research in the state with the UAS, Raichur as the sub-centre. The university has the distinction of releasing the first ELS hybrid Varalaxmi followed by several high yielding varieties of *hirsutum*, *arboreum*, *herbaceum* and H × B (ELS), H × H, A × A and h × A hybrids, besides a good number of agro technique recommendations for farmers.

**Tamil Nadu:** In the southern-most state of the South zone, the Tamil Nadu Agricultural University (TNAU), Coimbatore and the CICR regional station play a major role in cotton development. The two institutes have released some of the outstanding popular varieties and hybrids especially of ELS types. In addition, several recommendations for cultural practices have been made for farmers.

Thus, the public sector has so far released about 225 cotton varieties including about 50 hybrids. A brief list of some popular varieties/hybrids is given in Table 2.6. All these have been vigorously tested in varied agro-climatic conditions and stood the test of time. Many of these crossed state barriers and were popular in far flung states. Some of the varieties have been grown for fifty years in a row, indicating their versatility and sturdiness. Besides the varietal development, the public sector has contributed immensely in developing production technology which included nutrient requirement, crop geometry, sowing time, irrigation management, weed control, intercropping, micronutrients, use of PGRs. (plant growth regulators), integrated pest management for sucking pest and bollworms, pest and disease monitoring and forewarning and their management, evaluation of new chemistries, insecticide resistance management and many more, both for conventional cotton as well as Bt

**Table 2.6:** Some Top-performing Popular Varieties/Hybrids Released by Public Sector in India (1950-2011)

| *Zone* | *Category* | *Genotypes* |
|---|---|---|
| North | *Hirsutum* CV | F 414, H 777, LH 900, F 505, LH 886, LH 1134, RST 9, H 1226, F1378, F 846, F1861, H1098, H1236, RST 9 |
| | *Barbadence* CV | None |
| | Asiatic Cotton CV/h x A, AxA | G 27, LD 230, RG 8, LD 327, DS 5. CISA 310, CISA 614, CISAA 2, HD 123, HD 432, AAH 1 |
| | Hybrids (non Bt) | Fateh, Dhanlakshmi, Maruvikas, Om Shankar, CSHH 243, PAU 626, HHH 287MS, CSHH 238 |
| Central | *Hirsutum* CV | B 1007, Khandawa 2, DHy 286, SRT 1, G. Cot. 16, AKH 081, LRK 516, LRA 5166, CNH 36, G. Cot. 20, CNH120 MB, Suraj, AKH 8828, NH 615, NH 452, PH 348 |
| | *Barbadence* CV | None |
| | Asiatic Cotton CV (H x A) & (A x A) | AKH 4, AKA 5, AKA 7, G. Cot 15, AKA 8401, G. Cot 11, G. Cot 13, G. Cot. DH 7 & G. Cot. DH 9, G. Cot. 21, G. Cot. 23, AKA 8, AKDH 5, PA 402, PA 183, G. Cot. 25 |
| | Hybrids (non-Bt) | H 4, JKHY 1, H 6, PKV Hy 2, NHH 44, H 8, G. Cot. Hy-10, G. Cot. Hy-12, CICR HH 1, NHB 12, PKV Hy 4 (CMS), G Cot MDH 11, G Cot 102 |
| | Bt hybrids | G. Cot. Hy-6 (BG II), G. Cot. Hy-8 (BG II) |
| South | *Hirsutum* CV | MCU 5, MCU 6, MCU 7, MCU 9, MCU 10, Krishna, GS 23, Abhadita, Sharada, Sowbhagya, MCU 5 VT, Supriya, LPS 141, SVPR 2, Surabhi, LK 861, RAHS 100, KC 3, Anjali, Sahana, |
| | *Barbadence* CV | Sujatha and Suvin (100-120 HSC) |
| | Asiatic Cotton CV/H x A | K 8, K 9, K 10, Saraswati, Mahanandi, Aravinda, Raichur 51 (h) and DB 3-12 (h), DDH 2 (h × A), DDHC 11, |
| | Hybrids (non-Bt) | Savitha, Varalaxmi, DCH 32, TCHB 213, DHB 105, NHB 12, DHH 11, RAHH 95, RAHB 87. |

*Source*: Narayanan *et al.* 2011, Modified.

cotton. In fact, the public sector has been instrumental in bringing the Bt technology to this stage by testing and evaluating it at different levels and providing production technologies.

## 6. Cotton Biotechnology in Public Sector

Cotton crop is reported to be inhabited by about 1325 insects (Hargreaves 1948). By and large, 230 insect/pest species affect cotton, of which 46 have been identified as major insects in 32 countries. Around 10–15 principal pests in nine major countries cause significant losses to cotton. The bollworms, *Helicoverpa*, *Earias* and *Pectinophora* are the major damaging pests of cotton in India. To avoid losses and to control pests, large amounts of pesticides were used on cotton. It is said that of all the pesticides used on all crops, nearly 22 per cent are on cotton in the world. In India, pesticide use in cotton used to be around 55 per cent of total pesticide use in the country before the advent of Bt cotton in 2002. In spite of the huge quantities of different kind of pesticides used, control of pests in cotton continued to baffle the scientists and the farming community and the problem of pest control and its impact turned into a psycho-socio-economic tension for the farmers of India. New pesticides like synthetic pyrethroids introduced in the 1980s provided respite to the cotton farmers for a few years but due to indiscriminate use, turned into a nightmare and resulted in the development of resistance in bollworms and resurgence of whiteflies menace (Khadi 2005).

Efforts on cotton breeding for resistance/tolerance to insects in cotton were made during early 1980s to 1990s. These conventional approaches are still being used by incorporating characters responsible for non-preference, antibiosis and avoidance. However, these efforts never resulted in a perfect cultivar which exhibited resistance to bollworms.

It may be recalled here that use of *Bacillus thuringinesis kurstkii*—a soil borne bacteria powder was in vogue as pesticide since 1960s but had its own limitations like effectiveness only at certain temperature and humidity, non-use during rainy season, contact nature of action etc. Hence it could not gain enough ground amongst cultivators.

Meanwhile, biotechnological tools came in handy in transferring pest resistance in cotton. The crystal protein gene (Cry 1Ac) derived from *Bacillus thuringinesis* offered greatest resistance to lepidopterous tobacco budworm *Heliothis virescens*. This gene was transferred to cotton through *Agrobacterium tumifaciens* mediated transformation with CaMV 35S promoter (Perlak *et al.* 1991).

In India, although the use of biotechnology in agriculture particularly cotton is 15-17 years old now, so far, the research output of the public sector is yet to be transferred to farmers' fields. (The first such efforts by the University of Agricultural Sciences, Dharwad and the CICR, Nagpur relating to the variety Bikaneri Narma Bt in 2008, followed by hybrid NHH 44 Bt in 2009, known as BNLA-601 event, unfortunately ran into controversy and currently stands withdrawn). The Main Cotton Research Station, Surat of the Navsari Agricultural University under an agreement between Gujarat State Seed Corporation Ltd. Gandhinagar and Mahyco-Monsanto Biotech Limited, Mumbai recently released two Bt cotton hybrids, G.Cot.Hy.6 (BGII) and G.Cot.Hy.8 (BGII), an accomplishment by a public sector Corporation in India. Nevertheless, the efforts being put in by the public sector in this area shall bear fruits in the near future.

Currently, the following institutes are involved in cotton biotech research. The focus is not only on bollworms but also sucking pest resistance, drought/salinity tolerance, disease resistance, fibre quality and other important issues.

### *6.1 Central Institute for Cotton Research, Nagpur*

The Central Institute for Cotton Research (CICR) is spearheading the cotton biotechnology research programme in the country. Some of the major areas covered are as under.

*Bollworm resistance in diploid and tetraploid cotton:* The transformation of elite cotton varieties like LRK 516 and LRA 5166 of *G. hirsutum* and RG8, PA 255 and PA 402 of *G. arboreum* is in various stages of development.

*Genetic engineering for abiotic stress tolerance and identification of new genes for high water use efficiency:* LRA 516 and LRA 5166 have been subjected to transformation with pATRD29A::AtCBF$_3$/ DREB1A+POsMYB02::AtAVP1 gene construct by *Agrobacterium* mediated transformation.

*Development of disease resistance transgenic cotton:* Chitinase are known to hydrolyse chitin polymers and are effective against fungi. A 1.3kb full length noval class I *chitinase* gene has been amplified and cloned from *G. hirsutum* variety LRA 5166 and transformed through *Agrobacterium tumifaciens.*

*CLCuV*[13] *resistance transgenic cotton:* Three transgenic events one each involving Antisense Coat Protein (*ACP*) and Antisense Rep Gene (*ARep*) in *G. hirsutum* HS6 and Sense Coat Protein (*SCP*) in H777 proved tolerant against CLCuV. Fresh transformation events were generated with each of the gene constructs. Work is initiated to develop CLCuV resistance transgenic cotton through ds RNAi mediated silencing of viral genes.

*Improvement of fibre strength:* Attempts are being made to clone *cellulose synthase* gene (*Ath*A and *Ath*AB) from *Arabidopsis* which brings about rapid conversion of carbon to UDP[14] glucose to facilitate high fibre strength. Also, cDNA[15] clones were made.

*Drought Resistance Genes:* Intensive efforts are underway to clone and characterise native gene for cotton, viz. protease inhibitor involved in pest resistance, *Osmotin* and *dehydrin genes* that govern tolerance of cotton to drought.

*Nodule induction in cotton*: Cotton plants were subjected to root nodule induction by *Rhizobium fredii.* Although bacteriods were present in nodules, nitrogenase activity was not recorded but this will open up vistas of nitrogen fixation in non-leguminous plant like cotton.

In addition to these, the institute is undertaking molecular breeding, DNA fingerprinting on a large scale which would help in furthering crop improvement (Kranthi 2011).

### *6.2 National Botanical Research Institute, Lucknow*

The institute has been currently working on the development of high throughout marker assisted selection system for improvement of drought tolerant and fibre quality related traits in cotton. The institute has earlier patented *Cry1EC* gene which is effective against *Spodoptera* and the transformation of cotton lines in the private sector, Navsari Agricultural University, Surat and some other organisations are at various stages of development. The institute has a network collaborative programme for insect resistance, hybrid seed and yield heterosis with other public sector institutes in the country as well as private companies.[16]

### *6.3 National Research Centre for Plant Biotechnology (NRCPB), New Delhi*

The NRCPB, New Delhi is the leading public sector institute in the country involved in basic fundamental research in biotechnology and undertakes crop based problem oriented research programmes in biotechnology which includes gene discovery and characterisation, transformation and evaluation. The institute has network projects on gene stacking in *Bt* cotton and as well as targeted gene integration in cotton.

### *6.4 International Centre for Genetic Engineering and Biotechnology (ICGEB), New Delhi*

The ICGEB undertakes research on the development of abiotic stress (high salinity/drought/temperature) tolerant crops including cotton. Major contributions include the discovery and identification of the first stress-regulated helicases from plant system and the development of salinity stress tolerant plants.[17]

A programme is being carried out to understand genome involved in cotton fibre development and the influence of abiotic and biotic stress factors involved. The studies on the role of hormone signalling and transcription regulation in fibre initiation and differentiation showed that drought has led relatively less impact on fibre initiation but has profound impact on fibre elongation by down regulating genes involved in cell wall loosening and expansion process. This new information will be useful particularly for the development of drought tolerant cotton. Cotton has been transformed with several *Bt* genes in collaboration with Insect Resistance Group. Transgenic lines showing stable expression of *Bt* proteins offered protection to various pests. The transgenic lines are being evaluated with partners from industry and agriculture universities.

### 6.5 University of Agricultural Sciences, Dharwad

The university has programmes on development of cotton transgenic with improved fibre strength using *cellulose synthase* gene from *Arabidopsis,* gene stacking in *Bt* cotton, targeted gene integration in cotton with the NRCPB as the lead centre. Genetic improvement of cotton for insect resistance, hybrid seeds and yield heterosis, as also on boll and fibre development with the ICGEB at the lead centre are other programmes.[18]

### 6.6 Tamil Nadu Agricultural University, Coimbatore

Potent isolates of the native Bt have been selected by insect bioassay against *lepidopteran* pest of cotton. Indigenous Bt genes cloned from potent isolates of *Bt* and plasmid constructs were made to express codon optimised synthetic *Cry2A* genes. A chimeric *Cry2A* gene has been made using DNA sequences of two indigenous *cry2Ab* genes of *B. thuringensis*. They showed significantly higher toxicity to *H. armigera* (cotton bollworm) than the *Cry2Ab* protein which is currently used in the Bt-hybrids.[19]

### 6.7 Navsari Agricultural University, Navsari

The NAU (Navsari Agricultural University) undertakes cotton biotechnology programme at the Main Cotton Research Station, Surat. A *G. hirsutum* variety, G. Cot. 10, has been transformed with *cry1EC* gene (source NBRI, Lucknow) and is in BC4 $F_2$ generation. The research station has recently released two *Bt* cotton hybrids G.Cot.Hy-6 BG II and G.Cot.Hy-8 BG II encoding MON 15985 event sourced from Mahyco Monsanto Biotech Ltd. Other potential parental/inbred lines are at different stages of conversion.

In addition to these, other institutes and universities are also involved in biotech research especially marker assisted breeding programme, QTL (quantitative trait loci) mapping for various desired traits.

## 7. Cotton Research and Development: Future Perspectives

Before planning one has to visualise for the future—for example, what will be the cotton requirement for the growing population? What will be the productivity? What will be quality of cotton required by mills? and how will we achieve that? All these in a climate which is drastically changing and natural resources are fast shrinking. This needs strategic planning, sincere efforts and policy support from the government. Here four important areas have been identified for critical research in the future.

### *Genetic Enhancement/Modification of Cotton for Different Agro Ecological Situations*

- Conservation, cataloguing and utilisation of germplasm.
- Focus on development of desi cotton varieties having shorter duration, bigger bolls with strong and finer fibres.
- ELS cotton in *hirsutum* and *barbadense* with appropriate quality.
- Photosynthetically efficient plant types.
- Development of low input, high use efficiency plant types.
- Multiple resistances to abiotic stresses as drought, salinity, high temperature and water logging.
- Development of transgenic cotton for pest and disease and abiotic stress resistance and quality traits.
- Exploit possibility of $C_4$ carbon fixation pathway and N (nitrogen) fixation in cotton.
- Traits based marker assisted breeding programme.

### *Production and Management*

- Technology for efficient use of fertilisers and water.
- Reduce dependence on chemical inputs; instead, blend traditional knowledge with today's technology.
- Integrated weed management system, high density planting and organic farming.
- Enhance total output per unit area, per unit time, per unit investment in cotton based cropping system.
- Narrow gap between irrigated and rainfed yield.

- Technology that bridges gap in average yield and high yield realised especially in irrigated area.
- Use nano technology for quick and efficient delivery of inputs.
- Develop farm machinery for small and medium farmers particularly for picking.
- Use of secondary and micronutrients for higher productivity.
- Gearing up for climate change and strategies to mitigate harmful effects.

***Pest and Disease Management***

- Development and evaluation of new chemistries for pest management.
- Eco-friendly biopesticides.
- Monitoring and fore warning of pests and diseases.
- Identity tolerant/resistance varieties for further exploitation.
- Pest management for organic cotton.

***Social-economic Aspects***

- Efficient transfer of technology to the end users through mass media, NGOs and other entrepreneurs.
- Pest and disease forecasting and measures thereof.
- Advance information system in market demand and prices.
- Seed quality monitoring.
- Availability of seed and other inputs of right time and price.

To conclude, by 2030, Indian population will grow at an annual growth rate of around 1.1 per cent to reach 14.50 billion. Assuming per capita consumption of cloth at the current rate, the country would need 100 million bales besides the man-made fabrics. If recent trends in petroleum products availability are any indicators, the production of man-made fibre would decline as the sources are drying up. At the same time, pressure on land resources for food crop would immensely increase in view of the burgeoning population. Therefore, the area under cotton is unlikely to increase. Hence the country faces a challenging task to enhance cotton production roughly three times on the existing resources from the present 33.4 million bales. The current productivity shall have to be doubled. Modern technology with a blend of traditional wisdom will have to be judiciously used if we have to meet the goals without risking our natural resources. This would also call for an appropriate role for the public and private sector to take care of the long term interests of the different stakeholders particularly the farming community.

## Notes

1. Research Scientist Cotton (retired), Main Cotton Research Station, Navsari Agricultural University, Surat.
2. Unless stated otherwise, the abbreviations here and elsewhere in the chapter indicate the name of the cotton varieties.
3. Refers to a cross between *G. herbaceum* and *G. arboreum*.
4. In cotton parlance, the strength of the fibre is indicated by g/tex indicating gram per tex (Yarn. Tex is the term used for measurement).
5. Commercial Crops.
6. Chaudhary Charan Singh.
7. Refers to a cross between *G. hirsutum* × *G. hirsutum*.
8. Refers to a cross between *G. arboreum* × *G. arboreum*.
9. Chandra Shekhar Azad.
10. Maharani Vijaya Raje Scindia.
11. Refers to a cross between *G. hirsutum* × *G barbadence*.
12. Acharya N.G. Ranga.
13. Cotton leaf curl virus.
14. Uridine diphosphate.
15. Complimentary deoxyribose nucleic acid.
16. See www.nbri.res.in
17. See www.icgeb.org
18. See www.usad.edu
19. See www.tnau.ac.in

## References

Annual Report. 2012. *All India Coordinated Cotton Improvement Project*. CICR, Coimbatore.

Handbook of Agriculture. 2006. Cotton ICAR, New Delhi, pp. 1018–1036.

Hargreaves, H. 1948. *List of Recorded Cotton Insects of the World*. Commonwealth Institute of Entomology, UK.

Khadi, B.M. 2005. 'Bt Cotton Scenario in India: Problems and Prospects'. In *Souvenir*, National Workshop on Recent Advancement in Improvement of Cotton Seed Quality, November, Surat. pp. 1–12.

Kranthi, K.R. 2011. 'CICR: Spearheading Cotton Research in India'. In Souvenir, World Cotton Research Conference-5, pp. 16-25.

Kranthi, K.R., M.V. Venugopalan and M.S. Yadav. 2011. *Vision 2030*. Central Institute for Cotton Research, Nagpur, p. 50.

Kumar, V. 2011. 'Water Management Technology for Sustainable Cotton Production in Central India'. World Cotton Research Conference-5, Book of Papers, pp. 376–385.

Narayanan, S.S., C.D. Mayee and A.J. Shaikh. 2011. 'Post Independence Scenario on Cotton Genetic Improvement in India'. In *Souvenir*, World Cotton Research Conference-5, pp. 38–62.

Perlak, F.J., L.F. Roy, A.D. Duff, L.H. Sylvia and D.A. Fischhoff. 1991. 'Modification of Coding Sequences Enhances Plant Expression of Insect Control Protein Genes'. *Proceeding of National Academy Science*, US, pp. 3324–3328.

Singh, P. 2012. *Cotton Improvement in India*. New Vishal Publication, New Delhi, p. 200.

# 3

# Institutional Mechanisms to Ensure Biosafety among the Bt Cotton Adopters in India

*A. Indira*

India's tryst with agro-biotechnology[1] started in late 1980s to help the country reach the heights seen with the 'Green revolution', to ensure self-sufficiency in food, and make India a major global exporter of food. Bt cotton is the first, and so far the only genetically modified (GM) crop to be commercialised in India. Adoption of Bt cotton has made India the number one exporter of cotton globally as well as the second largest cotton producer in the world today. Between 2002 to 2008, Bt cotton has generated economic benefits for farmers valued at US$5.1 billion (Choudhary and Gaur 2010). Bt brinjal is the only other crop which was approved for commercial release in 2009, but soon after halted under pressure from various groups. As of 2011, research related to genetic modification was being carried out in 72 crops/plant species in India.[2]

Agriculture in India has seen a levelling off of productivity gains from the Ninth (1997–2002) and Tenth (2002–2007) Five Year Plan periods. The agricultural sectoral recorded a growth rate of 2.44 per cent and 2.30 per cent respectively during the Ninth and Tenth Five Year Plan period compared to 4.72 per cent during Eighth (1992–1997) Five Year Plan. Consultations prior to the Eleventh Plan identified that there was a stress on natural resources, technology fatigue, poor delivery systems in credit, extension and marketing, and insufficient agricultural planning at the local levels.

The Eleventh Plan sought to reverse the deceleration through various inputs, and managed to average 3.3 per cent per year in the plan period.

The Approach Paper to the Twelfth Plan (2012–17) emphasises the need to redouble efforts to tackle the weaknesses in the system and in particular focus on sustainable development. One of the most significant points of intervention that the Approach Paper has identified on the production side is that of technology as the prime mover of productivity in agriculture where natural resources are fixed. It states that significant breakthroughs are required to improve production technologies of predominantly rainfed crop groups. It clearly puts forward the case for keeping abreast with the latest advances in bio-technology. And, more importantly, it has emphasized that *an acceptable set of protocols and a clearly defined regulatory mechanism are required so that transgenic food crops can be cleared if advantages of doing so vastly outweigh any precautionary misgivings* (GOI 2011: 71).

There has been an enduring debate over the efficacy and safety of the use of agro-biotechnology in various countries. In the context of GM-crops, the concept of 'biosafety' broadly covers three areas: the health safety of humans and livestock, the safety of the environment (i.e. ecology and biodiversity) and socio-economic safety (i.e. the economic and social impact on farmers, consumers and different social classes, as well as on trade and the economy in general). While the biosafety regulations in force in industrialised countries (e.g. the USA, the European Union, Canada, Australia, Japan) address only the health and environmental risks and exclude socio-economic considerations, the regulations in some developing countries, including India tend to include all the three areas.

The nature of policy approach (permissive, precautionary or prohibitive) of the countries' governments to GM-crops has been influenced by the 'trade off' between potential benefits and risks in each case. Countries like the USA, Canada, Argentina and Australia, adopted a permissive attitude very early on, with lower production costs and greater export profits presumably outweighing other considerations. But the European Union, India, Brazil, Japan, the Philippines and others have taken a precautionary approach to the introduction, cultivation and development of GM-crops.[3]

This chapter sets the context of the biosafety regulatory regime in India and looks at the institutional framework for biosafety that has evolved over time. Part one gives a brief overview of the commercialisation of Bt cotton and the shortcomings in the regulatory system. Part two discusses the regulations and the regulatory mechanism in India. Part three lists the initiatives and reforms in progress in the Indian regulatory system. Part four briefly looks at some new experiences in other countries. And lastly, the conclusions.

### 1. Commercialisation of Bt Cotton in India

Bt cotton was approved by the Genetic Engineering Appraisal Committee[4] (GEAC) in March 2002 with the conditions that: (a) it was valid for the three-year period April 2002–March 2005, (b) the size of the pollen refuge belt surrounding a Bt cotton field shall be such as to accommodate at least five rows of non-Bt cotton or twenty per cent of the total sown area, whichever is more, (c) each packet of seeds of the approved varieties shall be accompanied by a separate packet of the seeds of the same non-Bt cotton variety which is sufficient for planting in the refuge prescribed as above, and (d) Mahyco (Maharashtra Hybrid Seeds Company) shall submit annual reports to the GEAC/Ministry of Environment and Forests (MoEF) by the end of March each year on the use of the approved Bt cotton hybrid varieties specifying, among other things, the dealers (seed selling firms), the acreage planted and the localities of planting (state and region). Further, the MoEF reserved the right to stipulate additional conditions during the period of validity of the approval, and the right to revoke the approval, if the implementation of the conditions were not satisfactory.

It is significant that the first GM-crop to have been approved by the biosafety regulatory authority GEAC for general release was the bollworm resistant Bt cotton (Event MON531) developed by the private sector India-USA joint venture Mahyco-Monsanto. Table 3.1 gives a brief chronology of the passage through the regulatory process.

Over the years, there has been a spate of studies looking at the economic impact of Bt cotton.[5] The studies confirm that there has been a 50 to 110 per cent increase in profits from Bt cotton. The yield increases have ranged from 30-60 per cent and there has been a reduction in the number of insecticide sprays. The adoption rate of Bt cotton in India 2009 was 87 per cent. The provisional estimates for 2011-12 showed that Maharashtra had the highest area of 40.97 lakh hectares under cotton cultivation, followed by Gujarat (30.23 lakh hectares) and Andhra Pradesh (18.54 hectares). In terms of productivity, Gujarat was the highest with 659 kgs per hectare, followed by Punjab (591 kgs per hectare) and Haryana (561 kgs per hectare). Gujarat also had the highest production of 114 lakh bales of 170 kgs, followed by Maharashtra (69 lakh bales of 170 kgs) and Andhra Pradesh (48 lakh bales of 170 kgs) (CICR 2012).

### 2. Biosafety Regulations in India

The Indian biosafety regulations are modelled closely on the OECD

**Table 3.1:** Chronology of Passage of Bt-cotton through the Regulatory Process

| | |
|---|---|
| 1991 | Monsanto approached GoI with a proposal to introduce its Bt-cotton in India. Reasons for not taking the offer: (i) a disagreement over the financial terms for the technology transfer and (ii) to promote the indigenous R&D, which was already underway in the public sector. |
| 1993 | Indian private company, Maharashtra Hybrid Seeds Company (Mahyco) began discussions with Monsanto about licensing its Bt-cotton technology. |
| 1995 | DBT granted permission to Mahyco to import Monsanto's Coker-312 seeds from the USA, under the existing regulations for GM-R&D. Mahyco crossed the transgenic trait into Indian elite cotton cultivars for six generations under controlled greenhouse conditions and then generated stable lines to breed hybrids. Eight hybrids containing the Cry 1Ac gene were developed. |
| 1996 | DBT permitted Mahyco to conduct limited field trials in 'open contained' plots of about 200 sq. mts each in 5 states (Andhra Pradesh, Haryana, Karnataka, Maharashtra and Tamil Nadu). |
| 1996 | Karnataka refuses, DBT intervenes – trials go on. |
| 1997 | Limited field trials begin in the five states. |
| 1998 | DBT permits multi-locational field trials in forty sites. |
| 1999 | Based on case filed by Civil Society Organisation (CSO) Research Foundation for Science, Technology and Environment, New Delhi, the 1998 Guidelines were amended to clarify the mandates and roles of the Review Committee on Genetic Manipulation (RCGM) and the Genetic Engineering Appraisal Committee (GEAC) with respect to small-scale and large-scale field trials, and to establish the Monitoring and Evaluation Committee (MEC). |
| 2000 | GEAC issued a preliminary approval for commercial release, based on the multi-locational field trials – a public interest petition was filed in the Delhi High Court by Gene Campaign, CSO based in New Delhi. |
| 2001 | An amendment to the regulations by which the Central government funded Indian Council for Agricultural Research under the Ministry of Agriculture was asked to conduct its own field trials of the Mahyco Bt cotton varieties. |
| 2001 | Illegal seeds in Gujarat found. |
| 2002 | GEAC approved three hybrid varieties – Bt MECH 12, Bt MECH 162, Bt MECH 184 for commercial releases in March 2002. |
| 2004 | Indian seeds company RASI produced Bt-cotton RCH-2A, obtained on license by Mahyco which was approved for release in the three southern and two central states. Mahyco produced four more hybrids. |
| 2005 | There were three companies producing 20 hybrids using the same event MON531. |
| 2006 | Three events approved - MON15985 featuring two genes cry1AC and cry2Ab was developed by Mahyco/Monsanto; Event -1 featuring cry1AC gene, sourced from Indian Institute of Technology-Kharagpur, India developed by JK Agri-Genetics; GFM Event featuring fused genes cry2Ab and cry1Ac sourced from China developed by Nath Seeds. |
| 2008-09 | Bikaneri Nerma (BN)- Indigenous Bt cotton event (BNLA-601) expressing the cry1AC was approved for commercial sale in 2008, but soon after s seed production was stopped in 2009 because of traces of contamination of the Mahyco's proprietary gene.* |
| 2009 | New event MLS-9124 developed indigenously by Metahelix Life Sciences was approved. This features a synthetic cry1C gene. |
| May 2012 | As of date, a total of 1128 varieties of Bt cotton hybrids released for commercial use by GEAC. |

*Note:* *The ICAR in 2011 was forced to close curtains on this event after it came to light that the commercial seeds were contaminated with Monsanto's proprietary Bt gene instead of the domestic one. This was brought to light through an RTI inquiry by two scientists which revealed how the University of Agriculture Sciences (UAS) at Dharwad went ahead after brushing aside all precautions to produce an indigenous variety working on a gene originally patented by Monsanto *(Indian Express 2011)*.

*Sources*: Indira *et al.* 2005; Karihaloo and Kumar 2009; Choudhary and Gaur 2010; GEAC 2012.

guidelines. They cover most of the safety aspects that are likely to arise in the context of R&D in the laboratory and greenhouse, contained tests on small plots (also referred to as 'contained small-scale field tests' or simply as 'contained field tests'), large-scale field trials out in the open, general release to farming and trading communities, commercialisation and trade. They are mandatory for all Indian institutions and companies that deal in Genetically Modified Organisms (GMOs). As with the regulations in force in the European Union (EU)[6], they are based on the precautionary principle[7], applied on a case-by-case[8] basis and use a step-by-step[9] approach. They are generally regarded as being one of the most comprehensive, detailed and stringent biosafety regulations in the world.

In India, the first national biosafety rules were issued by the Ministry of Environment and Forests (MoEF) as a Notification, and published in the Gazette of India, Extraordinary, Part II 3(i) on 5 December 1989. These 1989 rules are entitled *Rules for the Manufacture, Use, Import, Export and Storage of Hazardous Micro Organisms or Cells.* They were followed in 1990 by the DBT's *Recombinant DNA Safety Guidelines,* which were then revised, expanded and published in 1994 by DBT as the *Revised Guidelines for Safety in Biotechnology*. These three documents cover all the four main sectors of biotechnology: medical/pharmaceutical, agricultural, industrial and environmental. Four years later, the agricultural biosafety guidelines were separated out, expanded, further revised and published by the DBT in August 1998 as the *Revised Guidelines for Research in Transgenic Plants and Guidelines for Toxicity and Allergenicity Evaluation of Transgenic Seeds, Plants and Plant Parts.*

The 1989 Rules, as well as the Guidelines of 1990, 1994 and 1998, have been made mandatory for all practitioners in India of genetic modification technologies under the Environment (Protection) Act of 1986. These four documents together constitute the national biosafety regulations of India. The 1998 Guidelines were amended in 1999 to clarify the mandates and roles of the Review Committee on Genetic Manipulation (RCGM) and the Genetic Engineering Appraisal Committee (GEAC) with respect to small-scale and large-scale field trials, and to establish the Monitoring and Evaluation Committee (MEC).

The UN-sponsored international Cartagena Biosafety Protocol to the Convention on Biological Diversity governing the transboundary (i.e. inter-country) movement of living modified organisms (LMOs) was agreed in Montreal in January 2000. It came into force in September 2003, having been ratified by the requisite number of countries. India has ratified the Cartagena Protocol.

Further to this, a set of *Guidelines for Conduct of Field Trials of Regulated, GE Plants* have been approved by the RCGM and GEAC in June 2008. The Guidelines describe the application process and general requirements for confined field trials and the standard operating procedures (SOPs) for transport, storage, management, harvest/ termination and post-harvest management during the conduct of the trials.

The 1989 Rules classify R&D work involving GMOs in all sectors of biotechnology into four risk categories, and provide lists of bacterial, fungal, parasitic and viral agents that fall into each category. The 1998 Guidelines reaffirm this classification and specify the roles of the institution, the Institutional Biosafety Committee (IBSC) and the RCGM *vis-à-vis* the risk categories: Category I comprises *routine* recombinant DNA experiments conducted inside a laboratory; Category II consists of both laboratory and greenhouse experiments involving transgenes that combat biotic stresses through resistance to herbicides and pesticides; Categories III and IV comprise high risk experiments and field trials where the escape of transgenic traits into the open environment could cause significant alterations in the ecosystem.

The Government of India (GOI) has established a regulatory structure for implementation of the biosafety regulations. Two nodal agencies, Ministry of Environment and Forests and Department of Biotechnology, Ministry of Science and Technology are responsible for implementation of the regulations. GM-crops are regulated by the three tier structure (shown in Figure 3.1).

An application for a GM-crop after negotiating its passage through the three tiers of RCGM-GEAC-MEC, lands finally at the Ministry of Agriculture (MoA), whose responsibility it is to make sure that the application fulfils the provisions of the Seeds Act,[10] and if it does, to register the crop under the Seeds Act, whereupon it is allowed to enter the market.[11]

Adoption of the Bt cotton crop brought forth certain shortcomings in the system as elaborated below:

### *Shortcomings in the System*

In India, agriculture is a state subject. Under the Constitution of India, it is not the Union Government but the State Governments that exercise primary authority over agriculture. Thus, while the Government of India (GOI) may take the initiative in the policy arena and formulate policies concerning agricultural biotechnology and GM crops (in R&D as well as

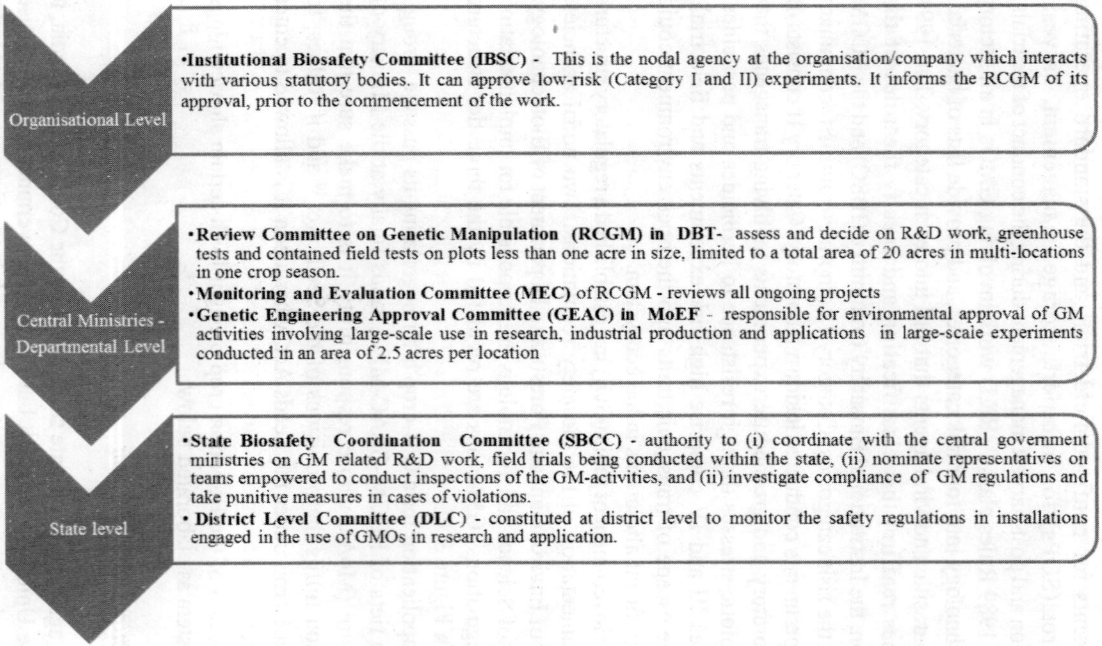


*Source*: Author.

**Figure 3.1:** The Three-Tier Structure

commercialisation), the agreement and active cooperation of state governments are indispensable for implementing the policies. The critical importance of consultation and coordination between the GOI and the state governments to the introduction and promotion of GM-crops was revealed from the very beginning (even prior to the commercial release). The Government of Karnataka objected to the perfunctory manner in which it had been informed of the limited field trials within its boundaries. The DBT got into active and serious consultations to sort out the matter and resolve the issue.[12]

GM-concerned CSOs, as well as researchers in agricultural universities, underlined the impossibility of small farmers (with holdings of one to two hectares) being able to comply with the biosafety directive that they must surround the field sown with GM-seeds by a non-GM sown area that is at least 20 per cent of the GM-acreage to act as a biodiversity *refuge* to ensure that GM-resistant pests and weeds do not, in the course of a few years, completely out-compete non-resistant ones.

The regulatory structure fell short in the implementation process. Early field research showed that the MEC directed monitoring process was limited to (i) the impact of GM-crops on some selected plants and insects in the trial plots and their immediate neighbourhood, (ii) documenting the frequency and intensity and sprayings of selected pesticides and herbicides, and (iii) recording the crop yields. It clearly came out that MEC lacked the capacity to ensure independent, extensive and intensive assessment of certain kinds of risks, e.g. (i) pollen and gene flow to other plant species in the near environment, (ii) the impact on the health of the livestock that were allowed to feed on the Bt cotton seeds by the farmers on whose the plots the trials were conducted,[13] and (iii) letting the plot owners sell the Bt cotton harvest and store the post-harvest residues for own use, instead of burning the harvest and the residues after the trials were over, as the regulations stipulated. All of these studies were undertaken by the company Mahyco itself.

The question of conflict of interest in the RCGM because of the simultaneous roles played by the representatives of R&D institutions as applicants for grants and as biosafety regulators has persisted since the inception of the committees. This has led to repeated questions being asked by GM-concerned CSOs about the credibility of the RCGM as an impartial and unbiased regulator and the legitimacy of the decisions it arrived at.

Quite a few researchers in public sector research institutions and private companies who were pro-GM were sceptical of the GEAC functions (Indira *et al.* 2005: 66–67). According to them, the GEAC did not have the scientific expertise to perform the tasks it was mandated to. Like its chairperson, the more influential of its members were senior civil servants with administrative experience but with no scientific credentials. The GEAC did not meet often enough to be able to deal expeditiously with the applications, which, in combination with its slow and inefficient way of working, led to great delays in processing the applications.

Very few states had established the SBCCs and DLCs. In those states where the SBCCs were established, there was nothing to indicate that there was any interaction with the various regulatory levels. Most officials in the Agriculture Department were unaware of the formation of the DLC. There seems to be no perceptible change in the role of the SBCC and DLC with regard to monitoring and ensuring the implementation of biosafety regulations within the states' boundaries.[14] As of 2007, SBCC and DLC were set up only in Andhra Pradesh, West Bengal, Maharashtra, Gujarat, Karnataka, and Tamil Nadu (GOI 2007).

The early studies of comparative performances of Bt and non-Bt cotton in the various states showed contradictory claims of yields. The cotton harvests, both in terms of yield and economic benefits at the farm level, were not higher than the non-Bt cases (Sahai and Rehman 2003).[15] The Bt cotton yields (average per acre of the MECH162 and MECH184 varieties) across all size categories of land holdings, were about 15 per cent lower than the non-Bt cotton varieties. While the Mahyco varieties were indeed lethal to the green bollworm, they were not effective against the pink bollworm. With competition from the green bollworm eliminated, the pink ones flourished, to the detriment of the crop yield. The official team charged by the state government in Andhra Pradesh to investigate complaints by local cotton farmers came to the conclusion that Bt-yields and Bt-fibre quality were indeed lower than the non-Bt ones. The farmers claimed compensation, for which the company was reported willing to 'compensate farmers only for (proven) failure of (Mahyco Bt-seeds) to germinate and (proven lack) of genetic purity of (Mahyco Bt-seeds), and not for yield losses' (Krishnakumar 2003).

The Parliamentary Standing Committee on Agriculture had asked the GOI to re-evaluate the economic viability of Bt cotton. On complaints from the state of Madhya Pradesh in central India, the GEAC/MoEF sent a team of seven scientists to evaluate the performance of the Mahyco hybrids sown by local cotton farmers. The team reported that while the Bt cotton crop had largely failed 'due to wilting and drying at the peak bolling stage', the traditional non-Bt crops had fared better. In a similar vein, a six-member team appointed by the Rajasthan state government came to the conclusion that Bt cotton was unfit for cultivation in the state and should be banned. This was in contrast to the research report of a few researchers from Mahyco (Barwale *et al.* 2004) which showed that the yield (weighted average over all the farms) was higher by about 60 per cent and the farm income had risen by approximately Rs. 18,000 per hectare and that the number of insecticide sprays had fallen from about three sprays to one spray.

Even before the approval of the Mahyco varieties, there were reports of unauthorized Bt cotton cultivation by farmers with the use of illegal seeds. The production and distribution of unapproved seeds was coordinated by a network of seed companies, seed producers and seed dealers. The government was in many cases unwilling to act against the errant companies or the farmers. It clearly showed that the seed producers did not give much cognizance to biosafety and hence had made no attempt to receive biosafety clearance (Lalitha *et al.* 2008).

A cause for greater concern was that leading public sector R&D institutions would at times break the rules with impunity. An example of this came to light when the Indian Agricultural Research Institute (IARI) in Delhi began GM-work on a native variety (*solanum melongen)* of brinjal. The IARI scientists introduced a Bt-toxin gene, Cry1A (b), obtained from Japan, into the native variety. Without first securing the legally obligatory approval of RCGM/DBT, the IARI team grew the transgenic brinjal in 1996, in an open experimental plot of 60 square metres within the IARI compound in Delhi. When the media uncovered this non-compliance of the biosafety regulations, the plot was burnt on the orders of the regulatory authorities. Not much is known about what disciplinary action was taken and how.

Despite repeated representations, CSOs remain excluded from the RCGM, the GEAC and the MEC and hence they have been wary of the entire regulatory system. In the early years, the system had not even acknowledged, let alone responded to, the queries, letters and submissions sent to them by civil society. The applications processed by the system, the proceedings and minutes relating to the processing, and the written 'output' (e.g. data, information, reports etc.) from the field trials, were not put into the public domain, despite repeated requests. The system had also neither imposed penalties nor taken punitive action against the institutions and companies that failed to comply with the biosafety regulations. The system was non-transparent and lacked public accountability.

It was not surprising, therefore, that activist CSOs repeatedly launched legal suits against the authorities at the highest levels (in the

High Courts at the state level and in the Supreme Court at the central level) to halt field trials, burnt GM crops on trial plots and mobilised farmers demonstrations. The main point being that the prevailing practice involved a clear conflict of interests where the test results were not available for public scrutiny and most of the testing was done by the biotech company that had a commercial interest in the approval of the GMO. In fact, fearing that the GEAC may grant approvals, in a petition admitted by the Supreme Court it was appealed that field trials for all GM products be stopped (*Frontline* 2010).[16]

### 3. Reforms in Progress

The first of the initiatives to reform the regulatory process was taken when the MoA (ICAR's parent ministry) set up an Agricultural Biotechnology Taskforce in 2003 under the chairmanship of Dr. M.S. Swaminathan[17] with the terms of reference to consult a wide range of stakeholders and on that basis make recommendations for reforming the regime. The Swaminathan Taskforce submitted its Interim and Final Reports to the Minister of Agriculture in April and June 2004, respectively. The Interim Report was sent to various stakeholders and the media.

The Swaminathan Taskforce Report (GOI 2004) recommended that:

- An autonomous Agricultural Biotechnology Regulatory Authority (ABRA) be set up.
- Pending the creation of an ABRA, (i) the GEAC should be split into two separate wings, one dealing with transgenic crops and the other with transgenic medical and pharmaceutical products, (ii) the GEAC's role be limited to 'environmental clearance', and the authority for deciding on the commercial release of GM-crops transferred to the ICAR and MoA, (iii) the RGCM should be authorised to grant approvals for open field testing for biosafety, while ICAR and the applicant company should be asked to conduct large-scale field trials, and (iv) the post-release monitoring of GM-crops be done by the MoA and ICAR.
- If a transgene has been declared bio-safe, its derivatives need not always be evaluated for bio-safety again.
- Farmers' and consumers' organisations be invited to participate in the process of assessing GM-crops (but not other GM-concerned NGOs/CSOs), with the farmers' and consumers' organisations given complete information on the benefits and risks associated with GM crops.

- Before embarking on the development of any specific GM-crop, the alternatives available for meeting the food and nutritional needs should be reviewed comprehensively and the export market should be kept in view.
- GM-foods and products be labelled as such, along the lines of the recommendations by the UN's Codex Alimentarius.
- Organic farming zones and agro-biodiversity sanctuaries be established to prevent GM-plants from cross-pollinating non-GM plants.
- The impact of GM-feeds and GM-vaccines on milk, meat and eggs be studied.
- The government create special insurance schemes, provide venture capital for and promote public-private partnerships for the production and marketing of GM-seeds.
- Government agencies be created at the state and district levels for preventing the proliferation of illegal GM-seeds.
- A 'single window information centre' be created for providing information on all aspects of bioethics and biosafety.

The two GM-concerned leading CSOs in India, Gene Campaign and Greenpeace India, both of which were consulted by the Taskforce, expressed misgivings on some aspects which were as follows (Sharma 2004):

- The proposed segregation of GM and non-GM crops by zones may not be feasible and the only way for protecting native germplasm from foreign genes with likely negative impact, like herbicide tolerance, is to disallow the GM version of that particular crop.
- Since GM technology has implications for life forms unlike other technologies and that there are social, economic and ethical concerns associated with it, it would be wise to set up a National Bioethics Commission to steward this technology with responsibility and sensitivity.
- It was a dangerous suggestion to assert that once a transgene is tested for biosafety in a particular crop it need not undergo tests while implanted in other crops.
- The suggested changes in modalities of operation did not clarify how the time taken for processing applications could be reduced without compromising the biosafety aspects. The Taskforce had

totally overlooked the needs of toxicological and other biosafety studies.

- The Taskforce had avoided the issue of (the government's and/ or the private sector's) accountability in case of failures in ensuring biosafety, in the event if Bt-resistant bollworms were to proliferate following the cultivation of Bt cotton.

The taskforce's report paved the way for further improvements in the regulatory system.[18] The report's recommendations regarding (1) an event based approval system; (2) that once an event has been declared 'bio-safe', its derivatives need not be subjected to extensive biosafety testing, and (3) a fast track approval process may be considered for notified varieties expressing the approved events were sought to be implemented. These are discussed later in the chapter.

In the meanwhile, as a first step, in 2004, the GEAC decided[19] to put on the MoEF website the status of projects pending consideration of the GEAC. It was also decided that all proposals (complete in terms of requisite documents) received in the ministry 30 days prior to the next GEAC meeting would be submitted for consideration in the subsequent GEAC meeting. The DBT also developed a website which shares data about biosafety to facilitate dissemination of all relevant information. The RCGM shares the minutes of the meetings giving an overview of discussion, list of participants and date for the next meeting.[20]

Meanwhile, in a writ petition[21], the Supreme Court gave significant interim orders. On 1 May, 2006, it directed that field trials of GMOs shall be conducted only with the approval of the GEAC. The Supreme Court also directed the GEAC to withhold approvals until it issued further directions after hearing all concerned. The Court asked the government to consider associating independent experts with the GEAC. Further, so as not to have a conflict of interests of members, it directed that before suggesting the names of such experts for the purpose, the government should ask prospective candidates whether they were being funded directly or indirectly by the biotech industry. The court acting on the application of Gene Campaign, in October 2006, directed the release of technical data relating to GM crops.

As a step towards comprehensive review of the regulatory processes the GEAC constituted a sub-committee on Bt cotton and related issues under the Chairmanship of Dr. C.D. Mayee, Chairman, Agricultural Scientists Recruitment Board and Co-Chair, GEAC.[22] The Mayee sub-committee recommended a new procedure for new hybrid/

variety with new gene for approval of transgenic crops that is in place (Figure 3.2).

The Mayee sub-committee recommended that extensive biosafety and agronomic testing was not necessary for an approved event and that once an event had been tested for its biosafety and approved for environmental release, it should be treated at par with the non-Bt hybrids. It was noted that an 'event based approval mechanism' (EBAM) instead

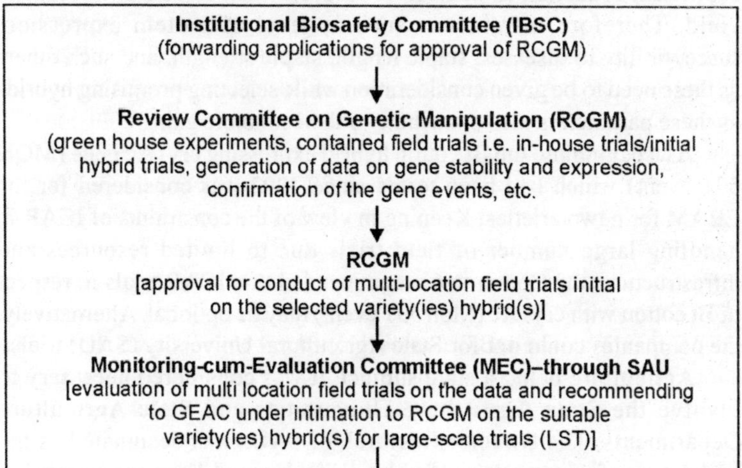


*Source*: Reproduced from http://dbtbiosafety.nic.in

**Figure 3.2:** Flow Diagram Showing the Recommended Procedure Based on the Recommendations of Sub-committee on Bt-cotton and Other Issues under the Chairmanship of Dr. C.D. Mayee

of the case by case approval process would speed up the introduction of new and diverse products for the Indian farmer, stimulate competition and offer a wider choice, without compromising biosafety and environmental safety.[23] The committee said that while due consideration for agronomic aspects of the GM-seeds should be given, the technology in no way increases the yield potential of a hybrid but because of the inherent protection to bollworms there is saving of bolls, and also significant reduction in number of sprays, which results in increase in yield. Therefore, parameters such as level of protein expression, susceptibility to diseases, staple length, staple strength, and such others as these need to be given consideration while selecting promising hybrids as these parameters also contribute to the economic gain.

As a beginning, the Bt cotton hybrid expressing cry1Ac gene (MON 531 event) which had been tested since 2002 was considered for the EBAM for new varieties. Keeping in view of the constraints of ICAR in handling large number of field trials due to limited resources and infrastructure, the Committee recommended that ICAR trials in respect of Bt cotton with cry1Ac (Mon 531 event) may be optional. Alternatively, the companies could opt for State Agricultural University (SAU) trials.

Agriculture being a state subject, it was considered necessary to involve the State Agriculture Universities and State Agriculture Departments. The Director Research of SAUs was designated as the nodal point for pre-release field monitoring and Director Extension of SAUs as the nodal point for post release monitoring mechanism, with representatives of State Agriculture Departments/District Agriculture officers in the committee. The cost of monitoring was to be borne by the applicant. A fee of Rs. 5000 per trial (per hybrid/location) under monitoring in multi-location trials (MLT) needed to be deposited with the Registrar/Controller of the University who in turn would make available funds to the Director of Research to meet the expenses for organizing and conducting the monitoring and report preparation as per the prescribed norms. If any large scale trials (LSTs) were conducted in the jurisdiction of a SAU, Rs. 500 per hybrid/per location was to be deposited by the applicant with the University for monitoring.

The state governments, however, represented that many of the hybrids approved by the GEAC were not suitable for the entire state. They suggested that it would be better if the GEAC followed a decentralising system which would empower the state government and SAUs for assessing the suitability of the genotype for the various regions of a state.

Following the recommendations of the sub-committee constituted for this purpose, a decentralised system involving the SAU[24] and State Department of Agriculture was put in place for Bt cotton hybrids with approved events.[25]

The decentralised system would follow the steps:

(a) Transfer of GM material to be approved by the RCGM through the IBSC.
(b) The strip trials to be conducted with the approval of the IBSC under intimation to the RCGM.
(c) Transformation and backcrossing under the mandate of the IBSC.
(d) In house trials/farmers field trials to be conducted by the Company.
(e) Request for SAU trials along with an affidavit confirming the following:
    (i) Confirmation of gene event through molecular characterisation;
    (ii) Level of protein expression in greenhouse and field trials;
    (iii) Morphological characters using DUS (distinctiveness, uniformity and stability) descriptors;
    (iv) Bio-efficacy data generated in labouratory conditions; and
    (v) Authorisation/no objection certificate (NOC) from technology provider in case of sub-licensee.
(f) Based on the SAU trial data and company data, the SAU and State Governments to identify the genotype suitable for commercial cultivation.
(g) In case the company likes to identify certain genotypes, as centrally notified varieties they should follow the prevailing AICCIP (All India Coordinated Cotton Improvement Project) procedure under the ICAR system.

The GEAC formally adopted the Event Based Approval Mechanism in respect of Bt cotton hybrids expressing approved events in its meeting held on 2 April, 2008. To further operationalise the event based approval mechanism, the GEAC decided to set up a 'Standing Committee'.[26]

The 'Standing Committee' is constituted by the Chairman, GEAC. It comprises of experts in the field of plant breeding, an entomologist, an agronomist and a plant pathologist. It was further agreed that the 'Standing Committee' would comprise of experts who are not members of the RCGM/GEAC. However, a representative from the GEAC/RCGM could be included to ensure continuity and also to interface with the Committee and the GEAC/RCGM. The 'Standing Committee' is serviced by the Department of Biotechnology.

The terms of reference of the Standing Committee are as follows:[27]

(a) Evaluate the duly submitted applications for commercial release of Bt cotton hybrids expressing approved events and verify that the affidavits are in order;

(b) Identify 10 per cent of the total applications received and send the samples to the referral labouratory at the Central Institute of Cotton Research (CICR), Nagpur for verification of the claims submitted along with the affidavit. Other labouratories including in the private sector having requisite capability will be identified for acting as referral labouratories for this purpose.

(c) Recommend the hybrid for commercial release in the given state, subject to the condition that punitive action may be initiated in case of providing any wrong information/ declaration irrespective of whether or not the application falls under the 10 per cent category;

(d) Maintain complete records of all the applications referred to it; and

(e) Lay down procedure for identifying 10 per cent of the applications for necessary verification.

To start with, the 'event based approval mechanism' was applicable for Bt cotton hybrids expressing four events namely: cry 1Ac (MON 531 event), cry 1 Ac and Cry 2 Ab (MON 15985 event, (cry1Ab + CryAc) GFM and cry1Ac (Event1). In respect of cry1AC (Dharwad) event developed by the Central Institute of Cotton Research (CICR), Nagpur, it was decided that the new procedure would be applicable only after the event completed three seasons of commercial cultivation.

Those applicants (private/public sector organisations) who are interested in commercial cultivation of Bt cotton hybrids expressing approved events need to file applications along with affidavits to the 'Standing Committee' with the required information.

Once an application passed the scrutiny of the Standing Committee, the release of Bt cotton hybrids in the respective state would follow the same process as that for conventional cotton seeds. Those applicants not satisfied with the decision of the 'Standing Committee' were free to approach the NEAA (National Environment Appellate Authority)[28] or represent their grievances in the appropriate court of law.[29] It was also decided that in addition to the CICR, Nagpur (which was notified by the Ministry of Agriculture as a referral lab), there was a need to identify NABL[30] accredited laboratories for verification of Bt gene in cotton seeds.

In the first meeting of 'Standing Committee' held in DBT on 17 April, 2009, a total of 25 companies submitted 120 applications for approval for commercial release of 217 Bt cotton hybrids expressing approved events under EBAM in northern, central and southern zones.

To further streamline the regulatory process, the 'standing committee' has also incorporated the requirements of (i) one year field functional data from concerned SAU of the particular hybrid; (ii) morphological characteristics using DUS descriptors as per PPVF&RA (Protection of Plant Varieties and Farmers Rights Authority)[31] guidelines as a mandatory requirement under the Event Based Approval Mechanism.[32]

However, during the GEAC discussions on EBAM, serious reservations were raised by the external expert, Dr. P.M. Bhargava[33]. They were as follows:

(i) Every event needed to be tested and reviewed for biosafety. He suggested that proteomics study and DNA finger printing should form an important component of the biosafety evaluation.
(ii) There was a need for an independent testing laboratory.
(iii) He called for a three to four-year total moratorium on GM crops and their products on the grounds that (a) the GEAC was relying on biosafety studies (pollen flow, seed germination study, soil microbial studies, toxicity and allergenicity studies etc.) generated by the applicant which was an interested party. There existed no mechanism to verify with the experimental and control groups, nor was the data reconfirmed by a third party. (b) Only acute toxicity studies were conducted. It was pointed out that in the soil microbial studies, it was not enough just to have the total number of organisms determined but also look at the bacterial profile and the effect on soil micronutrients. All toxicity studies must be done on the protein in the GMO (for example, appropriate extracts or whole parts of the plant). Toxicity studies done with the surrogate protein made, for example, in E.coli, should not be accepted. (c) The pollen flow studies in Bt cotton and Bt brinjal had indicated that pollen flow does take place up to 10 m and 20 m, respectively. This alone, even on the basis of unconfirmed studies of M/s Mahyco on pollen flow, should rule

out the use of Bt cotton by small and marginal farmers representing 84 per cent of the farming community.

iv) No GM crop should be released unless appropriate and reliable DNA finger printing, proteomics analysis and studies on reproductive interference in at least three mammalian species have been done by an accredited and independent laboratory with established expertise in the field.

With the 'event based approval mechanism' adopted by the GEAC, it was felt that there was an urgent need to suggest alternate and practical 'insect resistance management' (IRM) strategy suitable for the agricultural practices in the region. The GEAC also opined that until an alternate strategy was notified, the current practice of maintaining 20 per cent area as refuge should be strictly complied with.[34] However, in 2010, considering that there still remained the issue of non-compliance of refuge and the practical difficulty in getting farmers to plant the refuge, it was decided to form a network of technology providers and lead the resistance monitoring efforts in this country, involving SAUs specifically. Hence it was decided to put in place a network which would be actively involved in developing strategies to increase refuge compliance. Some of the suggestions for insect resistance management strategies included supplemental insecticidal spray, stubble destruction, pheromone traps and deep ploughing which would delay the development of insect resistance to cry1Ac gene. For refuge purposes, 'desi cotton' (local indigenous varieties of cotton) and late planted 'bhendi'(okra) were recommended.[35]

With regards to the issue of biosafety, the Supreme Court in its ruling dated 8 May, 2007 amended the order given on 22 September, 2006 directing the GEAC to stop all approvals to the extent that the GEAC may accord approval for commercial release of Bt cotton hybrids expressing approved gene events [cry 1Ac (MON 531 event), cry 1Ac and cry 2Ab (MON 15985) gene, cry 1Ac event 1 and cry 1Ab + cry 1Ac GFM]. However, the Supreme Court permitted conduct of field trials of GM crops (24 items approved by the GEAC during 1 May, 2007 and 22 September, 2007) expressing new gene events subject to the following conditions: (1) All trials should have a lead scientist's name with contact details who would be responsible for all aspects of the trials including regulatory requirements. (2) An isolation distance of 200 m would be maintained during field trials. (3) Prior to bringing out the GM material from the greenhouse for conduct of open field trial the company should submit a validated event specific test protocol at an LOD ('limit of detection') of at least 0.01 per cent to detect and confirm that there has been no contamination.

However, the GEAC[36] was of the view that the direction of the Supreme Court to maintain isolation distance of 200 m during GM crop field trials and a 0.01 per cent LOD cannot be uniformly applicable for all crops as the nature of pollen flow and level of cross pollination with related species would vary depending on the biology of the crop and the host environment.

Accordingly, the GEAC set up a Sub-Committee under the Chairmanship of Dr. Khadi, Director, CICR, Nagpur to examine the implications of the Supreme Court's order with respect to conduct of field trials and measures to be taken by the GEAC to ensure its compliance, as well as look at the feasibility of implementing the other two conditions.

With regard to a uniform isolation distance, the Khadi Sub-committee[37] said that it was scientifically not tenable as the nature of the pollen flow and level of cross pollination depends on the biology of the crop and the host environment in which it is being cultivated. Therefore, each crop has its own unique isolation distance. It was recommended that the isolation distance prescribed under the Indian Minimum Seed Certification Standards[38] manual may be followed. In addition to the prescribed isolation distance, it was suggested that additional biosafety measures may be prescribed on a case to case basis. The combination of physical and biological containment measures prescribed by the regulatory agencies was considered adequate to ensure the restriction of crosspollination to the highest possible levels.

With regard to the requirement for a validated protocol of 0.01 per cent LOD the sub-committee suggested it to be dispensed with because of the already stringent guidelines of maintaining scientifically prescribed isolation distance (which in case of several crops, is more than the 200 m isolation distance prescribed by the Supreme Court) and also that together with the additional biosafety measures prescribed during field testing of GM crops, the possibility of contamination is rare.

The sub-committee put forward that the arbitrary requirement for provision of a detection method with an LOD of 0.01 per cent at the time of field trial may be dispensed with, because it was suggested that it would be better if the LOD may be decided at levels that are practical to use prior to the commercial release/export requirements. This could be made compatible with the acceptable thresholds that may be defined for labelling purposes and based on the technical aspects of the analytical methods available in the international laboratories and the genome size of the crop.

The RCGM also met separately[39] and after deliberations on the Supreme Court directions gave the following recommendations (which were endorsed by the Khadi Sub-committee): (1) MLRT (multi-location research trials) for new events should not be conducted in the farmers' field. MLRT should be undertaken by the companies/institutions either in their own premises, research farms or long leased land or at the SAU/ICAR institutions. (2) Strip trials for new events should be undertaken by the company in their own premises/research farms. (3) Applicant should generate complete biosafety data along with MLRT. The applicant should submit a validated event specific test protocol before undertaking the trials. (4) For all new events, applicants would need to include plant parts, such as leaves/shoot/bolls in addition to seeds in the animal feeding studies for biosafety evaluation in the wake of certain reports indicating adverse effects of the leaves/shoot/bolls on animals. This was taken up in lieu of the reports of sheep deaths in Andhra Pradesh.[40] (5) MLRT may be conducted in minimum one location and maximum at two locations in each state.

It was also decided earlier by the GEAC[41] that applicants needed to obtain approval of the concerned Gram Panchayat prior to conduct of field trials. This had been done because of several representations that were received that the farmers were not aware of the GM crop trials in their field.

The GEAC brought out the *Guidelines for the Conduct of Confined Field Trials of Regulated, GE Plants* in 2008. These guidelines use specific terminology to differentiate between research conducted under 'contained' and 'confined' conditions. The guidelines are intended to provide guidance to applicants for the conduct of confined trials. The scope of the Guidelines clearly mentions that they are not intended to explicitly define all the requirements for the conduct of a confined field trial, as further terms and conditions/requirements may be identified during the review process by the regulatory authorities. This document covers all GE/transgenic plants modified through recombinant DNA (rDNA) technology.[42]

The Indian Council of Medical Research's (ICMR) *Guidelines for the Safety Assessment of Foods Derived from Genetically Engineered Plants in India,* 2008 has also been adopted. This document applies to all whole foods, food products, and foods used as ingredients that are derived from GE plant sources. Safety assessment is designed to identify whether a hazard, nutritional or other safety concern is present and if present, to collect and analyse information on its nature and severity following a structured and integrated approach performed on a case-by-case basis. A stepwise process is to be followed for the safety assessment of foods derived from GE plants. The guidelines provide guidance to both applicants and reviewers for regulatory purposes. It clearly states that further data requirements may be identified during the safety assessment process. Information on data submitted will be for the parts of the plant used as a food source.[43]

The GEAC has adopted the following criteria to address issues related to conflict of interest[44]: (1) If a member of the GEAC is either a team leader or member of a team that has developed a transgenic plant which has come up for consideration before the GEAC, that person shall not participate in the discussion regarding such an application. (2) If a member of the GEAC is a consultant for an industry/research foundation that has developed a transgenic plant which has come up for consideration before the GEAC, that person shall not participate in the discussion regarding the said application. (3) If a member of the GEAC involved in the development of transgenic plant constituting the same crop/trait of interest that is being considered by the GEAC or is involved in the development of a recombinant vaccine/drug against the same disease, that person shall not participate in the discussion regarding such an application. (4) Further, in all the three situations mentioned above, the GEAC member should not be involved in the conduct or monitoring of field trials/clinical trials with regard to an application being considered by the GEAC.

As per Supreme Court direction Biosafety Research Level-1 (BRL-1) trials require approval from the GEAC.[45] The GEAC has not accepted the suggestion made by the RCGM that the applications for conduct of all confined field trials of GE crops may be received and processed by the GEAC Secretariat including issuance of permission letter. The MoEF has filed an application for amendment of the order.[46]

The laboratory and greenhouse data is a prerequisite for consideration of the application for BRL-1 trials. The application for BRL-1 trials is to be submitted along with the laboratory and greenhouse trials data to the RCGM. The GEAC can consider the matter only on receipt of the RCGM recommendations. The applicants are advised to submit their proposal for BRL-1 at least 60–90 days before they intend to start the trials as the time frame for review of the protocols and procedure may vary for new events with novel traits that are under development.

The GEAC has decided that BRL-1 trials with direct imported GM seeds without carrying out R&D work in the country shall not be permitted.

Following the adoption of event based approval system and new set of guidelines and SOPs for confined field trials, the GEAC has brought out a new guidance document listing out studies to be conducted under BRL-1 and BRL-2 trials. The guidance includes that: (1) BRL-1 trials will be conducted for a minimum of two seasons in two to three locations; (2) BRL-2 trials will be conducted for a minimum of one season on the understanding that the data generated during BRL-1 trials will be generated during BRL-2 trials; (3) Additional trials may be called for based on the quality of data generated and/or new scientific evidence; and (4) Applicants may, for the purpose of research or otherwise may conduct BRL-1/BRL-2 trials for more than the prescribed seasons with the approval of the RCGM/GEAC.[47]

Even after so many years the debate about the efficacy of GM-crops continues. There still remain issues of transfer of illegal seeds and burning of fields in many states where GM-crop trials are taken up. The GEAC has repeatedly urged that the state governments should take cognizance of various complaints received, including strict vigilance of the ginning units, research farms, and fields where trials are done. The Ministry of Environment and Forests issued a Gazette Notification in September 2006 empowering the Seed Inspectors, Seed Analysts and Laboratories under the Seed Act, 1966 and Seed Control Order, 1983 also under the Environment Protection Act (EPA), 1986. The state government can initiate prosecution under EPA/Seeds Act against the erring companies/individuals.

The GEAC in its 100th meeting suggested that advance effective action against selected parties be taken well before the onset of the kharif season to send right signals for enforcement of law. To curb the illegal release of transgenic cotton seeds, the GEAC has mooted the idea of an annual audit report on the action taken by the state governments to be sent to the GEAC before the onset of the next sowing season. The GEAC has decided to prepare a proforma for submitting an annual audit report on the action taken. The issue had come in for more serious attention in view of the recent moratorium on commercialisation of Bt brinjal. The state governments now had more of a responsibility to not only monitor Bt cotton but also other crops in the pipeline.

The GEAC has decided that a scanned copy of all approval letters for field trials with GM crops will be sent to Director Agriculture/ Commissioner of Agriculture by e-mail. As per the conditions imposed in the approval letter, the applicants are required to inform the state governments about the field trials within 15 days of sowing. If this is not complied with, the state governments, through the SBCC/DLC, could take punitive action as they deemed fit and inform the GEAC.[48]

The GEAC has also deliberated on the slow process of getting a no objection certificate (NOC) from the state governments and has decided on the following options for obtaining state government approval for conducting GM crop field trials: (1) The GEAC may write to the state government on a case-by-case basis which have been approved for field trials and issue the approval letter only on receipt of consent from the state government. In case no communication is received in one month, the applicants may be allowed to initiate the trials. (2) The GEAC may accord approval to the applicant for field trials subject to the condition that sowing will be initiated only on receipt of consent from the state government. Copy of the consent from the state government and details of sowing should be submitted to the GEAC within seven days of receipt of consent from the state government. (3) The applicant may be advised to submit consent from the state government along with the application for which the GEAC may communicate to the state governments the new requirement.[49]

Just as in India, there has been a relook at regulatory mechanisms in the United States, Canada and E.U., which is reviewed next.

## 4. Experience in Other Countries

The U.S. government took to regulating GMOs under its existing laws in 1986. The three main agencies which are primarily responsible for regulation are the U.S. Department of Agriculture (USDA)[50], U.S. Environment Protection Agency (USEPA Regulation 2002), and the U.S. Food and Drug Administration (USFDA). The GMO products are viewed as science driven and are regulated according to their intended use. GM crops are known as pesticidal plants—those plants that produce their own pesticides, such as Bt corn and Bt cotton. The USEPA is the regulatory authority over pesticidal plants which are regulated much in the same way as chemical pesticides are regulated. The USDA is the regulatory authority over all other transgenic plants.

The principal statute used by the USFDA to regulate genetically engineered organisms is the U.S. Federal Food, Drug and Cosmetic Act. However, there is little formal regulation of genetically engineered foods by this agency. There is no pre-market testing required for engineered foods, no requirement for labelling, and no requirement for USFDA approval before an engineered food is placed on the market.

For many years the regulatory authorities USDA had not paid attention to the major issues of weediness, pest resistance, impacts on non-target organisms, gene flow and new weeds, as there were newer technologies being developed to resist them. However, there have been evaluation studies by the United States National Academy of Sciences and the Ecological Society of America which have been very critical of the regulatory system. The regulations have been shown as being opaque, not scientific in their approach and not applied uniformly.[51]

The USEPA is more concerned about environmental impacts. This has been spurred on by various studies. A study that showed impact of Bt corn pollen on Monarch butterflies, forced the USEPA to pay attention to non-target impacts. Similarly, hectic lobbying by various environmental and organic farm organisations has made the USEPA recognise the need for maintenance of refuge. The USEPA requires that farmers plant five per cent non-GM crops in Bt cotton fields. However, surveys have shown that only 80 per cent of farmers complied. In the state of Hawaii, the USEPA has in fact put precautionary measures in the growing of cotton to prevent gene flow from wild and weedy relatives. However, the USEPA has not stopped the commercial introduction of any GM-crop. The system continues to evolve on regulatory measures to handle issues as and when they arise.

Canada sought to promote development of biotechnology through the National Biotechnology Strategy in 1983. In 1993, the Federal Regulatory Framework for Biotechnology was put in place. This framework came about from an agreement among federal regulatory departments on six principles for an efficient, effective approach for regulating biotechnology products so as to balance practical benefits of biotechnology with need to protect human health, animal health and the environment. The six principles framework included—(a) maintaining Canada's high standards for the protection of the health of workers, the general public and the environment; (b) using existing legislation and regulatory institutions to clarify responsibilities and avoid duplication; (c) continuing to develop clear guidelines for evaluating products of biotechnology which are in harmony with national priorities and international standards; (d) providing a sound scientific database on which to assess risk and evaluate products; (e) assuring that both the development and enforcement of Canadian biotechnology regulations are open and include consultation; and (f) contributing to the prosperity and well-being of Canadians by fostering a favourable climate for investment, development, innovation and adoption of sustainable Canadian biotechnology products and processes. The three agencies central to regulatory mechanism are Canadian Food Inspection Agency, Health Canada and Environment Canada.[52]

In 2007, a new Science and Technology Strategy and governance structure replaced the Canadian Biotechnology Strategy, while the Canadian Regulatory System for Biotechnology remains in place. In the current biotechnology framework, the federal government acts as both the promoter and regulator of GM foods. In contrast, provinces primarily function independently in the promotional area. The federal government takes the leadership in regulation, and relations among the orders of government are generally characterised by independence and non-hierarchy. The collabourative nature of intergovernmental relations has served well, where the federal government works towards harmonization on international agreements and trade issues, while the provinces pursue economic development with the knowledge that the federal government ultimately takes responsibility for regulation. The main issues faced in the governance of biotechnology include accountability, transparency and public participation (Gabler 2008).

The European Union has a comprehensive legislation in the Directive 2001/18/EC of the European Council and the Parliament, which governs genetically modified foods and crops in its territories. It is an overarching law and covers both the experimental release of GMOs into the environment and the placing on the market of GMOs. EU regulation of GMOs aims to (a) Ensure strong protection of human life, health and welfare; (b) Protect the environment; (c) Defend consumer interests; (d) Ensure that the EU's internal market works effectively.[53] In 2009-2011, an independent evaluation of GMO legislation was carried out to address cultivation and GM food and feed. It brought out that the stakeholders supported the legislation's main goals in protecting health and environment, and creating an internal market. The stakeholders, however, opined that some adjustments were necessary which has already been acted upon. These include: (a) Proposal for increased flexibility on GMO cultivation, (b) Technical information on the socio-economic implications of GMO cultivation, (c) Reviewing and transforming the risk assessment guidelines for food and feed and environmental release into legal documents approved by EU countries; (d) Reinforcement of the environmental monitoring; (e) Harmonised sampling and testing for low level presence in food, (f) Assessment of new plant breeding techniques; and (g) Intensified communication on GMO issues.

What clearly comes out is that in United States issues of refuge, gene flow, and new weeds have gained attention. Canada has tried to look more closely into the issue of federal-province relations. Whereas E.U. is looking at how GMOs can be looked at in a more flexible manner so that it meets socio-economic needs, while being environmentally sustainable.

## 5. Conclusion

The adoption of Bt cotton has grown by leaps and bounds in India covering nearly 80 per cent of the total cultivated area of cotton. It has, however, evoked wide ranging reactions among the various stakeholders. It would not be wrong to say that the biosafety regulations in India have evolved with the growth and acceptance of the Bt cotton crop in the country. The main shortcomings to complete acceptance of the technology are in the nature of implementation of the regulatory measures—(a) ensuring planting of refuge; (b) stringent vigilance with regards to illegal seeds; (c) building infrastructure for testing facilities in terms of labouratories; (d) improving networking to better utilise human resources; (e) working cogently to build institutional relations between the Central and State regulatory mechanisms; and (f) being trustworthy in terms of accountability and transparency.

Efforts have been made to address the above concerns and streamline the processes. It needs to be recognised that the three tiered regulatory mechanism in place is not really hierarchical in nature. It is many layered and top-driven. The GEAC is responsible for the large scale field tests and commercialisation. It has followed upon the recommendations of the Taskforce Report, 2004. An event based approval system for Bt cotton hybrids expressing approved events has been put in place. A 'Standing Committee' has been set up to look at approvals in the states. Today the state government departments and the SAUs work together with GEAC. The state governments need to give a 'no objection certificate' before the field trials can be undertaken. The Gram Panchayat is to be informed well in advance before the trials begin. Efforts are being made to look at measures to improve better adherence in growing refugia. Scientists are also beginning to take a relook at the indigenous (*desi*) varieties of cotton which seem immune to drought and pests.

However, it clearly comes out that the responsibility of implementation of the regulating the GM-crops is decentralised to the various agencies such as the IBSC and SBCC. An important dimension that has not been addressed, but which needs to be recognised as playing an important and influential role in the system is that of agricultural extension. The performance of the extension services has grown weaker and contributed to the difficulties in keeping pace with the changing technologies. Strengthening of the regulatory system in the long run would clearly depend on improving the extension services such that it can cater to the needs of the farmers.

As Bt-technology has progressed, newer hybrids expressing new genes/events are in the offing. But the testing of the new genes seems to be recreating situations similar to what was seen right in the beginning with the coming of cry1AC gene. There have been reports of repeated defiance with regards to refugia and illegal seeds which the regulatory authorities have not been able to control.

There continue to be reports which undermine public trust and confidence in the regulatory system. Recall the first of the indigenous event developed by the CICR, Nagpur and the UAS, Dharwad that was commercialised in 2009. The ICAR in 2011 has been forced to close curtains on this event after it came to light that the commercial seeds were contaminated with Monsanto's proprietary Bt gene instead of the domestic one. An inquiry committee had been set up to look into the entire process. The members had largely opined that there may have been a case of conflict of interest which had compromised the regulatory process.

The GM-concerned civil society organisations have played a very crucial vigilant role in highlighting the deficiencies of the system and bringing forth a discussion on alternate methods and systems. In recent times the concerns include the preservation of bio-diversity and issue of biosafety. In case of Bt brinjal, the National Biodiversity Authority has decided to initiate action against Monsanto. It is alleged that the stated actors had illegally accessed and genetically modified six varieties of Indian brinjal to produce Bt Brinjal.

Then again the issue of biosafety of GM-crops in terms of food and feed continue to be unresolved. The case in discussion here is of soyameal for which the government (Empowered Group of Ministers) had allowed duty-free imports. Concerned organisations fear that such imports could result in GM material landing up in the food chain, as most of the soyameal produced overseas is of the GM variety. The missing role of the GEAC and the lack of coordination among the various agencies was once again brought to notice by the CSOs.

As seen elsewhere, there is a continuing and growing debate over the sustainability of the GM-crops. The most recent report tabled in the Lok Sabha is stated to have recommended study of socio-economic and health impact of Bt cotton seeds and a thorough probe into Bt brinjal, saying that adequate tests had not been carried out and the approval committee was under 'tremendous pressure' from the 'industry and a minister' to approve it. It is also said that while the ICAR has termed GM crops as being safe, some other departments under the Ministry of Science and Technology have expressed apprehensions about the technology and have apparently argued for more research before arriving at a final decision on introduction of GM food crops. The Council of Scientific and Industrial Research has been quoted as saying that the regulatory system in India was not strong enough to introduce GM crops as it would involve matters related to biodiversity, biosafety, environment, human health, for and fauna. Many GM-sceptics have hailed the report on GM crops as a historic, comprehensive and well-grounded document, and has thrown light on certain aspects of agrarian crisis that did not figure in any study or experts' reports earlier. This clearly shows that the controversy regards the Bt technology in India is far from settled.

## Notes

1. As per Cartagena Protocol on Biosafety to the Convention on Biological Diversity (Secretariat of the Convention on Biological Diversity 2000: 4) – 'Modern biotechnology', means the application of—a. In vitro nucleic acid techniques, including recombinant deoxyribonucleic acid (DNA) and direct injection of nucleic acid into cells or organelles, or b. Fusion of cells beyond the taxonomic family, that overcome natural physiological reproductive or recombination barriers and that are not techniques used in traditional breeding and selection.
2. See answers to Unstarred Question No. 4267, in the Lok Sabha (lower house of the Parliament) answered on 20 December, 2011 by Shri Harish Rawat, Minister of State in the Ministry of Agriculture, Food Processing Industries and Parliamentary Affairs, 'Research on GM Crops'.
3. There are recent studies debating whether there is growing regulatory polarisation between the US and EU biotechnology approaches. Falkner and Gupta (2009) while studying countries like Mexico, China and South Africa show that there is no convergence around either of these two international models but countries have in fact steered a course that suggests substantial regulatory diversity in their policies.
4. The Genetic Engineering Appraisal Committee is now called the Genetic Engineering Approval Committee vide the Gazette of India Notification of the Ministry of Environment and Forests, 16 July, 2010.
5. Choudhary and Gaur (2010) mention a total of 11 studies, all conducted by public sector institutes during the period 1999–2009, covering both pre and post-commercialisation of Bt cotton.

6. The European Union (EU) has established a stringent legal framework for regulating GM food and feed in the EU; several countries in the EU have taken a decision not to allow cultivation of GM food crops. Import and sale of GM commodity as food and feed are permitted as per the EU regulation subject to labelling laws. In the EU, maize is the only GM food crop approved for cultivation for use as food and feed. GM potato called Amflora has also been approved for cultivation for industrial use. See answers to Starred Question No. 360, in the Lok Sabha (lower house of Parliament) answered on 30 April, 2012 by Smt. Jayanthi Natarajan, Minister of State for Environment and Forests, 'Commercialisation of GM Crops'.
7. The precautionary principle is defined here as the principle that allows action on the suspicion of any potential damage to the environment and the health of humans and livestock, without first requiring full scientific proof.
8. Case-by-case assessment implies that as long as the experiences with GMOs are limited, each GMO case has to be assessed on an individual basis in relation to its potential impact on environment and health.
9. The step-by-step process implies that tests (greenhouse, contained and large-scale field) involving GMOs are done in a stepwise manner, building on experience from previous tests. This means that large-scale field trials have to be preceded by greenhouse and contained tests.
10. The Seeds Bill 2004 is still being debated. For a legislative brief on the Seeds Bill 2004, see http://www.prsindia.org.
11. ICAR (which falls under the MoA) administers the national registry of seeds that have been approved and certified for general release and marketing.
12. The DBT said that the situation was the result of a 'misunderstanding' by Karnataka, and that it had already communicated in 1990 to all the state governments that given the location-specific nature of agri-biotechnology, its development would be dependent on the cooperation and involvement of the state governments. (Indira *et al.* 2005: 86).
13. In 2005, sheep mortality was reported in the district of Warangal in Andhra Pradesh. A fact finding team of local CSOs who visited the area reported that a total of 1800 sheep and goats had died after grazing on Bt cotton fields (See *Frontline* 2007).
14. The following website reports that RTI queries regarding SBCC and GM crop field trials in Gujarat received responses from the Forests and Environment Department in Gujarat that the Agriculture and Cooperation Department is the nodal department in the case of Gujarat *vide* notification no. EPN-1099-GOI-64-P (Part-II) dated 4 February, 2004, while the Agriculture Department responded first by saying that it pertains to the Forests and Environment Department, and later amended that the work of SBCC meeting is not being done by this (Agriculture and Cooperation)

department. Available at http://indiagminfo.org/?p=376#more-376 (dated 28 February 2012).

15. Soon after, in 2004, Gene Campaign (a CSO whose Convenor is Suman Sahai) filed a writ petition seeking the court to direct the central government to bring in consonance with the Constitution the 1989 Rules.
16. Supreme Court admitted a writ petition by Aruna Rodrigues in 2005, from Mhow, Madhya Pradesh.
17. Dr. Swaminathan an eminent scientist, was the Head of Indian Agricultural Research Institute in Pusa in Delhi in the 1960s and 1970s. He later became the Head of the International Rice Research Institute in the Philippines, a CGIAR-institution. He is widely credited with the 'launching' of Green Revolution in India and he made decisive interventions at GoI levels that contributed to its success. On retirement, he set up the M.S. Swaminathan Science Research Foundation in Chennai.
18. The implementation of the taskforce recommendations was discussed in the 43rd GEAC meeting held on 9 June, 2004.
19. Decisions taken in the 40th meeting of GEAC dated 31 March, 2004. The GEAC meetings and decisions can be accessed at http://www.envfor.nic.in/divisions/csurv/geac.
20. All meetings of RCGM and MEC can be accessed at http://dbtsafety.nic.in. The biosafety data submitted by the companies is available at http://igmoris.nic.in.
21. Writ petition filed by Aruna Rodrigues in 2005 (Biosafety-info 2012).
22. Decisions taken in the 66th meeting of the GEAC held on 2 May, 2006. The Mayee Committee was set up by the MoEF, GoI *vide* OM No.10/13/2005-CS-GEAC dated 28 April, 2006. See also 69th meeting of the GEAC held on 30 April, 2006.
23. Decisions taken in the 69th meeting of GEAC held on 30 April, 2006.
24. Capacity building activities for efficient management of field trials of GM crops are being regularly undertaken. Training of 19 SAUs with respect to the use of new guidelines for confined field trials, safe operation practices, formats for monitoring and recording of data was completed during 2008–09. The training was funded jointly by the DBT and MoEF. Biotech Consortium India Limited (BCIL) organized 79 awareness workshops on GM crops with a focus on post release monitoring in all major cotton growing states during 2005–07 to 2008–09. Global Environment Facility (GEF) Phase I on biosafety was completed in June 2008 under MoEF with GEF-World Bank assistance of US $ 1 million of which approximately 60 per cent was utilized for training. Phase-II of Capacity Building Project is underway. The co-financing by the GOI is to be to the tune of US $ 6 million, of which 67 per cent has been earmarked for Human Resource Development (HRD) component. See answers to Unstarred Question No. 760, in the Lok Sabha (Lower House of Parliament) answered on 2 March, 2010 by Prof. K.V. Thomas, Minister

of State in the Ministry of Agriculture and Minister of State in the Ministry of Consumer Affairs, Food and Public Distribution, 'Transgenic Crops'.

25. Though approved, the adoption of EBAM had been deferred. But since there was a demand, a sub-committee was set up with the TOR to (a) review the need for continued case-by-case regulation of Bt cotton expressing cry1Ac (MON 531 event) and implication of deregulation if any; (b) rationalise the zones based on the agro climatic conditions; and (c) give recommendations regarding the suitability for release of Bt cotton in other cotton growing states. See decisions taken in the 82nd (11 January, 2008) and 83rd (2 April, 2008) meetings of the GEAC.
26. Decisions taken in the 90th and 91st meeting of the GEAC held on 12 November, 2008 and on 14 January, 2009 respectively.
27. F.No.13/39/2007.CS-II, Ministry of Environment and Forests, GoI. Office Memorandum—New Procedure for commercial release of Bt cotton hybrids expressing approved events. Available at www.dbtbiosafety.nic.in (accessed on 27 September, 2011).
28. The National Environment Appellate Authority Act, 1997 was enacted to provide for the establishment of a National Environment Appellate Authority to hear appeals with respect to restriction of areas in which any industries/operations or processes or class of industries, operations or processes shall not be carried out or shall be carried out subject to certain safeguards under the Environment (Protection) Act, 1986 and for matters connected therewith or incidental thereto.
29. Decisions taken in the 96th GEAC meeting held on 9 September, 2009.
30. National Accreditation Board for Testing and Calibration Labouratories, Department of Science and Technology, GoI.
31. PPV&FR Authority has been brought in order to provide for the establishment of an effective system for protection of plant varieties, the rights of farmers and plant breeders and to encourage the development of new varieties of plants (available at http://plantauthority.gov.in).
32. Decisions taken in the 111th GEAC meeting held on 6 July 2011.
33. Dr. P.M. Bhargava, is widely regarded as the architect of modern biology in India. He was the Founder-Director of Centre for Cellular and Molecular Biology, Hyderabad, currently a centre for excellence recognized by the UNESCO. He is currently the Chairman of March (Medically aware and responsible citizens of Hyderabad). He is a recipient of Padma Bhushan.
34. Decisions taken in the 84th GEAC meeting held on 2 May, 2008.
35. Decisions taken in the 100th meeting of the GEAC held on 12 May, 2010—meeting with the technology providers and concerned departments to discuss the development of insect resistance to Bt cotton in Gujarat (Monsanto Report in *Hindustan Times* dated 6 March, 2010).
36. Decisions taken in 76th meeting of the GEAC held on 11 May, 2007.
37. Decisions taken in the 78th meeting of the GEAC held on 22 June, 2007.
38. Indian Minimum Seed Certification Standards published by the Central Seed Certification Board sets Seed Standard which consist of: (i) The

minimum percentage of pure seeds and maximum permissible limits for inert matter, other crop seeds have been prescribed, (ii) The maximum permissible limits for objectionable weeds, seeds infected by seed borne diseases have been prescribed to ensure goods seed health. Available at www.indiaagronet.com/indiaagronet/seeds/contents/minimum_seed_certification_stand (accessed 28 July, 2012).

39. RCGM meeting held on 1 June, 2007 as referred to in the 78th meeting of GEAC held on 22 June, 2007.
40. The matter regarding the sheep deaths was examined by the Directorate of Animal Husbandry, Hyderabad and Veterinary Research Institute, Izatnagar, Uttar Pradesh. The study reports indicated that the sheep deaths might have occurred due to high content of nitrates/nitrites, residues of hydrocyanide (HCN) and organophosphates which are common constituents of pesticides used during cotton cultivation and not that of Bt toxin. It was further informed that data on toxicity of Bt protein to higher mammals is extensively available as Bt cotton was released globally about a decade ago. Further, prior to release of Bt cotton in India, a battery of studies to assess the safety of Bt toxin to the environment and animals have been conducted (83rd GEAC meeting held on 2 April, 2008).
41. Decisions taken in the 72nd meeting of the GEAC held on 13 December, 2006.
42. Decisions taken in the 84th meeting of GEAC held on 2 May, 2008. During the discussions on the 'Guidelines for the conduct of confined field trials of regulated, GE plants', the GEAC said it would look into the suggestions made by Dr. Bhargava, which may be considered for suitably amending the Guidelines in due course. These were (i) the need for extensive DNA fingerprinting and proteomic study; (ii) study of possible interaction with the commonly used drugs (especially pro-biotic interferences); and (iii) reproductive interference.
43. In the *Guidelines for the Safety Assessment of Foods derived from Genetically Engineered Plants in India,* information and data for the parts of the plant used as a food source is identified under section 6.2.1—a food may be considered to have a history of safe use if it has been commonly used in the diet for a number of generations in a large, genetically diverse human population where it has been used in ways and at levels that are similar to those expected or intended in India.
44. Decisions taken in the 105th meeting of the GEAC held on 8 December, 2010.
45. The multi-location field trials conducted under the purview of RCGM are called as Biosafety Research Level-1 trials. And the large-scale experiments conducted under GEAC are called the Biosafety Research Level-2 trials. As per Rules 1989, the mandate of the RCGM is to review all R&D activities involving GMOs including bringing out manuals of guidelines laying down the procedure and protocols for the trials whereas

the mandate of the GEAC is to accord approval for large scale trials and environmental release.

46. Decisions taken in the 96th meeting of the GEAC held on 9 September, 2009. Decisions regarding biosafety research trials were taken up during the 104th meeting held on 15 November, 2010.
47. Decisions regarding biosafety research trials were taken up during the 104th meeting held on 15 November, 2010.
48. Decisions taken in the 100th meeting of the GEAC held on 12 May, 2010.
49. Decisions taken in the 108th meeting of the GEAC held on 9 March, 2011. The discussions followed the withdrawal of permission for BRL-2 trials for transgenic maize in Bihar on the basis of the request received from the Chief Minister, Bihar.
50. Research reports pertaining to U.S. agriculture issues can be accessed at http://www.ars.usda.gov.
51. Stabinsky (Date Unknown), 'Critique of US Biosafety Regulation and Implementation'. Available at www.biosafety-info.net/file_dir/23669488571ff487c1.doc (accessed on 23 August, 2012).
52. The Office of the Auditor General of Canada produces a variety of reports and publications on behalf of the Auditor General and the Commissioner of the Environment and Sustainable Development which are available at www.oag-bvg.gc.ca.
53. Information relating to food and product safety, consumer rights, public health in the European Union is available at http://ec.europa.eu/food/plant/gmo/index_en.htm.

## References

Barwale, R.B., V.R. Gadwall, U. Zehr and B. Zehr. 2004. 'Prospects for Bt-cotton Technology in India'. *Agbioforum*, 7(1&2), 23–26. Available at http//www.agbiofroum.org.

Biosafety-info. 2012. Writ Petition (Civil) No.260/2005 (under Article 32 of the Constitution of India) in the Supreme Court of India – Aruna Rodrigues & Ors *versus* Union of India & Ors. Available at www.biosafety-info.net/file_dir/23669488571ff487c1.doc (accessed on 23 August, 2012).

CICR. 2012. *Cotton Database*. Central Institute of Cotton Research. Available at http://www.cicr.org.in/Database/dbcapp4.html (accessed on 20 July, 2012).

Choudhary, Bhagirath and Kadambini Gaur. 2010. *Bt Cotton in India: A Country Profile*, ISAAA Series of Biotech Crop Profiles, July.

USEPA Regulation. 2002. EPA's Regulation of *Bacillus thuringiensis* (*Bt*) Crops, 735-F-02-013, May 2002. Available at http://www.epa.gov/oppbppd1/biopesticides/pips/regofbtcrops.htm

Falkner, Robert and Aarti Gupta. 2009. 'The Limits of Regulatory Convergence:

Globalization and GMO Politics in the South'. *International Environmental Agreements*, Vol. 9, No.2: 113–133.

*Frontline*. 2007. 'Fatal Feed'. *Frontline* (magazine published by the national daily newspaper *The Hindu*), Vol. 24, Issue 8: 21 April–04 May.

*Frontline*. 2010. 'Legal Struggle'. *Frontline* (magazine published by the national daily newspaper *The Hindu*), Vol. 27, Issue 5: 27 February–12 March.

Gabler, Melissa. 2008. 'Intergovernmental Relations in Food Biotechnology Governance: Complementary Disentanglement in Regulation with Collaboration in Food Safety and Inspection'. In 'Special Series: The Role of Federalism in Protecting the Public's Health'. Institute of Intergovernmental Relations, School of Policy Studies, Queen's University, Public Health 2008(5). Available at http://www.queensu.ca/iigr/workingpapers/PublicHealthSeries/gablerpaperfinal.pdf

GEAC. 2012. Yearwise List of Commercially Released Varieties of Bt Cotton Hybrids by GEAC (Year 2002–upto May 2012). Available at http://igmoris.nic.in/Files2/Yearwise_list2002_May2012.pdf

GOI. 2004. 'Report of the Task Force on Application of Agricultural Biotechnology', M.S. Swaminathan, Chairman, Task Force on Agricultural Biotechnology, May 2004, Ministry of Agriculture, GOI. Available at http://agricoop.nic.in/Taskforce/tf.htm

GOI. 2007. 'Report on Training Workshops on Biosafety Issues for the Members of SBCCs, DLCs and IBSCs, April–June 2007'. Sponsored by the Ministry of Environment & Forests, Government of India and organized by Biotech Consortium of India Limited, New Delhi, 2007. Available at http://www.envfor.nic.in/divisions/csurv/biosafety/files/sbcc_report.pdf (accessed on 14 April, 2011).

GOI. 2011. 'Faster, Sustainable and More Inclusive Growth—An Approach to the Twelfth Five Year Plan (2012–17)', Government of India, Planning Commission, October.

Haq, Zia. 2010. 'Bt cotton flunks pest resistance test in Gujarat'. *Hindustan Times*, 6 March. Available at http://www.hindustantimes.com/StoryPage/Print/515648.aspx

Indian Express. 2011. 'Hyped "desi Bt" cotton has Monsanto gene, government stops production'. *Indian Express*, 29 December. Available at http://www.indianexpress.com/news/hyped-desi-bt-cotton-has-monsanto-gene-govt-stops-production/893287/

Indira, A., M.R. Bhagavan and I. Virgin. 2005. *Agricultural Biotechnology and Biosafety in India: Expectations, Outcomes and Lessons*, Centre for Budget and Policy Studies and Stockholm Environment Institute, April 2005. ISBN 91 975238 0 1. Available at www.esocialsciences.org.in.

ISAAA Brief 43. 2011. Executive Summary - Global Status of Commercialised Biotech/GM Crops. Available at http://www.isaaa.org/resources/publications/briefs/43/executivesummary/default.asp

Karihaloo, J.L. and P.A. Kumar. 2009. *Bt Cotton in India—A Status Report.*

Second Edition. Asia-Pacific Consortium on Agricultural Biotechnology and Asia-Pacific Association of Agricultural Research Institutions, New Delhi. Available at www.apaari.org.

Krishnakumar, Asha. 2003. 'A Lesson from the Field'. *Frontline*, 24 May–06 June, 2003, Vol. 20, Issue 11.

Lalitha, N., Carl E. Pray and Bharat Ramaswami. 2008. 'The Limits of Intellectual Property Rights –Lessons from the Spread of Illegal Transgenic Seeds in India'. Discussion Paper 08-06, March, Indian Statistical Institute, New Delhi (accessed on 26 October, 2011).

Sahai, Suman and Shakeekur Rehman. 2003. 'Performance of Bt-cotton'. *Economic and Political Weekly*, July 26, Vol. 28, No. 30.

Secretariat of the Convention on Biological Diversity. 2000. *Cartagena Protocol on Biosafety to the Convention on Biological Diversity: Text and Annexes.* Montreal: Secretariat of the Convention on Biological Diversity.

Sharma, Ashok B. 2004. 'India: Panel for Clipping GEAC Wings'. In the *Financial Express* (India), 27 April, 2004. Available at http://www.financialexpress.com/fe_full_story.php?content_id=57937

Stabinsky, Doreen. Date Unknown. 'Critique of US Biosafety Regulation and Implementation. PhD, Greenpeace International, College of the Atlantic, USA. Available at www.biosafety-info.net/file_dir/23669488571ff487c1.doc (accessed on 23 August, 2012).

# 4

# Economic and Environmental Performance of Bt Cotton in India

*K.R. Ashok, K. Uma, M. Prahadeeswaran,* and *H. Jeyanthi*

## 1. Introduction

India is the largest cotton cultivating country in terms of acreage and second largest in terms of production in the world. With more than ten million hectare under cotton, India accounted for 34 per cent of the cotton area and 20 per cent of cotton production in the world in 2009. India's cotton yield is one of the lowest amongst the leading cotton producing countries in the world and it ranks only 46th in terms of productivity, with an average yield of about one tonne per hectare. The other major cotton producing countries in the world are China, USA and Pakistan. Though cotton is mainly grown by nine states in India, two states, Maharashtra and Gujarat, alone account for nearly 59 per cent of the cotton area in the country in 2009–10. Out of the total cotton production, Gujarat alone accounts for nearly 33 per cent though it has only 24 per cent of the cotton area. It is because of high productivity of cotton in Gujarat and it ranks third in cotton productivity in the country after Punjab and Haryana. On the other hand, Maharashtra, with 35 per cent of the area, accounts only for 24 per cent of the cotton production because of low productivity. Cotton productivity in Maharashtra is one of the lowest in the country. In contrast, Tamil Nadu has a comparatively lower share in production but has the fifth largest productivity in India.

Over the last fifteen years, modern agricultural biotechnology has been adopted rapidly at the global level, including in several developing countries. This trend has been most apparent for genetically modified

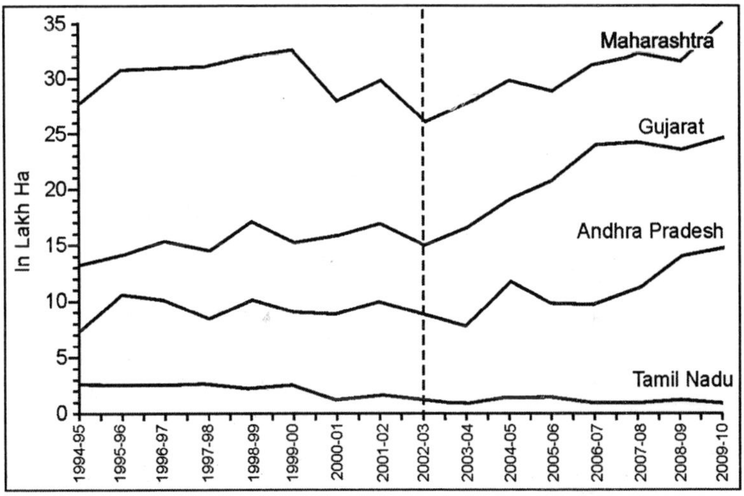


**Figure 4.1:** Area under Cotton before and after Introduction of Bt Technology

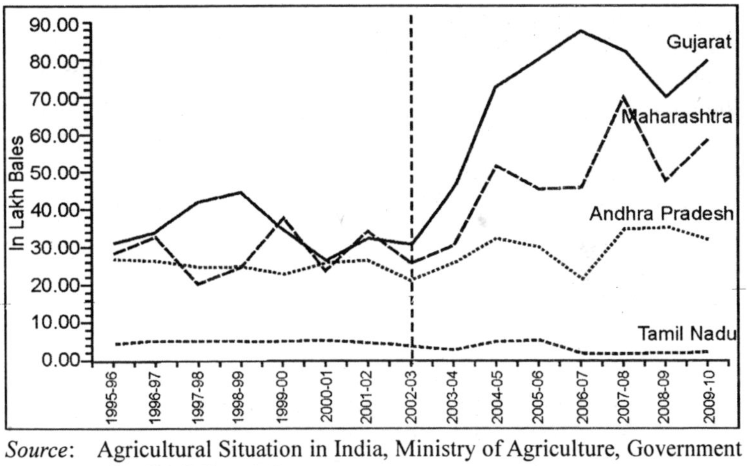


*Source*: Agricultural Situation in India, Ministry of Agriculture, Government of India, Relevant Issues.

**Figure 4.2:** Production of Cotton before and after Introduction of Bt Technology

(GM) crops. This is the fastest diffusion of any new crop technology in the history of humankind (Qaim 2005). The number of countries growing biotech crops has increased steadily from 6 in 1996, the first year of commercialisation, to 18 in 2003 and 29 in 2011. The growth rate between 1996 and 2010, the 15th anniversary of the commercialisation of biotech crops, was an unprecedented 87-fold increase, making it the fastest adopted crop technology in recent history. Reportedly, about 6.3 million small farmers benefited from planting 9.4 million hectares of Bt cotton in India, equivalent to a high adoption rate of 86 per cent (Clive 2010). The area and production of cotton in the major producing states were nearly stagnant or declining for nearly a decade before the introduction of Bt in 2002–03, showed significant increase except in Tamil Nadu (Fig. 4.1 and 4.2). But the latest figures show slight decline in production (Fig. 4.2).

In India efforts to harness genetic engineering technology for bollworm resistance in cotton began in the 1990s with the import of genetically modified cotton and initiation of research programmes in national laboratories (APCoAB 2006). The first approval of commercial cultivation of Bt cotton in India was granted in 2002 to three Bt cotton hybrids, MECH-12, MECH-162 and MECH-184, developed by the seed company Mahyco following a series of trials approved by the Government of India. In the first year of Bt cotton commercialisation in India, only three Bt hybrids had been approved, which were grown on large areas in different regions. Since these hybrids were not well adapted to all environments, the productivity advantage associated with the Bt gene was partly offset by general germplasm disadvantages in some locations (Qaim *et al.* 2006). The crop failure along with other concerns has led an intense scientific debate and public controversy. Much of this controversy surrounds ethical arguments, or concerns of multinational control of the world's seeds, human health, or environmental risks (Herring 2007a, 2007b). As the wider global society struggles to come to terms with the benefits and (unknown) risks of the biotechnology, better understanding of consumer interests and concerns is needed to formulate and implement effective private and public policies (Onyango *et al.*, 2004). Factors influencing consumer preferences for bioengineered crops include their perceptions of benefits and risks of bioengineered crops on human health and the environment, their ethical stance towards genetic engineering, and their trust in government regulations concerning risk assessment and management (OECD 2000). Analysing the socio-economic impacts of this new technology is important in the context of India, as it is emerging as one of the leading countries with large acreages under GM cotton. In 2011, India with 10.6 million hectares under Bt cotton was the fourth largest country in terms of acreage of biotech crops in the world after USA (69 million hectares), Brazil (30.3 million hectares) and Argentina (23.7 million hectares). Also out of the 16.7 million biotech farmers in the world, 7 million are in India. Bt cotton is the only GM crop approved for commercial cultivation in India and hence all the biotech farmers in India are cotton farmers. In the above scenario this study addressed issues like reduction in pesticide use, increase in yield and income, profitability, distributional aspects and environmental consequences of Bt cotton technology in India. The chapter is organised broadly into four sections, the first one on the context and objectives of the study, the second section describes the data source and analytical methods, the third section discuss the results and the fourth section concludes.

## 2. Data and Methods

The study was based on a multi-stage sample survey undertaken among farmers in four states, viz. Maharashtra, Gujarat, Andhra Pradesh and Tamil Nadu. The multi-stage sampling procedure considered the area, production, productivity and the spread of Bt cotton across the major districts and talukas within a district as well as blocks in each taluka. Accordingly, from each district, two major cotton growing talukas/blocks, and from each taluka/block three to five villages were identified considering the area under cotton and spread of Bt technology in consultation with the Agricultural Development Officers/Staff of State Agricultural Universities in the respective states. A total of 120 farmers were surveyed at random from each state comprising 80 Bt cotton growers and 40 non-Bt cotton growers, making an aggregate sample size of 480 from four states during 2007–08. The detailed sample distribution by state is given in Annexure 4.1.

### *2.1 Impact of Bt Technology*

Impacts of Bt technology on key variables like yield, pesticide use, profitability etc., were assessed through the Analysis-of-Variance (ANOVA) model (Gujarati 1988).

$$Y = \alpha + \beta C_{Bt} + u$$

where:

$Yi$ = Variable for which impact is estimated like yield, pesticide use, profitability

$C_{Bt}$ = 1 for adoption of Bt cotton

= 0 for non-adoption of Bt cotton

The model estimates whether Bt adoption makes any significant difference in the yield, pesticide use, profitability, etc., assuming that all other variables are held constant. For example, take the impact of Bt on yield. The intercept $\alpha$ gives mean yield of non-adopters and the slope $\beta$ tells by how much the mean yield of adopters differs from the mean yield of non-adopters. The term $\alpha + \beta$ gives the mean yield of the Bt adopter. The test of the null hypothesis that there is no difference between the yields of Bt adopters and non-adopters ($H_0$: $\beta = 0$) can be made by running the regression in the usual manner and finding out whether the estimated $\beta$ is statistically significant.

### *2.2 Yield Advantage in Bt Cotton: Technology Effect or Input Effect?*

Earlier studies have reported yield advantage in Bt cotton compared to non-Bt cotton (Morse *et al.* 2004, Qaim and Zilberman 2003, Qaim and de Janvry 2005, Qaim and Traxler 2005) though the Bt technology *per se* is meant for controlling the bollworms incidence in cotton. The higher yield in Bt cotton may be due to better plant growth in the absence of pest attack. It is also likely that farmers use higher inputs in Bt cotton cultivation. Hence the yield advantage in Bt may be due to the Bt

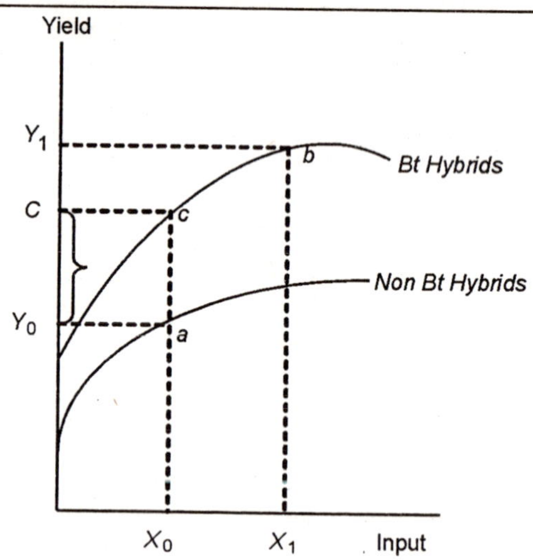


$Y_0C$ = Yield gain due to Technology
$CY_1$ = Yield gain due to additional inputs

**Figure 4.3:** Decomposition of the Yield Increase in Bt Cotton

technology and/or due to additional input use for Bt cultivation. Sources of the yield difference between Bt and non-Bt cotton was estimated by decomposing the productivity change into technology effect and input effect following the methodology of Bisaliah (1977).

Fig. 4.3 depicts the production technology of conventional hybrids and Bt hybrids. $Y_0$ is the yield from non-Bt hybrids with input use $X_0$. $Y_0C$ is the yield increase due to Bt technology embedded in seed, with the same level of input $X_0$, because technological progress shifts the production function upwards (Koutsoyiannis 1979). This shift shows that more yield equivalent to $Y_0C$, is obtained from the Bt hybrids (representing technological progress) with the same level of input, $X_0$ as that in non-Bt hybrids. But in practice, Bt farmers use more inputs than non-Bt farmers, equivalent to $X_0X_1$, which also contribute to increased yield, equivalent to $CY_1$. Thus the total increase in yield ($Y_0Y_1$) in Bt hybrids is due to the combined effect of technology ($Y_0C$) and higher input use ($CY_1$). The production technology is represented as:

The Production function for non-Bt cotton

$$\ln Y_0 = \ln \alpha_0 + \beta_{0i} \ln X_{0i} + u_0 \qquad ...(1)$$

The Production function for Bt cotton

$$\ln Y_1 = \ln \alpha_1 + \beta_{1i} \ln X_{1i} + u_1 \qquad ...(2)$$

Taking the difference of equation 1 and 2 and rearranging the terms, we get

$$\ln [Y_1/Y_0] = [\ln (\alpha_1/\alpha_0) + (\beta_{1i} - \beta_{0i}) \ln X_{0i}] + \beta_{1i} \ln (X_{1i}/X_{0i})$$

$\Delta$ in yield = $\Delta$ in yield due to Bt technology + $\Delta$ in yield due to additional inputs

In the present analysis, we use the Cobb-Douglas production functions for Bt and non-Bt cotton farmers as specified below:

$$LnY_B = Lnb_0 + b_1LnCHEM_B + b_2SEED_B + b_3LnFYM_B + b_4LnN_B + b_5LnP_B + b_6LnK_B + b_7LnLAB_B + b_8\,LnIRRI_B + u_B \qquad ...(3)$$

$$LnY_N = Lnb_9 + b_{10}LnCHEM_N + b_{11}SEED_N + b_{12}LnFYM_N + b_{13}LnN_N + b_{14}\,LnP_N + b_{15}LnK_N + b_{16}LnLAB_N + b_{17}LnIRRI_N + u_N \qquad ...(4)$$

Where,

CHEM = Cost of chemicals Rs. per acre
SEED = Seed rate in grams per acre
FYM = Manures in kg per acre
N = Total Nitrogen in kg per acre
P = Total phosphorus in kg per acre
K = Total Potash in kg per acre
LAB = Total labour in man days per acre
IRRI = Dummy for irrigated environment (I=1 otherwise 0)
$b_i$ = parameters to be estimated (i = 1 to 8 for Bt and i = 9 to 17 for Non-Bt)
u = random error
Subscript B = Bt cotton
Subscript N = non-Bt cotton

Taking differences between (3) and (4) and adding some terms and subtracting the same terms and by rearranging terms, the equation becomes:

$$Ln(Y_B/Y_N) = \{Ln(b_0/b_9)\} + \{(b_1 - b_{10}) LnCHEM_N + (b_2 - b_{11}) LnSEED_N + (b_3 - b_{12}) LnFYM_N + (b_4 - b_{13}) LnN_N (b_5 - b_{14}) LnP_N + (b_6 - b_{15}) LnK_N + (b_7 - b_{16}) LnLAB_N + (b_8 - b_{17}) LnIRRI_N\} + (b_9 (LnCHEM_B/CHEM_N) + b_{10} Ln (SEED_B/SEED_N) + b_{12} Ln(FYM_B/FYM_N) + b_{13} Ln(N_B/N_B) + B_{14} Ln(P_B/P_N) + b_{15} Ln (K_B/K_N) + b_{16} Ln(LAB_B/LAB_N) = b_{17} Ln(IRRI_B/IRRI_N)\} + \{u_B - u_N\} \quad ...(5)$$

The LHS (left hand side) of equation-5 denotes the difference in per acre productivity of Bt and non-Bt cotton cultivation. The RHS (right hand side) of the equation decomposes the difference in productivity into change due to technology and change due to input use. Equation-5 has three major terms within {}. The first term of the RHS refers to the gap attributable to neutral technological change, the second term refers to the gap attributable to non-neutral technological change, and the third term refers to change due to input use.

### *2.3 Distributional Impact of Bt Technology: Economic Surplus Model*

Economic surplus approach is widely used for estimating total benefits of a new technology and to evaluate its distributional impact. An economic approach to evaluating R&D begins with the basic commodity market model of research benefits which is depicted in Fig. 4.4. $S_0$ represents the supply function before the technical change, and $D_0$ represents the demand function. The initial price and quantity are $P_0$ and $Q_0$. Suppose research generates yield increasing or input saving technologies. These effects can be expressed as a per unit reduction in production costs, *K*, that are modelled as a parallel shift down of the supply function to $S_1$. This research-induced supply shift leads to an increase in production and consumption to $Q_1$ ($\Delta Q = Q1 - Q0$), and the market price falls to $P_1$ (by $\Delta P = P_0 - P_1$). Consumers are better off because the new technology enables them to consume more of the commodity at a lower price (Wood *et al.* 2001).

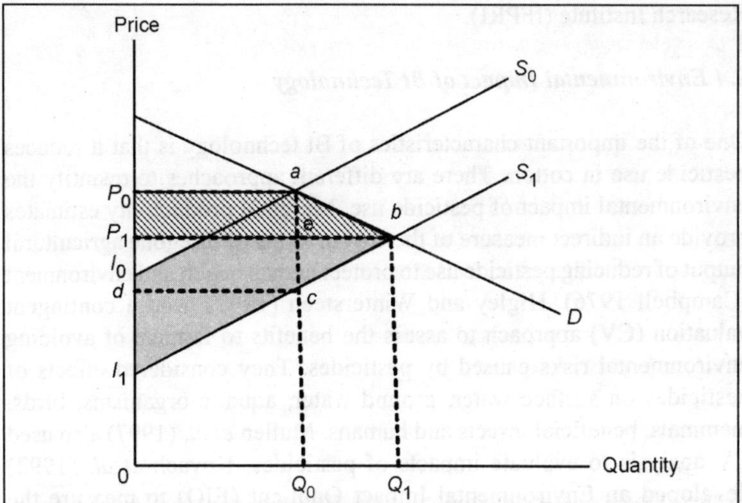


*Source*: Alston, Norton and Pardey 1988: 209.

**Figure 4.4:** The Economic Surplus Model

Although they receive a lower price per unit, producers who adopt the new technology are better off too because their unit costs have fallen by an amount, *K* per unit that is more than the fall in price. The change in consumer surplus which is the measure of the consumer benefit is equal to area $P_0abP_1$. The change in producer surplus which is the measure of the producer gain is equal to area $P_1 b I_1$ – area $P_0 a I_0$ in Fig. 4.4. Total benefit/the economic surplus is obtained as the sum of producer and consumer benefits. As an approximation, the cost-saving per unit multiplied by the initial quantity, $K.Q_0$, is often used. Thus, the size of the market, as indexed by the initial quantity $Q_0$, as well as the size of the research-induced savings in per unit cost of production. *K*, are critical factors in estimating the economic benefits from R&D. Fig. 4.4 is the basic static model for research evaluation. However, evaluations of the economic effects of research involve procedures to account for the timing of streams of benefits and costs, since there may be lengthy lag times between the initial investment in research, the eventual adoption of research results, and the flow of research benefits. The economic surplus model was estimated through DREAM (Dynamic Research Evaluation for Management) package, developed by International Food Policy Research Institute (IFPRI).

### *2.4 Environmental Impact of Bt Technology*

One of the important characteristics of Bt technology is that it reduces pesticide use in cotton. There are different approaches to quantify the environmental impact of pesticide use. Marginal productivity estimates provide an indirect measure of the cost in terms of foregone agricultural output of reducing pesticide use to protect human health and environment (Campbell 1976). Higley and Wintersteen (1992) used a contingent valuation (CV) approach to assess the benefits to farmers of avoiding environmental risks caused by pesticides. They considered effects of pesticides on surface water, ground water, aquatic organisms, birds, mammals, beneficial insects and humans. Mullen *et al.* (1997) also used CV analysis to evaluate impacts of pesticides. Kovach *et al.* (1992) developed an Environmental Impact Quotient (EIQ) to measure the environmental impact of different pesticides. The EIQ value of individual pesticide refers to the average of its effect on farm worker, consumer and ecology. Later Oskam and Vijftigschild (1999) modified the EIQ field rating method to study the environmental impact of pesticides in agricultural use. In this study, EIQ values were adopted from Kovach *et al.* (1992).

$$\begin{aligned} EIQ = \{ & C\,[(DT*5) + (DT*P)] \\ & + [(C*((S + P)/2)*SY) + (L)] \\ & + [(F*R) + (D*((S+P)/2)*3) + (Z*P*3) + (B*P*5)]\}/3 \end{aligned}$$

Where,

DT = dermal toxicity,
C = chronic toxicity,
Sy = systemicity,
F = fish toxicity,
L = leaching potential,
R = surface loss potential,
D = bird toxicity,
S = soil half-life,
Z = bee toxicity,
B = beneficial arthropod toxicity,
P = plant surface half life.

To account for different formulations of the same active ingredient and different use patterns, a simple equation called EIQ Field Use Rating was developed:

EIQ Field Use Rating = EIQ × Per cent of active ingredient × Rate

With this method, comparisons of environmental impact between pesticides and different pest management programmes can be made. Oskam and Vijftigchild (1999) developed an indicator to measure the risk of pesticides to the environment and other impacts, RM using EIQ values as:

$$\text{RM} = \frac{\sum_{k=1}^{n}\left(\text{use of a.i.}_k \times \text{EIQ}_k\right) \text{ for crop j}}{\text{area of crop j}}$$

Where

RM = risk measure,
a.*i* = active ingredient of pesticides,
EIQ = Environmental Impact Quotient,
$k = 1, \ldots, n$, insecticides and
$j = 1, \ldots m$ crops.

**3. Results and Discussion**

The results of the study analyses the impact of Bt technology on key variables in the production of Bt cotton like on yield, pesticide application, seed cost, cost of cultivation and profitability across the states. The impact of Bt technology on the key variables of cotton production in this study was estimated through the Analysis-of-Variance (ANOVA) model. The intercept of the model gives the mean yield of Non-Bt and the slope coefficient shows the difference in yield between Bt and non-Bt. Discussion across states facilitates comparison of key variables and help in understanding how the production environment in each state is different and accordingly how the impact of Bt cultivation varies across the states in terms of the key variables. The detailed results are presented in Annexure 4.2.

### *3.1 Impact of Bt Technology on Cotton Cultivation*

One of the important advantages of Bt technology on cotton is higher yield. This may be due to the prevention of yield loss due to pest attack. For example, analyses of several years of Indian field trial data demonstrated that Bt technology can significantly reduce pesticide applications and increase effective yields under experimental conditions (Qaim and Zilberman 2003, Qaim 2003, ICAR 2002, Naik 2001). According to Clive (2008), yield increased by 31 per cent on conservative estimates for small farmers. In this study, the estimated mean yield of non-Bt was 1903 kg per ha in Andhra Pradesh, 1740 in Tamil Nadu, 1686 in Gujarat and 829 in Maharashtra (Fig. 4.5). The slope coefficient shows the difference in the yield of Bt and non-Bt. The mean difference in yield ranged from nearly 338 kg per ha in Maharashtra to 874 kg per ha in Gujarat. At an all-India level, mean difference in yield was 631 kg per ha. The difference in mean yield was statistically significant in all the states.

The estimated mean expenditure on pesticides in non-Bt was Rs. 4707 per ha in Andhra Pradesh, Rs. 4262 in Gujarat, Rs. 2713 in Tamil Nadu and Rs. 2391 in Maharashtra (Fig. 4.6). The expenditure on pesticides in Bt was less than non-Bt and statistically significant in all states. The reduction in expenditure on pesticides ranged from Rs. 1718 per ha in Andhra Pradesh, Rs. 1713 in Gujarat, Rs. 1442 in Tamil Nadu and Rs. 1057 in Maharashtra. Earlier studies about the agronomic and economic impacts of Bt cotton cultivation demonstrated that the adopting farmers benefit from income increase through reduced pest control costs and higher effective yields (Morse *et al.* 2004, Qaim and Zilberman 2003, Qaim and Traxler 2005). Benefits will vary according to varying pest infestation levels in different years and locations. In China, for instance,

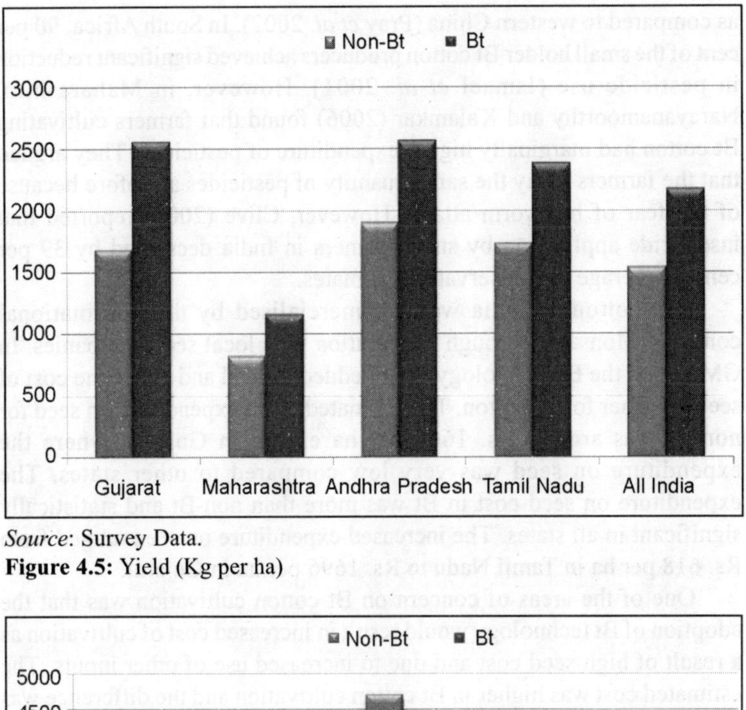


*Source*: Survey Data.

**Figure 4.5:** Yield (Kg per ha)

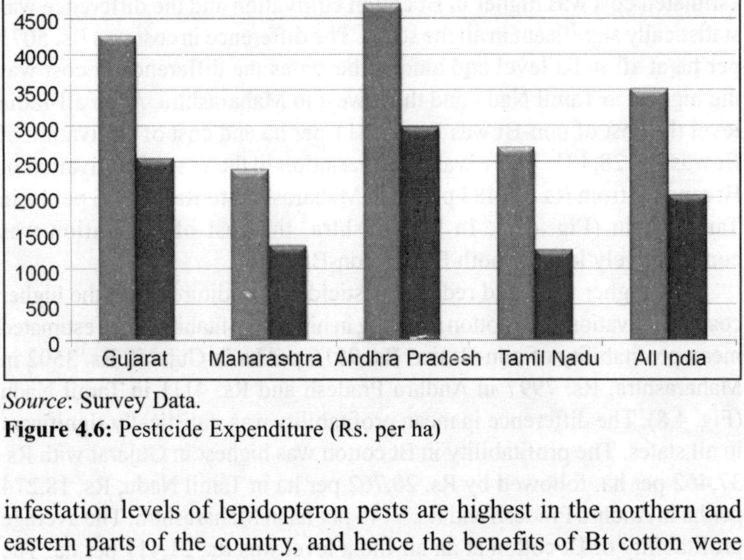


*Source*: Survey Data.

**Figure 4.6:** Pesticide Expenditure (Rs. per ha)

infestation levels of lepidopteron pests are highest in the northern and eastern parts of the country, and hence the benefits of Bt cotton were most pronounced there. This was reflected in much higher adoption rates, as compared to western China (Pray *et al.* 2002). In South Africa, 90 per cent of the small holder Bt cotton producers achieved significant reduction in pesticide use (Ismael *et al.* 2001). However, in Maharashtra, Narayanamoorthy and Kalamkar (2006) found that farmers cultivating Bt cotton had marginally higher expenditure of pesticides. They argued that the farmers spray the same quantity of pesticides as before because of the fear of bollworm attack. However, Clive (2008) reported that insecticide application by small farmers in India decreased by 39 per cent on average by conservative estimates.

GM cotton in India was commercialised by the multinational company Monsanto through cooperation with local seed companies. In GM cotton the Bt technology is embedded in seed and hence the cost of seed is higher for Bt cotton. The estimated mean expenditure on seed for non-Bt was around Rs. 1600 per ha except in Gujarat, where the expenditure on seed was very low compared to other states. The expenditure on seed cost in Bt was more than non-Bt and statistically significant in all states. The increased expenditure on seed ranged from Rs. 618 per ha in Tamil Nadu to Rs. 1696 per ha in Gujarat.

One of the areas of concern on Bt cotton cultivation was that the adoption of Bt technology would result in increased cost of cultivation as a result of high seed cost and due to increased use of other inputs. The estimated cost was higher in Bt cotton cultivation and the difference was statistically significant in all the states. The difference in cost was Rs. 5071 per ha at all India level and among the states the difference in cost was the highest in Tamil Nadu and the lowest in Maharashtra. At an all-India level the cost of non-Bt was Rs. 23,411 per ha and cost of cultivation of Bt was Rs. 28,481. There was wide variation in the cost of cultivation of Bt ranging from Rs. 15,483 per ha in Maharashtra to Rs. 37,435 per ha in Tamil Nadu (Fig. 4.7). In Maharashtra, the cost of cultivation was comparatively low for both Bt and non-Bt cotton.

The higher yield and reduced pesticide expenditure offset the higher cost of cultivation in Bt cotton resulting in higher profitability. The estimated mean profitability of non-Bt was Rs. 9516 per ha in Gujarat, Rs. 3502 in Maharashtra, Rs. 7997 in Andhra Pradesh and Rs. 4111 in Tamil Nadu (Fig. 4.8). The difference in mean profitability was statistically significant in all states. The profitability in Bt cotton was highest in Gujarat with Rs. 37,462 per ha, followed by Rs. 20,762 per ha in Tamil Nadu, Rs. 18,274 per ha in Andhra Pradesh and Rs. 9146 per ha in Maharashtra. The average profitability of Bt cotton at an all-India level was Rs. 21,411 per ha. The profitability was relatively very low in Maharashtra compared to other states.

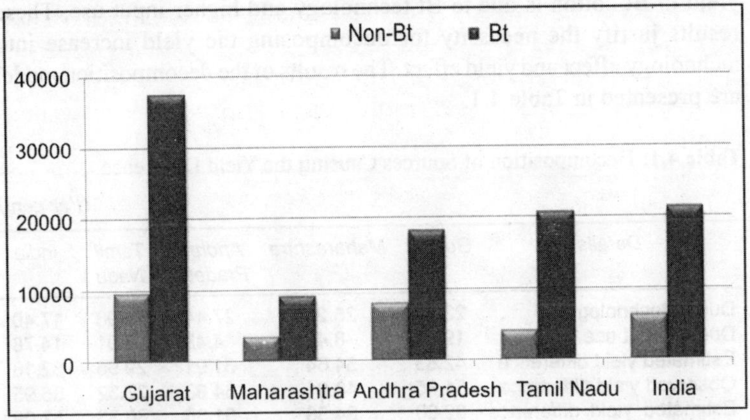


*Source*: Survey Data.

**Figure 4.7:** Cost of Cultivation (Rs. per ha)

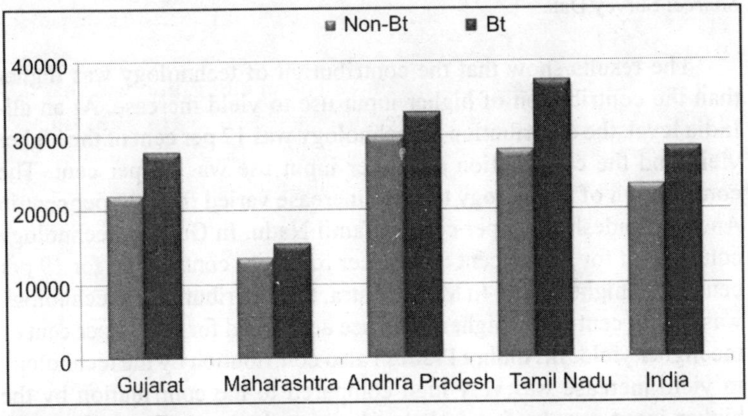


*Source*: Survey Data.

**Figure 4.8:** Profitability (Rs. per ha)

### *3.2 Decomposition of Yield Increase into Technology Effect and Input Effect*

It is evident that the higher profitability of Bt cotton was mainly due to higher yield and reduced expenditure on pesticides though the Bt technology *per se* is meant for controlling the bollworms incidence in cotton. But it is also important to note the high cost of cultivation of Bt cotton despite the reduced expenditure on pesticides which implies higher use of other inputs in Bt cotton cultivation. This indicates the higher yield in Bt cotton is due to Bt technology and higher input use. These results justify the necessity for decomposing the yield increase into technology effect and yield effect. The results of the decomposition model are presented in Table 4.1.

**Table 4.1:** Decomposition of Sources Causing the Yield Difference

*(Per cent)*

| Details | Gujarat | Maharashtra | Andhra Pradesh | Tamil Nadu | India |
|---|---|---|---|---|---|
| Due to technology | 23.79 | 26.22 | 27.44 | 14.96 | 17.40 |
| Due to input use | 19.05 | 8.42 | 4.48 | 15.01 | 14.78 |
| Estimated yield difference | 42.83 | 34.64 | 31.91 | 29.98 | 32.18 |
| Observed yield difference | 51.85 | 40.81 | 34.93 | 37.32 | 36.95 |
| Estimated yield difference as per cent of observed yield difference | 82.60 | 84.88 | 91.35 | 80.33 | 87.09 |

*Source*: Survey Data.

The results show that the contribution of technology was higher than the contribution of higher input use to yield increase. At an all-India level, the contribution of technology was 17 per cent of the higher yield and the contribution of higher input use was 15 per cent. The contribution of technology to yield increase varied from 27 per cent in Andhra Pradesh to 15 per cent in Tamil Nadu. In Gujarat, technology contributed for 24 per cent and higher input use contributed for 19 per cent of the higher yield. In Maharashtra, the contribution by technology was 26 per cent while higher input use accounted for only 8 per cent of the higher yield. In Andhra Pradesh also contribution by the technology to yield increase was very high compared to the contribution by the higher input use. In Tamil Nadu, the contribution by the technology and higher input use contributed 15 per cent each. The decomposition model accounted for about 80 to 90 per cent of the observed yield difference.

### *3.3 The Distributional Impact of Bt Technology: Economic Surplus Approach*

The results in the foregoing sections indicate the gains from Bt technology. The nationwide impacts of higher production of cotton influence the macro supply and consequently the gains of producers and consumers. To understand this, the distributional impact was analysed through economic surplus approach and the results are presented in Table. 4.2.

**Table 4.2:** Estimated Economic Surplus due to Bt Cotton

*(Rs. in Billion)*

| | *Supply Elasticity** | | |
|---|---|---|---|
| | 1.2 | 0.31 | 0.13 |
| Producer surplus | 147.29<br>(40.00) | 254.53<br>(72.07) | 292.36<br>(86.02) |
| Consumer surplus | 220.93<br>(60.00) | 98.63<br>(27.93) | 47.51<br>(13.98) |
| Economic Surplus | 368.22<br>(100.00) | 353.16<br>(100.00) | 339.87<br>(100.00) |

*Source*: Survey Data.

*Note: Supply elasticities were taken from Poonyth et al. (2004),* Shepherd (2006) and Sumner (2003).

The producer surplus and consumer surplus were estimated under three supply elasticity scenarios of cotton. Under relatively inelastic supply assumption, 0.31 and 0.13, producer benefits more from the technology than the consumers and at elastic supply consumer benefits more than the producers.

### *3.4 Environmental Consequences*

Under the Environment Protection Act (1986), the Ministry of Environment & Forests, has notified the Rules for the Manufacture, Use, Import, Export and Storage of Hazardous Microorganisms/Genetically Engineered Organisms or Cells. Bt cotton being a transgenic crop requires environmental clearance under Rule 7-10 of the 1989 'Rules for Manufacture, Use, Import, Export and Storage of Hazardous Microorganisms/Genetically Engineered Organisms or Cells' notified under the Environment (Protection) Act, 1986. The bio-safety and environmental issues related to this GM crop includes molecular characterisation of induced gene, biochemical characterisation of the expressed protein, estimation of the level of the expressed proteins in cotton, proteins in cotton products, safety of the expressed proteins to non-target organisms, environmental fate of the Bt protein, and agronomic, compositional and food and feed safety evaluation of Bt cotton compared to non-Bt cotton seed. The environmental analysis in this paper is based on the cross-section data collected from the farmers during 2007–08 and hence refers to the short term impacts. The short term impacts may have long term implications, and any comment on the long term impact on environment could be made only after evaluating technology over years. The different aspects of environmental consequences based on the survey data is presented below.

### *3.4.1 Incidence of Bollworm Attack in Bt and Non-Bt Cotton*

Bollworm attack is the major cause of indiscriminate use of pesticides in cotton. Bt technology is specifically targeted against bollworm incidence in cotton. Farmers' experience with bollworm incidence in different states is given in Table 4.3.

**Table 4.3:** Severity of Bollworm Attack in Bt and Non-Bt cotton

*(Per cent of farmers)*

| *Pests* | *Severity* | *Gujarat* | | *Maharashtra* | | *Andhra Pradesh* | | *Tamil Nadu* | |
|---|---|---|---|---|---|---|---|---|---|
| | | *Bt* | *Non-Bt* | *Bt* | *Non-Bt* | *Bt* | *Non-Bt* | *Bt* | *Non-Bt* |
| American BW | No attack | 83.8 | 70.0 | 35.0 | 22.5 | 88.8 | — | 100.0 | 20.0 |
| | Mild attack & no pp measures needed | 6.3 | 2.5 | 51.3 | 45.0 | 11.3 | 5.0 | — | 2.5 |
| | Moderate | 10.0 | 20.0 | 13.8 | 20.0 | — | 95.0 | — | 67.5 |
| | Severe | — | 7.5 | — | 12.5 | — | — | — | 10.0 |
| Pink BW | No attack | 96.3 | 42.5 | 50.0 | 30.0 | 37.5 | — | 95.0 | 40.0 |
| | Mild attack & no pp measures needed | 1.3 | 2.5 | 25.0 | 37.5 | 45.0 | 5.0 | 1.3 | 2.5 |
| | Moderate | 2.5 | 55.0 | 20.0 | 20.0 | 17.5 | 95.0 | 3.8 | 37.5 |
| | Severe | — | — | 5.0 | 12.5 | — | — | — | 20.0 |
| Spotted BW | No attack | 57.5 | 32.5 | 22.5 | 20.0 | 88.8 | — | 100.0 | 67.5 |
| | Mild attack & no pp measures needed | 7.5 | 2.5 | 23.8 | 32.5 | 11.3 | 5.0 | — | 5.0 |
| | Moderate | 35.0 | 62.5 | 53.8 | 30.0 | — | 95.0 | — | 27.5 |
| | Severe | — | 2.5 | — | 17.5 | — | — | — | — |

*Source*: Survey Data.

More than 80 per cent of the farmers in Gujarat, Maharashtra and Tamil Nadu did not report any incidence of American bollworm in Bt cotton. In Maharashtra majority of the farmers reported mild attack of American bollworm. In non-Bt, majority of the farmers reported moderate attack of American bollworm in Andhra Pradesh and Tamil Nadu and no attack in Gujarat. As far as pink bollworm is concerned, majority of the farmers reported no incidence in Bt cotton in Gujarat and Tamil Nadu while farmers in Maharashtra and Andhra Pradesh reported no incidence to mild attack. In non-Bt, there was no incidence to mild incidence of pink bollworm in Gujarat, Maharashtra and Tamil Nadu where as moderate attack in Andhra Pradesh. The spotted bollworm attack was also considerably less in Bt cotton in all the states.

### *3.4.2 Cultivation of Refuge Crops*

Pest susceptibility (resistance) is an environmental good (bad). Pests will initially be susceptible to pesticides, but over time, with variations in their genetic makeup, become resistant to pesticides. Similar to this phenomenon, bollworms which are now susceptible to Bt cotton, may develop resistance to with the repeated exposure to Bt cotton. To avoid/ delay the development of resistance, growing of refugee crops (non-Bt cotton which are susceptible to pests) along with Bt cotton are recommended. The refuge crop supply non-resistant insects to mate with possible resistant insects in Bt cotton to produce non-resistant insects. Thus refuge crop is meant to prevent the emergence of resistant pests which may require even higher dose of chemicals. Details of refuge cotton cultivation are given in Table 4.4.

**Table 4.4:** Details on Refuge Crop Cultivation

*(Per cent)*

| *Details* | *Tamil Nadu* | *Maha-rashtra* | *Andhra Pradesh* | *Gujarat* |
|---|---|---|---|---|
| Awareness of refuge seed supplied | 80 | 98 | 100 | 86 |
| Use of refuge seeds | 6 | 96 | 28 | 29 |
| Awareness of susceptibility of refuge crop | 100 | 53 | 36 | 43 |
| Spraying of refuge crops | 40 | 73 | 13 | 20 |

*Source*: Survey Data.

The Bt seed companies provide a small packet of refuge seeds along with Bt seeds to cultivate around the Bt fields. Most of the farmers in all the four states were aware of these refuge seeds. But only less than 30 per cent of the farmers used the refuge seeds for the purpose for which it was given. In Maharashtra, nearly all the farmers used the refuge seeds as normal seeds and also sprayed the refuge crop to save from the bollworms, thus defeating the very purpose for which it was supplied. Awareness of the susceptibility of refuge crop was also very low except in Tamil Nadu.

The reasons for not using the refuge seeds were analysed and presented in Fig. 4.9.

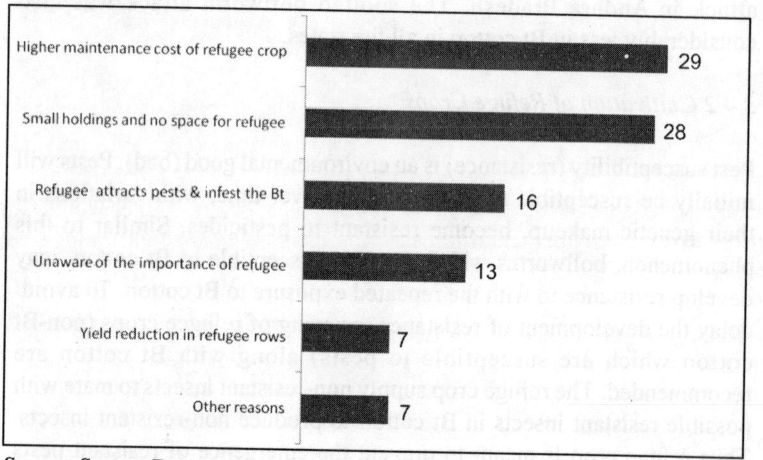


*Source*: Survey Data.

**Figure 4.9:** Reasons for Not Using Refuge Crop

The most important reason for not using refuge seeds was that farmers feared severe pest attack and more expenditure for pest control and reduced yield in refuge crop. In 28 per cent of the cases farmers found it difficult to comply with this measure because of small land holdings. The other important reasons were fear of pest attack in Bt cotton from refuge cotton, lack of awareness about the importance of growing refuge crop and fear of yield reduction. Some of the other cultivated crops near the Bt fields may also serve as a refuge crop, since these crops were also susceptible to bollworms. The details of the other important crops grown near Bt cotton fields are given in Table 4.5 .

**Table 4.5:** Other Major Crops Grown Near the Bt Cotton Field

| *Tamil Nadu* | *Maharashtra* | *Andhra Pradesh* | *Gujarat* |
|---|---|---|---|
| Maize | Soybean | Chillies | Brinjal |
| Red gram | Mung | Maize | Okra |
| Black gram | Red gram | Red gram | Chillies |
| Castor | Jowar | Paddy | Groundnut |

*Source*: Survey Data.

### *3.4.3 Environment Impact Quotient (EIQ) of Cultivating Bt and Non-Bt Cotton*

Cotton is one of the crops in which traditionally pesticide use is very high. The introduction of Bt cotton reduced the quantity as well as the type of pesticides used in cotton. This shift in pesticide use has a definite impact on environment. Environmental Impact Quotient (EIQ) measures the adverse impact of individual pesticides on farm worker, consumer and ecology. EIQ field rating is an aggregate of the EIQ values of different pesticides used per unit area. Higher the EIQ field rating, higher would be the adverse environmental impact. The estimated EIQ field ratings are given in Table 4.6.

**Table 4.6:** EIQ Field Rating/Ha for Bt and Non-Bt Cotton

| *State* | *Bt cotton* | *Non Bt* | *Difference* | *Per cent change over non-Bt* |
|---|---|---|---|---|
| Gujarat | 43.23 | 50.64 | –7.41 | –14.63 |
| Maharashtra | 34.50 | 68.72 | –34.22 | –49.80 |
| Andhra Pradesh | 58.33 | 71.74 | –13.41 | –18.69 |
| Tamil Nadu | 26.58 | 82.31 | –55.73 | –67.71 |

*Source*: Survey Data.

The value of EIQ field rating for Bt cotton was lower than that of non-Bt cotton in all the states implying lower environmental impact due to pesticides. The decline in the EIQ field rating varied from 15 per cent in Gujarat to 68 per cent in Tamil Nadu.

## 4. Conclusion

Over the last ten years, modern agricultural biotechnology has been adopted rapidly at the global level, including in several developing countries. In India, this trend has been most apparent in the case of Bt cotton in the last few years. But the introduction of Bt cotton in the country in 2002 led to an intense scientific debate and public controversy. The study revealed that the yield of Bt cotton was higher ranging from nearly 338 kg per ha in Maharashtra to 874 kg per ha in Gujarat and the expenditure on pesticides was lower, ranging from Rs. 1718 per ha in Andhra Pradesh to Rs. 1057 in Maharashtra, resulting in higher net profit. The cost of seed was one of the contentious issues in the introduction of Bt in India. The expenditure on seed cost in Bt was more to the tune of

Rs. 618 per ha in Tamil Nadu to Rs. 1696 per ha in Gujarat. The total cost of cultivation of Bt was higher on an average by Rs. 5071 reflecting higher seed cost and more use of other inputs. The farmers in the major cotton growing states in India thus benefited significantly from adopting Bt technology though the Bt technology *per se* is meant for controlling the bollworm incidence in cotton. It is evident that the higher profitability of Bt cotton was mainly due to higher yield, reduced expenditure on pesticides and increased use of inputs. This necessitates the decomposition of the source of higher yield and the decomposition revealed that at the aggregate level, Bt technology contributed 17 per cent and higher input use contributed 15 per cent of the increased yield. Cotton is one of the highest pesticide consuming crops which impact the environment through polluting land and water and poisoning humans, animals and insects. Hence, an attempt was made to study the environmental impact of Bt cotton based on the cross-section data collected from the farmers. More than 80 per cent of the farmers in Gujarat, Maharashtra and Tamil Nadu did not report any incidence of American bollworm in Bt cotton while majority of farmers in Maharashtra reported mild attack. The environmental impact quotient was significantly lower for Bt cotton because of reduced pesticide consumption. But only less than 30 per cent of the farmers raised the refuge crop meant to prevent the emergence of resistant pests to Bt cotton. The state departments of agriculture and seed companies need to create awareness among the farmers on the importance of raising the refugee crop. The study did not address any long term impact of the technology on environment and bio-safety aspects.

**Acknowledgement**

The authors are grateful to Dr. C. Ramasamy, former Vice Chancellor, Tamil Nadu Agricultural University, Coimbatore-641003, for his erudite guidance and constant encouragement and to the anonymous referee for the meticulous editing and valuable comments.

## References

Alston, Julian M., George W. Norton and Philip G. Pardey. 1988. 'Science under Scarcity: Principles and Practices for Agricultural Research Evaluation and Priority Setting'. CAB International (CABI), Wallingford, U.K.

APCoAB. 2006. *Bt Cotton in India—A Status Report*. Asia-Pacific Consortium of Agricultural Biotechnology, New Delhi, India, p.34.

Bisaliah, S. 1977. 'Decomposition Analysis of Output Change under New Production Technology in Wheat Farming: Some Implications to Returns on Investment'. *Indian Journal of Agricultural Economics*, 32(3): 193–201.

Campbell, H.E. 1976. 'Estimating the Marginal Productivity of Agricultural Pesticides: The Case of Tree Fruit Farms in the Okangan Valley'. *Canadian Journal of Agricultural Economics*, 24: 23–30.

Clive, James. 2008. *Global Status of Commercialized Biotech/GM Crops: 2008*. ISAAA Brief No. 39. ISAAA: Ithaca, NY.

Clive, James. 2010. *Global Status of Commercialized Biotech/GM Crops: 2010*. ISAAA Brief No. 42. ISAAA: Ithaca, NY.

Das, Kasturi. 2006. 'GM Crops in India: Is the Government's Policy Stance Justified?'. *Economic Political Weekly*, 4 March, 2006.

Gujarati, D.N. 1988. *Basic Econometrics*. New York: McGraw-Hill Book Company, p. 432.

Herring, R.J. 2007a. 'The Genomics Revolution and Development Studies: Science, Politics and Poverty,' *Journal of Development Studies*, Vol. 43 No. 1, pp. 1–30.

Herring, R.J. (ed). 2007b. *Transgenics and the Poor: Biotechnology and Development Studies*. Oxford: Routledge.

Higley, L.G. and W. K. Wintersteen. 1992. 'A Novel Approach to Environmental Risk Assessment of Pesticides as a Basis for Incorporating Environmental Costs into Economic Injury Level'. *American Entomologist*, 38: 34–39.

ICAR. 2002. 'All India Coordinated Cotton Improvement Project: Annual Report 2001–02'. Central Institute for Cotton Research, Indian Council for Agricultural Research, Coimbatore.

Ismael Y. R., R. Bennet and S. Morse. 2001. 'Farm Level Impact of Bt Cotton in South Africa'. *Biotechnology and Development Monitor* 48: 15–19.

Koutsoyiannis, A. 1979. *Modern Microeconomics*. Hong Kong: MacMillan Press Ltd. p. 85.

Kovach J., C. Petzoldt, J. Degni and J. Tette. 1992. 'A Method to Measure the Environmental Impact of Pesticides'. *New York's Food and Life Sciences Bulletin*, No.139, New York State Agriculture Experiment Station, Cornell University.

Morse, S., R. Bennett and Y. Ismael. 2004. 'Why Bt Cotton Pays for Small-Scale Producers in South Africa.' *Nature Biotechnology*, 22: 379–80.

Mullen J.D., G.W. Norton and D.W. Reaves. 1997. 'Economic Analysis of Environmental Benefits of Integrated Pest Management'. *Journal of Agricultural and Applied Economics*, 29(2): 243–253.

Naik, G. 2001. 'An Analysis of Socio-Economic Impact of Bt Technology on Indian Cotton Farmers'. Mimeo, Centre for Management in Agriculture, Indian Institute of Management, Ahmedabad.

Narayanamoorthy, A. and S.S. Kalamkar. 2006. 'Is Bt Cotton Cultivation Economically Viable for Indian Farmers? An Empirical Analysis'. *Economic and Political Weekly*, June 30.

Onyango, B., G. Ramu, G. Hallman, Jang Ho-Min and V.S. Puduri. 2004. 'Consumer Acceptance of Genetically Modified Foods in Korea: Factor and Cluster Analysis'. Food Policy Institute, Working Paper No.WP1104–015, Rutgers University.

Organization for Economic Co-operation and Development. 2000. 'Modern Biotechnology and Agricultural Markets: A Discussion of Selected Issues'. Working Party on Agricultural Policies and Markets. December 2000.

Oskam A. and Rob Vijftigs child. 1999. 'Towards Environmental Pressure Indicators for Pesticide Impacts'. In M. Brouwer and B. Crabtree (eds), *Environmental Indicators and Agricultural Policy*, CAB International.

Poonyth, Daneswar, Alexander Sarris, Ramesh Sharma and Shangnan Shui. 2004. FAO Commodity and Trade Policy Research Working Paper-8.

Pray, C.E., J. Huang, R. Hu and S. Rozelle. 2002. 'Five Years of Bt Cotton in China—The Benefits Continue'. *The Plant Journal*, 31(4).

Qaim, M. 2003. 'Bt Cotton in India: Field Trial Results and Economic Projections'. *World Development* 31: 2115–27.

Qaim, M. and A. de Janvry. 2005. 'Bt cotton and Pesticide Use in Argentina: Economic and Environmental Effects'. *Environment and Development Economics*, 10: 179–200 .

Qaim, M. and D. Zilberman 2003. 'Yield Effects of Genetically Modified Crops in Developing Countries'. *Science* 299: 900–1002.

Qaim, M. 2005. 'Agricultural Biotechnology Adoption in Developing Countries'. *American Journal of Agricultural Economics*, 87 (5): 1317–1324.

Qaim, M., A. Subramanian, G Naik, and D. Zilberman. 2006. 'Adoption of Bt Cotton and Impact Variability: Insights from India'. *Review of Agricultural Economics*, in press.

Qaim, M., and G. Traxler. 2005. 'Roundup Ready Soybeans in Argentina: Farm Level and AggregateWelfare Effects.' *Agricultural Economics* 32: 73–86.

Shepherd, Ben. 2006. 'Estimating Price Elasticities of Supply for Cotton: A Structural Time-Series Approach'. FAO Commodity and Trade Policy Research Working Paper-21.

Sumner, Daniel. 2003. 'A Quantitative Simulation Analysis of the Impacts of U.S. Cotton Subsidies on Cotton Prices and Quantities', Mimeo.

Wood Stanley, Liangzhi You and Wilfred Baitx. 2001. 'DREAM Version 3'. User Manual. International Food Policy Research Institute, Washington, D.C.

## ANNEXURE 4.1

Distribution of Sample Farmers

| *State* | *District* | *Blocks* | *Villages* | *No. of farmers* |
|---|---|---|---|---|
| Gujarat | Surendranagar | Halvad, Vadhvan | Mayurnagar, Juna Devalia, Nava Devalia, Moolchand, Bhadresh, Bakarthali | 60 |
| | Rajkot | Doraji, Upleta | Torania, Naniparabadi, Motiparabadi, Motipaneta, Hariyasan, Kharachiya | 60 |
| Maharashtra | Amravathi | Amravathy, Chandoor Bazar | Aasrigav poorna, Nanori, Javla Shapur, New Akola, Sukhali, Nandorai | 60 |
| | Yawatmal | Ner, Yawatmal | Watfali, Manikwada, Dhanaj, Vagapur, Jambavadi, Laasina | 60 |
| Andhra Pradesh | Guntur | Dachapalli, Thadikonda, Guntur rural | Srinivasapuram, Dachepalli, Nidumukkula, Tonnalagadda | 60 |
| | Warangal | Bachennapet, Devaruppala | Itkyapalli, Nakkavarigudam, Thamidapalla, Chandrapuram, Devaruppala, Sitaramapuram, Kadavendi, Nirmala | 60 |
| Tamil Nadu | Salem | Kolathur, Mecheri | Ellaraimathikadu, Kannamoochi, Settiyur, Aandikavundanur, Virudhasampatti, Mallikundham | 60 |
| | Perambalur | Kunnam, Veppanthattai | Andhur, Chitthali, Anukur, Nergunam, Kudikadu, Nergunam | 60 |
| Total | | | | 480 |

*Source*: Survey Data.

## ANNEXURE 4.2

Impact of Bt Technology on Cotton Cultivation

| | | *Gujarat* | *Maharashtra* | *Andhra Pradesh* | *Tamil Nadu* | *All India* |
|---|---|---|---|---|---|---|
| Yield (kg/ha) | Non-Bt | 1685.59*** | 828.50*** | 1902.67*** | 1739.50*** | 1539.06*** |
| | | (14.446) | (14.022) | (24.456) | (25.69) | (25.997) |
| | Bt# | 873.95*** | 338.08*** | 664.58*** | 649.23*** | 631.46*** |
| | | (6.115) | (4.672) | (6.975) | (7.828) | (8.709) |
| PP chemicals (Rs/ha) | Non-Bt | 4262.323*** | 2391.014*** | 4707.203*** | 2713.036*** | 3518.394*** |
| | | (9.388) | (15.128) | (24.824) | (22.45) | (23.847) |
| | Bt# | -1713.82*** | -1057.041*** | -1718.05*** | -1441.73*** | -1482.662*** |
| | | (-3.082) | (-5.461) | (-7.398) | (-9.74) | (-8.205) |
| Seed cost (Rs/ha) | Non-Bt | 413.59*** | 1574.93*** | 1621.25*** | 1695.04*** | 1630.40*** |
| | | (6.715) | (13.325) | (15.515) | (24.95) | (28.628) |
| | Bt# | 1696.43*** | 889.05*** | 1119.84*** | 618.03*** | 776.63*** |
| | | (22.49) | (6.141) | (8.750) | (7.430) | (11.630) |
| Cost of Cultivation (Rs/ha) | Non-Bt | 22088.90*** | 13652.15*** | 29855.04*** | 28046.82*** | 23410.73*** |
| | | (13.492) | (23.62) | (42.442) | (7.146) | (18.900) |
| | Bt# | 5810.67*** | 1830.82*** | 3251.93*** | 9388.65** | 5070.52*** |
| | | (2.898) | (2.587) | (3.775) | (1.953) | (3.342) |
| Profitability (Rs/ha) | Non-Bt | 9516.44*** | 3502.03*** | 7997.32*** | 4110.73 | 6281.63*** |
| | | (3.308) | (2.789) | (4.383) | (0.864) | (3.851) |
| | Bt# | 27945.76*** | 5644.43*** | 10276.33*** | 16651.80*** | 15129.58*** |
| | | (7.931) | (3.671) | (4.599) | (2.858) | (7.573) |

*Source*: Survey Data.

Note: # *Difference between Bt and Non-Bt; *** Significant at 1 per cent level; ** Significant at 5 per cent level*

# 5

# To Bt or Not to Bt? Risk and Uncertainty Considerations in Technology Assessment[1]

*Sarthak Gaurav*[2] and *Srijit Mishra*[3]

## 1. Introduction

Since its legal introduction in 2002, there has been a remarkable diffusion of Bt cotton technology in India, with the genetically modified seeds being used to cultivate nearly 90 per cent of the total area under cotton (Arora and Bansal 2012; Choudhary and Gaur 2010; Government of India 2010).[4] During this period, the yields and overall production has increased on an average, and India, which is the second largest producer of cotton in the world, became a net exporter of cotton from being a net importer. Should we accept these trends as evidence of a successful adoption of the technology? Or, is there something more than what meets the eye—like the dominance of the QWERTY technology for keyboard layouts (Liebowitz and Margolis 1990)? In other words, can the increasing acreage under Bt cotton be solely attributed to inherent benefits of the transgenic seed technology or are there other mechanisms at work that demand explanation? More importantly, are we missing out on critical dimensions of assessing the technology amidst the 'deceptive rhetoric, spin, and soundbite science portraying the wonders or horrors of the new technology' (Stone 2002: 611; also see Stone 2012) propounded by the competing discourses of the pro and anti Bt camps?

The evidence on the performance of Bt cotton in India has been mixed. Several studies provide evidence on the positive and significant impacts of Bt cotton through increased crop productivity (yield), higher net returns, saved management time and savings on account of reduced insecticides cost (e.g. Basu and Qaim 2007; Gruère and Sengupta 2011; James 2009; Kathage and Qaim 2012; Morse *et al.* 2004; Naik *et al.* 2005; Qaim 2005; Qaim and Zilberman 2003; Rao and Dev 2009a, 2009b; Ramasundaram *et al.* 2007; and Subramanian and Qaim 2009, 2010; among others). The central argument of the proponents of the new seed technology in being instrumental in yield enhancement is that reduced pest attack on account of Bt seeds usage has a concomitant impact on improving yield. This, with no foreseeable price shocks (both input and output) and cost savings due to lower insecticide requirements have been argued to lead to greater economic returns. Large welfare gains for farmers and consumers, positive effects on the environment and human health have also been argued (e.g. Qaim and Zilberman 2003; James 2009).

On the other hand, the opponents of the technology mention about loss of seed sovereignty, adverse impact on health of livestock and people because of increased exposure to plant toxicity, insecticide resistance, emergence of new pests, and increased costs on account of seeds and additional requirements on account of the new technology (e.g. Kurungati 2009; Qayum and Sakkhari 2005; among others). Also, there has been variability in the impact of Bt cotton given the range of its performance across space and time because of differences in levels of pest pressure, pesticide usage patterns and 'germplasm effects' (Herring, 2008; Qaim *et al.* 2006).[5] Moreover, for the Bt cotton technology there is considerable lack of knowledge regarding the uncertainties associated with the technology as also the regulatory and institutional frameworks (e.g. Joseph 2007; Lalitha 2004; Pray *et al.* 2005); factors which could have far reaching implications for safe introduction and distribution of the transgenic seeds in the marketplace (e.g. Basu and Qaim 2007; Murugkar *et al.* 2007).[6]

It should be borne in mind that there are several methodological differences in the analysis of Bt cotton impacts that explains the spectrum of conclusions in the debate (Smale *et al.* 2006; Smale *et al.* 2009). However, notwithstanding the conflicting findings in the discourse, there has been a perceptible bias in the narrative of genetically modified (GM) agricultural technologies. Glover (2009; 2010) critiques the 'pro-poor narrative' of the GM technology benefits on the grounds that a range of technical, socio-economic and institutional factors which influence the performance of GM crops are ignored. Similarly, arguments against Bt seeds tend to ignore the positive outcomes even though they are based on some restrictive assumptions and conditions. This lack of accommodating plural and competing explanations among the pro and anti Bt camps has rendered the Bt cotton debate reductionist, which motivates our analysis.

It can be argued that while some farmers, in some regions and/or in some seasons, may have experienced benefits in some particular sense (higher production, or yield and income) from the use of Bt seeds, the genetic attribution of Bt seeds, that is, the Bt trait which causes a particular protein to be expressed, does not guarantee any effects beyond protection against specific bollworm (like American bollworm) infestation under certain conditions.[7] Cotton being a risky cash crop, varieties with Bt gene are as susceptible to all the risks in cotton cultivation that non-Bt varieties are (Herring 2007), and ignoring such risks is a serious analytical flaw. Moreover, owing to the complex interactions in crop production, there are inherent uncertainties in the effects of any agricultural technology, and transgenic seeds are not an exception.

For a scientific evaluation of the hypothesised superiority of Bt cotton seeds in the causal sense, the use of these seeds should be the differentiating factor in attainment of higher yields and profitability in cotton production, *ceteris paribus*. Also, it should be borne in mind that in the absence of appropriate counterfactuals and adequate controls, any evidence that rejects the null hypothesis of Bt seeds not being superior to some non-Bt seeds, does not necessarily imply the superiority of Bt seeds, and *vice versa*. In the case of Bt cotton, the co-existence of scores of varieties with Bt genes and non-Bt hybrids (e.g. Karihaloo and Kumar 2009) complicates the separability of the effects of the Bt strain. Unfortunately, the Bt cotton discourse has failed to acknowledge this fundamental logic while attempting to establish the positive or negative impacts.

In this study, we argue that an improved understanding of risk and uncertainty is central to the assessment of Bt cotton technology by analysing outcomes in two drought periods, across two cross-sections. First, we use unit-level data from the Situation Assessment Survey of Farmers from the 59th round of the National Sample Survey (NSS), for the period 2002–03 across specific regions of Gujarat (Saurashtra and Southern Plains) and Maharashtra (Western Vidarbha).[8] Then we employ data from our surveys and ethnographic inquiries in these agro-ecological zones of the two states for the period 2009-10. Furthermore, the Gujarat sample provides interesting comparisons across Bt and non-Bt growers and ethnographic findings from the field in Wardha district of Vidarbha allow us to draw critical inferences on the Bt technology.

By taking up the analysis in two drought periods—one at a time when the legal (official) commercial transgenic varieties were introduced, and the other at a time when the technological diffusion can be assumed to have played out well, this study attempts to sensitise Bt cotton analysis to issues of risk and technological uncertainty. We believe that this study has two important contributions to improving our understanding of the Bt cotton debate. First, we argue for the need to consider the fundamental aspect of risk in the evaluation of the impact of Bt seeds. This assumes significance at a time of crisis in the Indian agriculture and a spate of farmers' suicides amongst the predominant cotton growers (e.g. Reddy and Mishra 2009). Second, our study is a contribution in mixed-methods research paradigm (Johnson and Onwuegbuzie 2004; Tashakkori and Teddlie 2003). In this context, we introduce a methodological innovation that combines quantitative and qualitative findings from the field with secondary data from a nationally representative sample that provides a rough baseline approximation in the context of Bt cotton evaluation for the two major cotton growing regions of the country, namely Saurashtra and Southern Plains in Gujarat and Western Vidarbha in Maharashtra. It is quite likely that some farmers in the nationally representative sample would have already used illegal Bt seeds even before the legal commercialisation began (Roy 2012). Nevertheless, the analysis can be safely assumed to be representative of non-Bt usage scenario and hence a valid benchmark for conditions of cotton production in the relevant regions. For the current purpose we compare and contrast results from the NSS survey for the crop season of 2002–03 with that of some location specific and somewhat independent surveys in the above-mentioned two cotton growing regions during 2009-10.

This chapter is structured as follows. In section 2, we discuss the data sources and method employed in the study. Section 3 discusses how risk has been treated in the context of agricultural technology evaluation and presents a theoretical framework for assessment of risk in Bt cotton. Section 4.1 discusses the results of secondary data analysis for Saurashtra and Southern Plains in Gujarat, and Western Vidarbha in Maharashtra. This is followed by discussion of findings from the field: an analysis of the field data from the two regions by focusing on riskiness for Bt and non-Bt growers in Gujarat (Section 4.2.1); and problems of uncertainty, deskilling and fads in Western Vidarbha (Section 4.2.2). In Section 4.3, we discuss the aspects of risks in light of our empirical findings and ethnographic evidence from the field. Section 5 concludes the chapter.

## 2. Data and Methods

We conduct our analysis using two datasets. First, we use unit-level data from the Situation Assessment Survey (SAS) of Farmers in the 59th Round of the National Sample Survey (NSS), for the period 2002–03.[9] This survey is a rich source of information on income, expenditure and investments by the farmer households. We confine our analysis to two specific sub-samples of the larger dataset and focus on cotton farmers of two agro ecological zones of Gujarat (Saurashtra and Southern Plains) as well as the Western Vidarbha region of Maharashtra—the two leading cotton producing states of India. Second, we use relevant evidence from our field surveys (micro data) and ethnographic inquiries among farmers of cotton growing villages of Saurashtra and Southern Plains in Gujarat (277 farmers across 15 villages) and in Wardha district of Western Vidarbha in Maharashtra (120 farmers across five contiguous villages) in 2009-10, which could be considered to be fair approximations for the regions analysed using the secondary dataset mentioned above.[10]

The purpose of this methodological innovation is twofold. Firstly, in the absence of appropriate counterfactuals and inter-temporal variations, we approximate the production conditions in the pre-Bt period in our study regions by using the NSS dataset as the baseline approximations (corresponding to the period 2002-03). Both these periods being drought years, the findings provide an interesting approximation of conditions in years when the production conditions can be considered to be limiting and detrimental for crop production in comparison with normal years. Secondly, given the lack of relevant datasets, we have attempted to understand the potential benefits and riskiness of the seed technology based on ethnographic (qualitative) and limited quantitative evidence from the field at a time when the technology effects can be assumed to have played out. This combination of secondary and primary data, and quantitative and qualitative information, can be considered as a methodological departure.

As a caveat, although a cross-sectional analysis is not appropriate for addressing these concerns, given the lack of relevant longitudinal datasets (e.g. Sadashivappa and Qaim 2009) for the evaluation of Bt Cotton technology, we attempt at an analysis that is sensitive to the fact that farm and farmer fortunes fluctuate across time and space, and technological evaluations should incorporate these possibilities and pluralism of arguments and evidence rather than give a verdict based on reductionist ideologies and limited information.

Before discussing the results of our analysis, we explain the need for considering risk and uncertainty in agricultural technology evaluation in general, and Bt cotton in particular. Our treatment of risk and uncertainty in the context of Bt cotton is based upon the theoretical foundations discussed in the next section.

### 3. Risk in Bt Technology Evaluation: Theoretical Frameworks

Risk is integral to the performance of any agricultural technology due to the inherent uncertainties in crop production and the stochastic nature of crop outcomes. Since there are a host of factors that determine crop yield (e.g. rainfall, availability and quality of inputs, credit, capital, labour, farm characteristics, farming practices, pest attacks, crop disease, soil health), and a host of others that influence farm profitability for a given level of yield (e.g. input and output price volatility, storage conditions, policy uncertainty), it is not possible to net out the exact effects of specific factors on variability in yield or net returns. Risk can emerge from each of the components associated with the production process, rendering *ex-ante* decision making under risk and uncertainty (e.g. Kurosaki 1998; Feder 1979) extremely complex. Moreover, downside risks have potential adverse effects on farmers' welfare, necessitating an improved understanding of the riskiness of the inputs in a stochastic production process.

Surprisingly, studies on the impact of Bt cotton provide only ex-post estimates of the effects of Bt cotton and remain silent on the aspects of risk and *ex-ante* expected values from adoption of the new technology.[11] Studies on Bt cotton adoption at the farm level are limited by their ignorance of preference driven demand side factors. Most importantly, they have not looked into farmers' risk attitudes (e.g. Just and Pope 2003) and cognitive biases (e.g. Cole *et al.* 2009) in adoption of Bt cotton, which are central to technology adoption decisions in agriculture (e.g. Just and Zilberman 1983). A difficulty with the inability to analyse such perspectives is that the increased acreage of Bt cotton, might well be driven by network effects, bandwagon effects or farmers preferring Bt to non-Bt varieties because of fads, rather than by way of well-informed rational choices or experiential factors.

Given the high adoption rate of Bt cotton among India's cotton farmers and a lack of consensus on the effects of the technology, it is imperative to assess the risks associated with Bt cotton technology and weigh the identified benefits, not only against the associated costs but also with the unmitigated risks embedded in the new technology. In this context, a fundamental question that remains to be answered is whether Bt cotton increases or decreases risk?[12]

The agricultural production literature provides some clues on whether an input (e.g. pest control) reduces or increases riskiness of the production process, depending on how the uncertainty associated with the input's potential (e.g. pest uncertainty) enters the production function (Horowitz and Lichtenberg 1994; Pannell 1991); and there is empirical evidence on both cases (e.g. Farnsworth and Moffitt 1981, Horowitz and Lichtenberg 1993).

Though there is no discussion of risk in the context of Bt cotton, a study by Hurley *et al.* (2004) challenges the risk reduction attribute of Bt corn and argues that in some circumstances it is advisable to 'subtract a risk premium from the "mean benefits" of Bt corn, because farmers will use it to increase, not decrease, risk' (Hurley *et al.* 2004: 346). They derive the theoretical conditions under which Bt corn increases/decreases risk or acreage with farmer's welfare. They highlight the importance of distinguishing between marginal and aggregate risk effects in evaluating the effect of Bt corn on risk and demonstrate that the effect of Bt corn on risk depends on the price paid for the technology. An interesting finding of Hurley *et al.* (2004) is that Bt corn is not universally risk increasing or decreasing, rather, conditional on the price paid for the technology and the expected value of loss, it can be risk increasing even when its effect on risk improves farmer welfare. Their empirical findings show that conditional on the price, Bt corn can be marginally risk increasing or decreasing and can either increase or decrease corn acreage. Applying these perspectives to the case of Bt cotton, it can be argued that if planting cotton is optimal for farmers, Bt cotton can be considered risk increasing if the expected loss (by way of protection from crop losses due to pest infestation) it eliminates exceeds its price, but not if its price exceeds the expected loss.[13] Interestingly, understanding the riskiness of Bt cotton could also help explain the acreage decisions of farmers, namely what share of their cultivable lands would they allocate to Bt and non-Bt hybrids.

Another simplistic yet intuitive theory of risk assessment developed in Mishra (2008) can be used to scrutinise the success of agricultural production techniques against the alternatives since some technologies while being successful in increasing production as well as productivity, may also add to the risk simultaneously. The framework is as follows:

Let technique $T_i$ be the one where inputs $X_i$ lead to output $Y_i$; where $i$ = 0, 1 indicates two possible techniques. In our case $i$ = 0 could be assumed to reflect the traditional (non-Bt) technique and $i = 1$ be the new and improved Bt technology. According to this theory, we would consider $T_1$ as an improvement over $T_0$ if either $X_1 < X_0$ or $Y_1 > Y_0$. What these inequalities imply is that a technique would be preferred over another if either it is 'input-saving' (i.e. it uses less inputs for producing the same output) or 'output-enhancing' (i.e. it uses same inputs to *p* more output), or both.[14]

The arguments in favour of Bt-cotton indicate the condition of $Y_1 > Y_0$ but what remains to be factored into the technology assessment is the concomitant condition of $X_1 > X_0$. There could have been more production and productivity (yield) due to Bt cotton, but the technique associated with it may be using more inputs to produce more output, this enhancement being indicated by a shift in the production function, *ceteris paribus*.

From a risk perspective, it is important to understand that it may so happen that the output is higher but the rate of increase in output could well be lower than the rate of increase in input, i.e. $(Y_1/Y_0) < (X_1/X_0)$, indicating the adoption of the Bt technology as increasing the riskiness of cotton production. Our understanding of risk in this article will revolve around the empirical validation of this theory of risk using the data on production costs and net returns in cotton production for two drought years (one being relatively good compared to the other), for the two leading cotton producing states of India. Interestingly, this framework fits into Scott's (1976) 'subsistence ethic' argument in evaluating modern agricultural technology *vis-à-vis* traditional technology as it factors in farmers' preferences for reliability and stability in choice of technology. Moreover, considering risk perceptions of the competing technology (as long as farmers have an option to choose from alternative technology) is also crucial for a proper evaluation of agricultural technologies. We incorporate these philosophies into our analysis by the use of stochastic dominance analysis (Hazell and Norton 1986; Anderson 1974), which has been developed in the literature to evaluate a modern technology *vis-à-vis* a traditional alternative. We now present the main results of our investigation and discuss the relevance of the findings.

## 4. Results and Discussion

### *4.1 Findings from NSS data for the Period 2002–03: Baseline Approximations*

In order to understand the conditions prevailing at a time when the Bt cotton seeds were legally introduced into farming systems, we look at the yield, production costs and net returns per acre, for farmers in the NSS sample for the year 2002-03. As evident from Table 5.1, seeds and fertilizer expenses are substantial in cotton production. While seeds comprise of 25.4 per cent of total cost in Gujarat and 25.9 per cent in Vidarbha, share of fertilizer expenses in production costs were higher than seed costs in both the regions, at 26.3 per cent and 31.6 per cent for Saurashtra and Southern Plains and Western Vidarbha respectively. While the share of pesticide costs are more or less same in both the regions—11.6 per cent in Saurashtra and Southern Plains and 10.5 per cent in Western Vidarbha—the share of casual labour expenses in Western Vidarbha are relatively higher at around 22 per cent compared to 14 per cent in Gujarat. The yield in Gujarat (2.12 quintals per acre) is more than double that of the Western Vidarbha sample, which can be attributed to higher irrigation in cotton production in Saurashtra and Southern Plains in comparison with Western Vidarbha.[15]

**Table 5.1:** Production Conditions for Cotton Farmers in Study Regions of Gujarat and Maharashtra, NSS, 2002-03

| *Items* | *Gujarat (N = 369)* | | *Maharashtra (N = 1392)* | |
|---|---|---|---|---|
| | *Mean* | *Share of total costs, %* | *Mean* | *Share of total costs, %* |
| Seed | 540.4 | 25.4 | 395.6 | 25.9 |
| Pesticide | 299.2 | 11.6 | 195.0 | 10.5 |
| Fertilizer | 526.8 | 26.3 | 537.5 | 31.6 |
| Irrigation | 274.4 | 8.6 | 77.6 | 3.0 |
| Regular Labour | 101.6 | 2.4 | 31.6 | 1.4 |
| Casual Labour | 346.3 | 14.4 | 408.8 | 22.0 |
| Others | 239.0 | 11.2 | 109.1 | 5.6 |
| Total Cost | 2327.5 | 100.0 | 1755.2 | 100.0 |
| Net Returns | 7139.4 | | 3724.5 | |
| Yield | 2.12 | | 1.03 | |

*Notes:* Seed, Pesticide, Fertilizer, Irrigation, Regular Labour, Casual Labour, Others, Total Cost and Net Returns are expressed in rupees per acre; reported means are unweighted averages.

*Source:* Authors' computation using data from the Situation Assessment Survey (SAS) of the 59th Round of NSS, 2002–03.

The average net returns per acre in Saurashtra and Southern Plains of Gujarat is more than 90 per cent that in Western Vidarbha of Maharashtra even though the cost of production is only 30 per cent higher. It is of concern that in Western Vidarbha total costs is nearly half that of the net returns. It should be borne in mind that the period under study being a drought year, farmers could have had a higher cost if they went in for a second or third sowing or a lower cost if they reduced their investment in variable inputs given expectations of a bad crop. It is impossible to decompose such effects from the current data. Nevertheless, the figures here could be considered as representative of a drought year, and not a normal year. The scenario can also be used to explain a baseline condition with little or no Bt cotton usage.

The observations from 2002-03 can be contrasted with field level observation in the period 2009-10, another drought year. This was also a time when the use of Bt seeds had become ubiquitous, which could provide important insights into the assessment of Bt cotton's impact. Before, discussing the findings from the field in Saurashtra and Southern Plains of Gujarat and in Western Vidarbha of Maharashtra in the next section, we would like to highlight a fundamental methodological point that has surprisingly been missed out in the Bt cotton debate, and which is integral to our interpretation of risk in agricultural production systems.

From a methodological viewpoint, for a causal interpretation of the impact of Bt cotton on profitability of cotton farming, it should be understood that in a pre-Bt environment, higher yield and higher prices could enhance net returns, given production costs. Hence, the channel through which Bt cotton could increase net returns is through savings by way of specific pesticides and pesticides spraying costs (labour) given the frequency and intensity of pest pressure. The narrative of the pro-Bt camp that the Bt technology provides yield gains (e.g. Qaim 2003; 2005; Rao and Dev 2009a; 2009b) should at best be interpreted as a reduction in yield gap from a potential because what the technology offers is mere protection against yield losses on account of specific pest incidence (cotton bollworms), rather than yield gains for a given pest pressure. This differentiation is crucial for understanding the true effects of Bt cotton because protection from yield losses is not same as yield increment in the sense that the Bt cotton seed technology can be considered as mitigating the risk of bollworm induced losses, instead of increasing yields in the presence or absence of bollworms. Keeping this distinction in mind, we discuss our findings from the field in Gujarat and Maharashtra in the following section.

## 4.2 *Findings from the Field in Gujarat and Maharashtra*

As discussed earlier, for a proper evaluation of a new agricultural technology, it is imperative to factor in the dynamics of farmers' experiences with the new technology; the prevailing environmental, ecological and economic conditions of crop production; and aspects of risk and uncertainty associated with the new technology. In this section we discuss aspects of risk by delving into the differences in conditions of cotton production; and compare crop outcomes of two relevant groups—Bt users and non-users from the Gujarat sample, and discuss evidence from the field in Wardha district of Western Vidarbha, to understand the uncertainty around the Bt technology and farmers' adjustment problems.

### *4.2.1 Risk in Cotton Production: Comparing Bt and Non-Bt Users in Gujarat*

Our field survey from Gujarat, for the period 2009-10 allows us to compare the crop economics and outcomes of those farmers who did not grow Bt hybrids with those who did, thus providing a good counterfactual to usage of Bt cotton. This dataset, though primarily generated with the objective of understanding farmer's participation in financial markets, is nevertheless unique in the sense that it lends us to understand crop economics of some non-Bt growers as well, providing a reasonable estimate of the variations across a sample of Bt and non-Bt users. However, the Maharashtra sample does not provide us with this opportunity. Furthermore, the Maharashtra sample will help us understand a complete Bt scenario, with the non-Bt growers in the Gujarat sample mimicking the counterfactuals. In light of the discussion of results of the analysis using NSS data for 2002-03, it is worth noting that those findings can be considered as baseline scenarios for Bt cotton evaluation in the study regions as it pertains to a period when both Gujarat and Maharashtra farmers could be assumed to be non-Bt growers.

The random sample of 277 cotton growers from three districts of Gujarat, namely Amreli, Bharuch and Bhavnagar, for the period 2009-10 enables us to compare crop outcomes of Bt and non-Bt hybrid growers. Interestingly, 56 per cent of the farmers in our sample are non-Bt growers, quite representative of the general patterns of relatively lower adoption of Bt technology among Gujarat farmers (Arora and Bansal 2012). However, there are three caveats which should be noted for a better contextualisation of our research setting and prevalence of Bt growers in our sample in Gujarat.

First, while Gujarat has the second largest acreage of cotton (second to Maharashtra), it is the largest producer of cotton in the country and has the second highest productivity (after Tamil Nadu) among the major cotton growing states of India. The state is idiosyncratic in the Bt cotton story in India because even before the official transgenic seeds were released into the market in March 2002, unofficial 'loose seeds' or 'stealth cultivars' which were being supplied to farmers had their origins in Gujarat (Roy *et al.* 2007).[16] We did not delve further in getting details whether the Bt grown is loose or illegal (unapproved) or legal (approved); a factor which could explain the relatively low figure of around 56 per cent non-Bt growers than that reported in three studies by the Gujarat Institute of Development Research (GIDR), (Lalitha *et al.* 2009; Lalitha and Viswanathan 2011; Viswanathan and Lalitha 2010). As the GIDR surveys have specifically included this point in the questionnaire, the adoption rate that GIDR surveys get is almost 80 per cent, with marginal difference across varieties. We are unable to make such a distinction in our dataset. Second, Gujarati cotton farmers have difficulties in reporting the seed variety used on account of problems with identification of the varieties and differences in perceptions about the varieties used. Roy (2010; 2012) shows that reports of self-identified Bt growers should be taken with a pinch of salt given the prevalence of various types of seeds (as reported in Lalitha *et al.* 2009; Roy *et al.* 2007). Third, the study year being a drought year, the usage of Bt hybrids may not be in line with what it would have been in a normal year given the likelihood of losing out on the seed costs in the eventuality of crop failure.

**Table 5.2:** Economics of Production: Bt and Non-Bt Growers, Gujarat, 2009-10

| *Item* | *Non Bt (N=154)* | | *Bt (N=123)* | | *All (N=277)* | |
|---|---|---|---|---|---|---|
| | *Mean* | *CV* | *Mean* | *CV* | *Mean* | *CV* |
| Yield | 2.89 | 0.69 | 3.01 | 0.78 | 2.94 | 0.73 |
| Revenue | 11798.4 | 1.99 | 9651.4 | 0.91 | 10845.02 | 1.7 |
| Cost | 6539.73 | 1.41 | 12747.5 | 0.75 | 9296.22 | 1.06 |
| Net Returns | 5258.64 | 4.84 | –3096.1 | –4.08 | 1548.79 | 13.64 |

*Source:* Data from the first author's doctoral field work in Gujarat (Gaurav 2014).

Table 5.2 reports the crop economics of Bt and non-Bt growers.[17] Our cross-sectional findings bring out the relatively high cost of production for Bt growers. The non-Bt hybrids growers in the sample have significantly lower cost of production and higher net returns, while the yield advantage of Bt over non-Bt is not statistically significant (p-value 0.3356). The relatively higher net returns per acre can be attributed to lower production costs for the non-Bt farmers. Furthermore, the cost of production for Bt growers is almost double that of the non-Bt growers and this substantial difference is statistically significant (p-value 0.0000). From a risk perspective, lower input costs have relevance because the only factor that a farmer has control over, is the choice of inputs (e.g. Dercon 1996). For a given level of revenue, the net returns and the associated risk that a farmer has on her (his) portfolio in any season is a decreasing function of the costs incurred.

An interesting observation from the perspective of risk is that the yield variability of Bt cotton is higher. Assuming this as an indicator of inter-temporal variability in yield, farmers who have not yet chosen Bt cotton, may be inferred to have a preference for reliability of the yields which the non-Bt varieties could be offering. If farmers chose stability and lower fluctuations in yield to higher expected returns (yield), their risk-avoidance behaviour in terms of not choosing Bt cotton seeds is indeed rational. This risk-aversion is also reasonable from Scott's (1976) 'subsistence-ethics' perspective as it would be very difficult for farmers with lower asset-base to recover from adverse production shocks which the use of new technology could be associated with. The losses experienced by the Bt cotton users as against the gains of the non-Bt users (Table 5.2) and farmers' views from focus group discussions (FGDs) corroborates this argument.

Having had an idea of the general differences in crop outcomes, a look at the components of the cost of production should be informative about which variable inputs weigh heavy in the production costs. Table 5.3 brings out the differences among input costs of Bt and non-Bt growers. There is a clear pattern of a costly external input system of production being undertaken by the Bt growers *vis-à-vis* their non-Bt counterparts, which is a matter of grave concern given the unsustainability of returns to cultivation in India (Gaurav and Mishra 2014). Except for draught (bullock) costs, the difference between variable input costs for Bt and non-Bt hybrid growers is statistically significant at all levels, corroborating the earlier finding of relatively higher production costs of Bt technology. The seed costs, pesticide costs (on account of sucking pests) and labour costs of the Bt growers is more than twice that of the non-Bt growers which is reflected in average production costs which are almost double that of the non-Bt growers.

Concerns about high input costs in Bt cotton production have been raised in earlier studies. Sahai and Rahman (2003) found Bt seed costs to

**Table 5.3:** Production Costs Per Acre for Bt and Non-Bt Growers, Gujarat (2009)

| *Farmer Type* | *Seed* | | | *Fertilizer* | *Manure* | *Pesticide* | *Irrigation* | *Labour* | *Draught* | *Tractor* | *Total* |
|---|---|---|---|---|---|---|---|---|---|---|---|
| | *Non-Bt* | *Bt* | *Varieties* | | | | | | | | |
| Non Bt (N = 154) | | | | | | | | | | | |
| Mean | 741.3 | – | 18.7 | 763.0 | 864.6 | 911.7 | 479.3 | 1540.3 | 577.7 | 643.2 | 6539.7 |
| % Share | 11.34 | – | 0.29 | 11.67 | 13.22 | 13.94 | 7.33 | 23.55 | 8.83 | 9.84 | 100.00 |
| Bt (N = 123) | | | | | | | | | | | |
| Mean | 27.5 | 1358.9 | – | 1441.4 | 1811.8 | 1978.3 | 1105.0 | 3154.9 | 747.8 | 1122.0 | 12747.5 |
| % Share | 0.22 | 10.66 | – | 11.31 | 14.21 | 15.52 | 8.67 | 24.75 | 5.87 | 8.80 | 100.00 |
| All (N = 277) | | | | | | | | | | | |
| Mean | 424.3 | 603.4 | 10.4 | 1064.2 | 1285.2 | 1385.3 | 757.2 | 2257.2 | 653.3 | 855.8 | 9296.2 |
| % Share | 4.56 | 6.49 | 0.11 | 11.45 | 13.82 | 14.90 | 8.14 | 24.28 | 7.03 | 9.21 | 100.00 |

*Notes*: Non-Bt/Bt refer to non-Bt/Bt hybrids; Varieties refers to straight varieties.
*Source:* Data from the first author's doctoral fieldwork in Gujarat (Gaurav 2013).

be four times the traditional varieties, and that the higher seed costs failed to be compensated by corresponding increments in yield or savings on account of reduced pesticide use, resulting in lower net profits for Bt-cotton than non-Bt cotton. Swaminathan and Rawal (2011) found that the Bt cotton varieties had higher expenditure on chemical pesticides and higher variability as in our Gujarat field sample. Bennet *et al.* (2004; 2006) identified the problem of higher seeds costs and overall cost of production associated with the Bt adopters.

From an agricultural technology perspective, it should be borne in mind that no matter how effective or superior it might be, there would be times in agricultural production process when realised crop outcomes will fall short of (or go beyond) farmers' expectation. In other words, it will have either adverse or favourable deviations from the normal values. In seasons when there are adverse realisations, the adopters of the costlier technology stand to lose more than their counterparts who did not adopt. At times of losses, the benefits or merits of the technology cannot come to their rescue as the loss event has occurred, rather it may push them down into deeper financial crisis on account of having incurred higher cost (also higher opportunity costs). Glover (2010) argues on similar lines and discusses how a Bt adopter is potentially even more vulnerable than a non-adopter in the face of such eventualities.

To further substantiate our argument of a costlier production technique being riskier, especially in bad crop seasons, we look at the cumulative distribution functions (CDF) of net returns from the two technologies. Risk perceptions in alternative technology evaluation are embodied in the CDF, which indicates the cumulative probability that the net returns fall below a given level. The lower cumulative density plot first order stochastically dominates (e.g. Anderson 1974) the higher plot, indicating the lower CDF being superior in the sense that, for a given cumulative probability, we can always obtain a higher level of net returns. As depicted in Figure 5.1, we see that the Bt technology is first order stochastically dominated by the non-Bt technology, which implies that for rational decision makers (farmers here) guided by the principle of local non-satiation (more is better), the non-Bt technology is more risk-efficient and it is clearly preferred to Bt technology[18].

What this finding implies is that the farmers' present practices of using non-Bt hybrids dominates the Bt technology in a risk-prone environment and this behaviour is reasonable as it would not be appropriate to experiment with a costlier and uncertain technology like Bt for farmers who are credit and liquidity constrained. This result of a

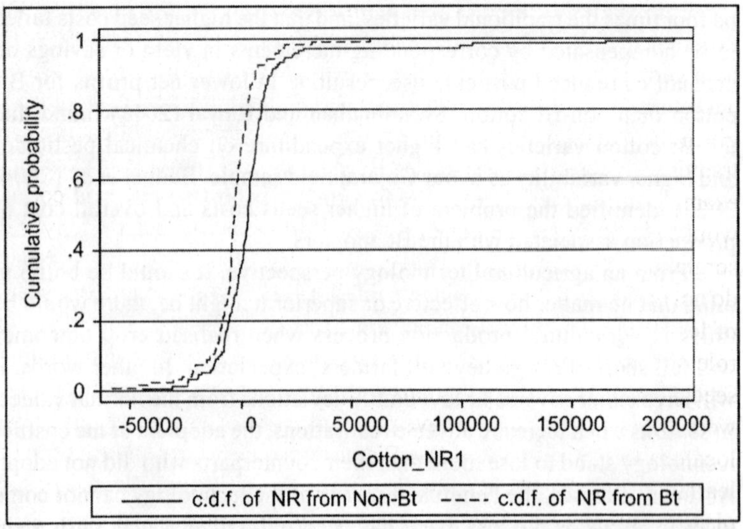


*Note:* c.d.f. stands for cumulative distribution function; NR stands for net returns expressed in rupees per acre; Bt and Non-Bt stand for Bt and non-Bt hybrids; Cotton_NR1 represents net returns from cotton in rupees.

*Source:* Data from the first author's doctoral field work in Gujarat (Gaurav 2014).

**Figure 5.1:** Stochastic Dominance between Alternative Seed Technologies: Bt and Non-Bt Hybrids, Gujarat Sample: 2009

traditional technology being perceived as better than a new technology that is claimed as 'improved' is similar to that of a study by Dvorak (cited in Walker and Ryan 1990) where applying inorganic fertilizer to post-rainy season sorghum grown in Sholapur (ICRISAT village level studies) is worse than no fertilizer applications as it adds to the production cost and risks, without increments in potential benefits. It should be borne in mind that if the CDF plots crossed once (or multiple times), we would have had to check for second (or higher) order stochastic dominance, which are scenarios where risk-aversion of the farmers become potentially important. Moreover, crossing does not necessarily indicate a trade-off between risk and expected profitability as one technology may be preferred to another in low probability and low returns events. If it is likely that persistently low or even negative returns is realised at the time of adversities, farmers may not prefer the new technology like Bt, as the expected profitability may be dominated by concerns of stability of returns (e.g. Perrin and Winkelman 1976).

This analysis suggests that farmers who do not have substantial experience of the returns from a particular cash-intensive technology in both good and bad crop years (i.e. subjective probability distribution of net returns from the new technology unlike the traditional technology), given their local conditions of farming, can be expected to choose the traditional technology. Walker and Ryan (1990: 225) provide an excellent example of this complexity: 'In abnormally good years, hybrid cotton would significantly outperform local cotton, *mung* bean, or hybrid sorghum, but those years of ideal growing conditions are so few that they do not make up for hybrid cotton's inferiority in the more numerous years of less than ideal growing conditions'. Implicit in this argument is the role of experimentation or familiarity with a seed variety in determining seed choices. From conversations with farmers during our fieldwork, we were informed that farmers have preference of particular traits given the localised conditions of farming. Either through own experimentation or learning from others (social learning), farmers decide which varieties to plant. It can be conjectured that the longer the acquaintance with a specific variety then the greater is the accumulated knowledge on what works and what does not. This local knowledge is an important element of production in general and cotton production, in particular, at a time when the markets are infested with a large variety of seeds.

### *4.2.2 Technology Uncertainty, Deskilling and Fads: Insights from Vidarbha Villages*

The cash-crop oriented production systems in the five contiguous villages in Wardha district are representative of rainfed production in other parts of Western Vidarbha—cotton or soybean as main crops intercropped with *tur* (pigeon pea/red gram) as intercrop during the Kharif season; and wheat and *chana* (Bengal gram/chickpeas) in the Rabi season, for those who have access to irrigation.[19] Though some farmers grow food crops like *juar* (sorghum), there has been a significant reduction in the acreage of food crops, particularly sorghum, with the growth of soybean acreage over the last three decades.

Table 5.4 reports the condition of crop production in 2009-10. We will contrast this information on costs and returns with the findings from the NSS sample for Western Vidarbha (2002-03) in the following section to get an idea of the riskiness in the production system. A similar calculation will also be done for the two Gujarat samples.

**Table 5.4:** Cotton production in Five Wardha Villages, 2009-10

| Items | Mean |
|---|---|
| Yield (quintal per acre) | 3.32 |
| Revenue (Rs. per acre) | 10056.96 |
| Cost (Rs. per acre) | 3578.28 |
| Net Returns (Rs. per acre) | 6478.68 |
| Seed Cost (Rs. per acre) | 1038.28 |
| Seed Cost % | 29.02 |

*Note*: Costs reported are paid out costs. Though the full sample comprises 120 farmers, in 2009-10 only 104 farmers took up cotton production; All the cotton plots were a mix of Bt cotton hybrids intercropped with *tur* (pigeon pea). Seed Cost % refers to percentage share of seed costs in total costs.

*Source*: Data from the first author's doctoral fieldwork in Gujarat (Gaurav 2013).

For the farmers in our five study villages, 2009-10 was one of the worst crop years they had faced in the last five years given a delayed onset of monsoon and prolonged dry-spells during the germination phase of cotton. In spite of it being a relatively bad drought year, the average cotton yield of 3.3 quintals per acre fetched positive net returns (an average of Rs. 6500) due to better price realisations in comparison to the previous year (an average of Rs. 3029 per quintal as against Rs. 2752 per quintal in 2008-09, which was also a drought year with a marginally higher yield of 3.4 quintals per acre and almost similar (lower) costs that gave average net returns of Rs. 5876 per acre). The fact that farmers can receive positive returns despite poor yields and high costs brings out the role of output prices and price risk (in both inputs and output) in the determination of net returns from cotton production.[20] Such decomposition of price effects and yield effects are vital for a sound evaluation of the effects of Bt cotton, which has not been attempted here (see Ranganathan and Gaurav 2013, for an analysis in the context of Vidarbha region).

Related to this issue is the role of intercrops in influencing input decisions as well as yield and returns given inherent complementarities and substitutabilities among multiple crops in an intercropped system. Swaminathan and Rawal (2011) discuss the importance of differentiating between mono-cropped and inter-cropped cotton in the context of Bt cotton evaluation in a village study in Vidarbha. Accounting for this distinction is critical because the inter-crops could confer positive as well as negative externalities on yield and costs. For example, growing *tur* (pigeon pea) as an inter-crop with cotton not only works as a 'pest distracter' for cotton plants, but also plays an important role in nitrogen fixation and providing ecological benefits to the cotton production. Farmers' own labour allocation and use of hired labour for farming activities is also dictated by the cropping patterns.[21] The presence of intercrops also influences the crop management practices and influences variables like availability of 'refuge land' for non-Bt varieties, for instance, which play a role in determining crop yields. From a price risk perspective (e.g. Newbery and Stiglitz 1981), growing intercrops also works as a natural hedge to the downside risk in cotton prices and ensures nutritional security for the farm households (e.g. Parsuraman and Rajaratnam 2011).

Farmers in our sample complained of the relatively higher seed costs. As shown in Table 5.4, seed expenses comprised of around 29 per cent of the total production costs, as against 26 per cent in 2002-03 (Table 5.1). Though this seed inflation may be attributed to the higher prices of Bt cotton seeds on an average, it becomes disproportionately penalizing for those farmers who incur substantial losses in bad crop years.[22] With around one-third of the farmers in our sample reporting substantial losses in crop production in 2009-10 compared to one-tenth in 2002-03, it is for these farmers that such high costs are symptomatic of higher risk in the production system as the relative increase in costs is likely to be higher than that of the net returns. Moreover, significantly large negative income shocks could also increase the likelihood of farm households falling into poverty traps (e.g. Carter and Barrett 2006).

Another problem that complicates farmers' decision making process is the proliferation of seed varieties with Bt traits.[23] In recent years, there has been an increasing trend to adopt multiple genes (mostly two genes). The first two-gene event (second generation) MON15985, commonly known as Bollgard®II (BG®II) featured the two genes 'cry1Ac' and 'cry2Ab', and was approved for sale for the first time in 2006—four years after the approval of the single gene event (BG®I) MON531 Bt cotton hybrids in 2002-03. Surprisingly, all the farmers in our sample were ignorant of the basic distinction between the single gene and multi-gene technology and reported expectations about double yield from the costlier BG®II hybrids based on anecdotal evidence from some farmers in their social network or news of some progressive farmers in a neighbouring village having experienced supranormal profits by its adoption. To further complicate farmers' seed purchase decisions, with episodes of higher weeding costs (labour costs) and pest resistance being reported from many regions which saw the higher adoption of hybrids with BG®II trait, the farmers reported rumours of a likely emergence of next generation Bollgard® seeds which could overcome such problems.

Closely associated with the proliferation of multiple gene seeds is the conspicuous unavailability of straight varieties and alternative cultivars in the sowing season. This has been reported to be a major concern for farmers in our study villages because the lack of affordable alternatives and inexperience with the new arrivals force them to make ill-informed guesses while purchasing the seed mix every growing season. Moreover, a major source of uncertainty with the Bt seeds is the farmers' ignorance about the effect of Bt seeds on pest ecology, and hence the effect on optimal insecticide spray decisions of farmers.[24] This could be a result of the drastic transformation of cotton seed technology to Bt seeds. This transformation occurred at a time when the market for seeds had been undergoing unprecedented changes with the proprietary hybrid seeds flooding the market and crowding out the more regulated public hybrid seeds and straight varieties (Lalitha *et al.* 2009). The proliferation of the predominantly intra-*hirsutum* (intra-specific) proprietary hybrids had a complex interaction with the dynamics of pest ecology and crop's growth requirements, which resulted in the farmers being increasingly dependent on the market for their seeds (Stone 2011).[25] Furthermore, with new varieties of seeds, there has been a concomitant emergence of new kinds of insecticides with declining natural resistance to many pests, which have been attributed to ecological imbalances on account of higher dosage of synthetic sprays (Kranthi *et al.* 2002).[26]

This market transformation also renders the evaluation of Bt genes difficult. The claim of significant yield increments and the role of Bt genes in the hybrids is problematic. It is impossible to account for the share of yield increments attributable directly to the Bt trait in the Bt hybrids. Also, attributing this causality to Bt cotton, without controlling for access to irrigation, weather variations and socio-economic characteristics is unscientific.[27]

Many Bt growers in our study villages revealed that they applied fewer insecticides to control sucking pests than non-Bt growers in the early seasons assuming the Bt seeds to take care of the sucking pests as well. However, owing to increasing sucking pest pressure as well as resistance from bollworms they eventually resorted to relatively more insecticides sprays and experienced losses or lower profits. Accounting for such differentiation in pesticide expenses is important for the evaluation of Bt cotton because what matters for the economic viability of the production process is how the savings due to reduced insecticides usage exceeds the higher seed cost. If there are instances of pest resistance or other effects on the cropping system, the positive effects of Bt cotton could be reduced and there could well be negative effects.

Hence, the evidence from the field in Vidarbha suggests that, left on their own to wade through the myriad of uncertainties looming large in the absence of effective extension and information (e.g. Fitzgerald 1993), the informational gaps were filled in by local input dealers with vested interests of profiteering.[28] The farmers rely heavily on the advice of input dealers who promote the costlier seeds and unnecessary chemical inputs by a narrative that their use would lead to higher profits through cost savings associated with fewer sprays for controlling pests like *Spodoptera* as well as increasing yield significantly over single gene Bt hybrids. Since, farmers rely on social learning and their preferences and decision making was shaped more by the opinions and actions of others, rather than experiential learning, their behaviour regarding adoption of new seeds and insecticides appears to be dictated by 'highly localised seed fads' driven by marketing and happenstance rather than local agro-ecology (Stone 2011). The observed behaviour of the cotton farmers could be thought of as 'informational cascades' or 'observational learning' (Bikhchandani *et al.* 1992) or 'herd behaviour' (Banerjee 1992) as prominent in the theory of psychology, economics and marketing. From a Bt cotton assessment perspective, this is crucial because herd mentality, information cascades or fads do not necessarily imply the superiority of the adopted technology.

In some sense, Bt cotton is akin to an insurance product, and hence, the premium the farmers are paying by way of higher seed costs to avert the losses on account of unanticipated bollworm attacks should also be actuarially appropriate or the reasons for the higher prices be rationalised.[29] The price of Bt cotton seeds should reflect the risk that is inherent in the technology as it is not proper for the farmers to be paying a premium on account of trait fees or charges of royalties without the knowledge of the scenarios under which the technology could fail or perform worse than expected. This implies that informational asymmetries on account of the lack of knowledge of the loss events and corresponding probabilities amplify the uncertainty associated with the technology. Moreover, the germination rates and conditions for their success or failure, as well as side-effects should be properly communicated to the farmers like the 'red herring' or 'exclusion clauses' in insurance contracts. This would enable the farmers to properly evaluate the technology given their credit and liquidity constraints and then make a well informed buying (investment in seeds) decision.

The observed high production costs in both Gujarat and Maharashtra samples could also be a consequence of the problem of deskilling and unreasonable decision making by farmers that makes it a riskier proposition. In the following section, we discuss the riskiness of cotton production by combining data from the NSS for the period 2002-03 and from the field in 2009-10 for both Gujarat and Vidarbha.

### *4.3 Riskiness of Cotton Production*

A formidable test of sustainability of an agricultural technology and its ability to protect farmers against downside risks is how it fares in the eventuality of bad crop years or worse seasons. If Bt cotton does not perform well *vis-à-vis* non-Bt hybrids in drought years, when the expected yields and returns are lower than normal years, then the utility of the technology should be brought under the scanner.

As discussed in Section 3, our perspective on risk in cotton production is motivated by the dynamics of costs and net returns associated with Bt cotton and its alternative production technology. With Y representing net returns, X representing input costs and the subscripts 0 and 1 representing non-Bt and Bt technology respectively, a look at the relative change in net returns with relative changes in production costs should be indicative of the risk in the system. We undertake a methodological jump here assuming that 2002-03 represents a non-Bt scenario as it was formally legalised that year and the NSS data would be capturing baseline conditions of cotton production where Bt adoption was near zero.[30] This is contrasted with micro-data from the field for 2009-10 when the Bt diffusion had reached a high level in the study regions. In the absence of richer datasets for the relevant analysis, we focus on just two time periods and try to understand the relative change of costs and returns in cotton production given our assumption.

Using the notation developed earlier, it implies that even if $Y_1 > Y_0$ could hold, the rate of increase in net returns (gross value of output less costs) could well be lower than the rate of increase in production costs (input), i.e. $(Y_1/Y_0) < (X_1/X_0)$, indicating an increment in risk.[31]

From Table 5.5, it is clear that there has been an increase in riskiness of cotton production in both Gujarat and Maharashtra. In Gujarat, production costs have almost quadrupled and net returns have become one-fifth, resulting in an eighteen fold increase in riskiness. In the

**Table 5.5:** Riskiness of Cotton Technology, Gujarat and Maharashtra

| *Item* | *Gujarat* | *Maharashtra* |
|---|---|---|
| $X_0/Y_0$ | 0.33 | 0.47 |
| $X_1/Y_1$ | 6.00 | 0.55 |
| $X_1/X_0$ | 3.99 | 2.04 |
| $Y_1/Y_0$ | 0.22 | 1.74 |
| $Y_0$ | 7139.40 | 3724.50 |
| $Y_1$ | 1548.79 | 6478.68 |

*Note:* *X* and *Y* denote input costs and net returns respectively. The subscripts 0 and 1 denote non-Bt and Bt scenario respectively.

*Source:* Authors' computation using data from the Situation Assessment Survey (SAS) of the 59th Round of NSS, 2002-03.

meantime, Maharashtra saw a relatively lower increase in riskiness as costs doubled while net returns increased by around three-fourth. Even if both the periods are drought years, in 2009-10, Gujarat farmers experienced costs that were six times more than that of the net returns; in fact, the net returns of the Gujarat farmers in our sample of 2009-10 is about one-fifth of that in 2002-03. In Maharashtra, the net returns were higher in 2009-10 compared to 2002-03, but the rate of increase in net returns was lower than the rate of increase in input (i.e. $Y_1/Y_0 < X_1/X_0$), as in Gujarat, thereby, increasing the risk in cotton production. In other words, having input costs as the basis for deriving a price index that the farmer faces, the net returns for cotton farmers of Gujarat and Maharashtra in 2009-10 are lower than that of 2002-03 in real terms.

In any case, the absolute value of the observed net returns in both Gujarat and Maharashtra is not sustainable from a livelihood perspective (Gaurav and Mishra 2014), yet the case of Gujarat exemplifies the problem of a costly production system. In 2009-10, even $Y_1<Y_0$ (net returns is lower in absolute terms when compared with 2002-03) which implies that the riskiness of the cotton technology has increased since the introduction of Bt cotton to such an extent that in bad drought years, there are relatively greater losses on an average for farmers in the region. Such situations could affect the welfare of farmer households adversely and it is imperative to include such possibilities while assessing the relative costs and benefits of Bt cotton technology.

## 5. Concluding Remarks

In this chapter, we argue that the perceived benefits and costs of introduction of the genetically modified seed technology namely Bt cotton, needs to be re-evaluated from the perspective that by being a substantial addition to cost in the farmers input vector, it also has risk considerations that need to be analysed. Its dynamic interaction with a host of other factors like changing pest ecology, price risks, multi-crop systems and market dynamics needs attention.

Using data from the Situation Assessment Survey of Farmers of the NSS 59th round, for the drought period 2002-03, we derive baseline scenarios for regions in Gujarat and Maharashtra. Data from the field and ethnographic evidence in similar regions provide changes in conditions of cotton production in the drought year 2009-10. First, we show that Bt technology is first order stochastically dominated by non-Bt technology for our sample of cotton growers in Gujarat and the Bt technology adds substantial input costs to the production system.

Second, findings from the village level study in Maharashtra provide evidence of technology uncertainty, rampant deskilling and sub-optimal production decision making by farmers with limited information. Widespread institutional failures in the market for seeds emerge as a major concern in the farmers' adjustment process and demands immediate attention.

Last but not the least, we provide the first evidence on the riskiness of Bt cotton technology by employing a simple risk framework (Mishra 2008) which takes the ratio of costs to that of net returns from cultivation of a specific crop. In this theoretical framework, Bt technology can be both risk increasing or risk decreasing given the actual conditions under which the production occurs, but empirical evidence presented in this analysis indicate that Bt cotton has been risk increasing.

Even though our data are limited in terms of external validity and we employed a mixed-methods framework to address our core research question, we get substantive qualitative and quantitative confirmation on the Bt technology being associated with increasing risk in a drought year. This perspective is critical for technology evaluation because the farmers stand to lose more in bad crop years than they would have in the case of traditional varieties. Moreover, if the net gains from Bt (at higher expected yields) are offset by the high intensity losses (frequent or otherwise), the yield advantage promised by Bt seeds should be taken with a pinch of salt because the absolute returns are so low that an increase in returns in such years would not lead to an increase in savings that can be used to compensate in a bad year. The riskiness of this technology becomes even more important from the fact that around 65 per cent of India's cotton is produced in rainfed conditions as against 35 per cent under irrigated conditions.

We would like to emphasise that though this study is not an addition to the existing literature on the productivity or socio-economic impact of Bt cotton technology, it is an attempt to introduce a new perspective in evaluating the technology by analysing farm level data from a nationally representative sample and field level data from one of the most challenging socio-economic and institutional contexts where the merit of the technology could be put to test. Notably, the paper introduces an alternative perspective on how the inherent riskiness of a so-called innovation like Bt cotton should be identified, irrespective of its potential benefits or observed merits across different parts of the country and among different groups of farmers. We demonstrate how a neglect of the simple economics of crop production and the associated risk factors can play out against the farmers who make a conscious decision to adopt a particular technology given multiple constraints, with the hope that it will pay off well.

There are certain methodological issues in Bt cotton evaluation that have not been dealt with in this paper but could be taken up as future research. Bt-hybrids outperforming the earlier non-Bt varieties should not be taken as a positive increment in yield on account of Bt alone, and the separability of effects of the hybrid impacts from Bt traits in a seed should be strived at. Variability of yield and other dimensions like price risks need to be assessed rigorously. Moreover, Bt technology could also induce moral hazard in farming practices as the farmer would expect the superior technology to be causing higher yields. An analysis of the dynamic performance of the technology over an appropriate period of time across different agroclimatic conditions could throw more light and also help us study the associated effects on the ecology and environment. Despite these limitations, our analysis does suggest that the proposition—to Bt or not to Bt—demands more non-reductionist analyses that take cognizance of the localised variations and institutional realities.

## Notes

1. The authors would like to thank the editors of the volume and to participants in the National Seminar on Bt cotton held at the Gujarat Institute of Development Research (GIDR), Ahmedabad for their comments. Special thanks are in place to P.K. Viswanathan for presenting

a preliminary version of this paper on our behalf at the seminar and conveying to us the comments. Sarthak Gaurav (SG) has benefited from discussions with Vijay Barapatre, Chandrasekhar Dorlikar, Nirupam Datta, Vijay Jawandhia, Thiagu Ranganathan, Atul Sharma, Ashish Singh, Joseph Stiglitz and farmers in the study regions. Field survey was facilitated through the support of the staff from the Centre for Sustainable Agriculture, Wardha and Sajjata Sangh, Development Support Centre, Ahmedabad. The study in Gujarat was made possible through the ILO Micro-insurance Innovation Facility Research Grant to SG, as Principal Investigator with Shawn Cole, Harvard Business School. A version of this article is available as an IGIDR Working Paper, WP-2012-001.

2. Post-doctoral Researcher at Asia Research Centre, London School of Economics and Political Science (LSE). E-mail: s.gaurav@lse.ac.uk.
3. Associate Professor, Indira Gandhi Institute of Development Research (IGIDR), Mumbai. E-mail: srijit@igidr.ac.in.
4. Bt cotton is genetically engineered to produce a protein found in the soil bacterium Bacillus thuringiensis (Bt). The protein is toxic to lepidopterous insects, especially bollworms like *Helicoverpa armigera* (popularly known as American bollworm) and *H.zea*, and the trait promises to protect farmers from losses due to pest attacks that damage cotton production and affect quality.
5. Qaim *et al.* (2006) argue that farmers in low pest-pressure areas experience relatively lower benefits from Bt and in some cases the benefit could even be smaller than the mark-up on the seed price—in which cases the seed price caps could indeed lead to significant increases in adoption.
6. Several studies demonstrate the potentially harmful health effects of the GM technology in general (e.g. Losey *et al.* 1999; Ewen and Pusztai 1999; Stanley-Horn *et al.* 2001), bringing the issue of GM foods and crops to the forefront of public consciousness apart from resulting in protracted protests against the GM technology by civil society groups and environmental organisations.
7. Expression of insect-resistant cotton encoding the 'cry1Ac' toxin gene from Bacillus thuringiensis (Bt).
8. Western Vidarbha region in the NSS dataset comprises the districts Akola, Amravati, Buldhana, Nagpur, Wardha, Washim and Yavatmal.
9. The SAS dataset provides detailed farm and farmer level information from 51,770 households spread over 6,638 villages across the country for the period 2002–03.
10. The sample in the five villages of Wardha district in Vidarbha is part of an ongoing panel data designed by the first author's doctoral research and focuses on issues of risk and vulnerability of farm households in the region.
11. Fernandez-Cornejo and Jans (1999) discuss this aspect in the context of US Agriculture.
12. An input is defined as risk decreasing (increasing) if a risk averse farmer

uses more (less) than a risk neutral farmer at a given price (Ramaswami 1992).

13. Following Hurley *et al.* (2004), a testable hypothesis in this context could be: Bt cotton increases (decreases) cotton acreage if the change in the expected marginal revenue of cotton acreage is greater (less) than the change in the marginal risk premium (Ramaswami 1992; Kurosaki 1998).
14. Inputs can be measured by input costs (expenditures) while gross value of output (revenue) or net returns could be used as measures of output.
15. Though both regions fall under the central zone of cotton production in India, which is predominantly rainfed and is under black soil (vertisols) subjected to runoff, soil erosion and nutrient losses, tracts in Gujarat have a higher irrigation percentage than in Maharashtra.
16. Mostly F2 hybrid seeds packaged as the popular *Navbharat 151* (NB 151) seed and illegal brands like *Rakshak.* See Roy *et al.* (2007) for a discussion on the multiple seed varieties available to Gujarati farmers when the Bt seeds were legally introduced.
17. See Gandhi and Namboodiri (2004); Roy *et al.* (2007); Sadashivappa and Qaim (2009) which also surveyed both Bt and non-Bt growers.
18. However, one should not read too much into an analysis of stochastic dominance based on a cross-sectional data in a drought year. Nevertheless, in the absence of appropriate time series data for the region, this evidence becomes relevant from a worst-case scenario perspective.
19. Though the villages are characteristic of the risky cultivation under rainfed conditions, these villages are seeing active participation in the community based sustainable agriculture (CMSA) practices of a civil-society group. The randomly selected study sample can be considered to be representative of local conditions in the five villages.
20. See Baffes (2011) for the emergence of Bt cotton and the evolution of cotton output prices in India and other leading cotton producers in the world.
21. This includes animal draught, human and mechanized labour.
22. Arora and Bansal (2012) discuss the implications of the high prices of Bt cotton seeds and the effect of price ceilings on account of legislation in some Indian states.
23. The problem of proliferation of Bt cotton hybrids and dwindling alternatives remains unaddressed. Choudhary and Gaur (2010) mention that a total of 522 Bt cotton hybrids were approved for planting in 2009 compared with 274 Bt cotton hybrids in 2008, 131 in 2007, 62 in 2006, 20 in 2005 and only four Bt cotton hybrids in 2004.
24. See Pannel (1991) for the role of uncertainty about some variables, such as pest density and pest mortality, on optimal pesticide use under risk aversion.
25. Interestingly, India is the only country to grow all four species of cultivated cotton *Gossypium arboreum* and *G. herbaceum* (Asian cottons), *G.*

*barbadense* (Egyptian cotton) and *G. Hirsutum* (American upland cotton). *Gossypium hirsutum* represents more than 90 per cent of the hybrid cotton production in India and all the current Bt cotton hybrids are *G. hirsutum*.

26. Synthetic insecticides have been observed to be rendered ineffective against most Lepidopterans (e.g. bollworms) and *Hemipteran* sucking pests like cotton aphid (*Aphis gossypii*), mealy bug (*Phenacoccus sp.*, *Maconellicoccus sp.*) and mirid (*Creontiades sp.*), whitefly (*Bemisia tabaci*) and other aphids (Stone 2011).
27. This is akin to asking: Would the non-Bt cotton not have registered the yield increments during the similar period, *ceteris paribus*?
28. Incidences of supplier-induced demand were also observed in a study on farmers' suicides in Vidarbha by Mishra (2006a, 2006b), but the scenario depicts a situation when Bt seeds were not used much.
29. An added advantage of such a perspective is that it necessitates the knowledge of loss distributions of the Bt cotton seed technology, which could be used to communicate to farmers, the intensity of losses and associated probabilities of the loss events.
30. Baffes (2011) confirms that this is a valid assumption given the relatively low adoption of Bt-cotton in India in 2002-03 compared to other cotton growing countries and subsequent time periods.
31. For a proper assessment of risk X should be cost in a bad year and Y be the net average savings from good years (net returns less consumption). The value, X/Y, indicates that we need so many good years to make good a bad year. In the absence of consumption/net savings we take net returns as proxy and the measure being relevant only for Y>0 implies that for zero or negative net returns farmers cannot make good the loss.

## References

Anderson, J.R. 1974. 'Risk Efficiency in the Interpretation of Agricultural Production Research'. *Review of Marketing and Agricultural Economics*, 42(3): 131–184.

Arora, A. and S. Bansal. 2012. 'Diffusion of Bt-cotton in India: Impact of Seed Prices and Varietal Approval'. *Applied Economic Perspectives and Policy*, 34 (1): 102-118.

Baffes, J. 2011. 'Cotton, Biotechnology and Economic Development'. Policy Research Working Paper 5896, Washington D.C.: The World Bank.

Banerjee, A.V. 1992. 'A Simple Model of Herd Behaviour'. *Quarterly Journal of Economics*, 107 (3): 797–817.

Basu, A.K. and M. Qaim. 2007. 'On the Adoption of Genetically Modified Seeds in Developing Countries and the Optimal Types of Government Intervention'. *American Journal of Agricultural Economics*, 89(3): 784-804.

Bennett, R.M., Y. Ismael, U. Kambhampati and S. Morse. 2004. 'Economic

Impact of Genetically Modified Cotton in India'. *AgBioForum*, 7(3): 96-100.

Bennett, R.M., U. Kambhampati, S. Morse and Y. Ismael. 2006. 'Farm-level Economic Performance of Genetically Modified Cotton in Maharashtra, India'. *Review of Agricultural Economics*, *28*(1): 59-71.

Bikhchandani, S., D. Hirshleifer and I. Welch. 1992. 'A Theory of Fads, Fashion, Custom, and Cultural Change in Informational Cascades'. *Journal of Political Economy*, 100(5): 992-1026.

Carter, M.R. and C.B. Barrett. 2006. 'The Economics of Poverty Traps and Persistent Poverty: An Asset-based Approach'. *Journal of Development Studies*, 42(2): 178–199.

Choudhary, B. and K. Gaur. 2010. *Bt Cotton in India: A Country Profile*. New Delhi: The International Service for the Acquisition of Agri-biotech Applications (ISAAA).

Cole, S.A., X. Gine, J. Tobacman, P. Topalova, R. Townsend and J. Vickery. 2009. 'Barriers to Household Risk Management: Evidence from India'. Harvard Business School Working Paper No. 09–116. Boston: Harvard Business School.

Dercon, S. 1996. 'Risk, Crop Choice, and Savings: Evidence from Tanzania'. *Economic Development and Cultural Change*, 44(3): 485–513.

Ewen, S.W.B. and A. Pusztai. 1999. 'Effect of Diets Containing Genetically Modified Potatoes Expressing Galanthus Nivalis Lectin on Rat Small Intestine'. *The Lancet*, 354 (9191): 1353-1354.

Farnsworth, R.L. and J.L. Moffitt. 1981. 'Cotton Production Under Risk: An Analysis of Input Effects on Yield Variability and Factor Demand'. *Western Journal of Agricultural Economics*, 6(2): 155–163.

Feder, G. 1979. 'Pesticides, Information, and Pest Management under Uncertainty'. *American Journal of Agricultural Economics*, 61: 97–103.

Fernandez-Cornejo, J. and S. Jans. 1999. 'Pest Management in US Agriculture'. Agricultural Handbook No. 717, Washington D.C.: U.S. Department of Agriculture.

Fitzgerald, D. (1993). 'Farmers Deskilled: Hybrid Corn and Farmers Work'. *Technology and Culture*, 34 (2), 324–343.

Gandhi, V.P. and N.V. Namboodiri. 2006. 'The Adoption and Economics of Bt-cotton in India: Preliminary Results from a Study'. Working Paper No. 2006–09–04, Ahmedabad: Indian Institute of Management.

Gaurav, S. and S. Mishra. 2011. 'Size-class and Returns to Cultivation in India: Revisiting an Old Debate. *Oxford Development Studies*, forthcoming.

Gaurav, S. 2014. *Risk and Vulnerability of Agricultural Households in India.* Unpublished Ph.D. Dissertation. Mumbai: Indira Gandhi Institute of Development Research.

Glover, D. 2009. 'Undying Promise: Agricultural Biotechnology's Pro-poor Narrative, Ten Years on'. STEPS Working Paper 15, Brighton: STEPS Centre.

Glover, D. 2010. 'Exploring the Resilience of Bt-cotton's "Pro-poor Success Story"'. *Development and Change*, 41(6): 955–981.

Government of India. 2010. Press release No.11011/2/2010-Trade. Ministry of Agriculture, New Delhi: Department of Agriculture and Cooperation.

Gruère, G. and D. Sengupta. 2011. 'Bt Cotton and Farmer Suicides in India: An Evidence-based Assessment', *Journal of Development Studies*, 47 (2): 316–337.

Hazell, P.B.R. and R.D. Norton. 1986. *Mathematical Programming for Economic Analysis of Agriculture.* New York: Macmillan.

Herring, R.J. 2008. 'Whose Numbers Count? Probing Discrepant Evidence on Transgenic Cotton in the Warangal District of India', *International Journal of Multiple Research Approaches*, 2: 145–159.

Herring, R.J. 2007. 'The Genomics Revolution and Development Studies: Science, Poverty and Politics'. *Journal of Development Studies,* 43 (1): 1–30. http://government.arts.cornell.edu/assets/faculty/docs/herring/JDS_HerringIntro.pdf (accessed on 20 October, 2010).

Horowitz, J.K. and E. Lichtenberg. 1993. 'Insurance, Moral Hazard, and Chemical Use in Agriculture'. *American Journal of Agricultural Economics*, 75 (4): 926-35.

Horowitz, J.K. and E. Lichtenberg. 1994. 'Risk-reducing and Risk-increasing Effects of Pesticides'. *Journal of Agricultural Economics*, 45 (1): 82–89.

Hurley, M.T., P.D. Mitchell and M.E. Rice. 2004. 'Risk and the Value of Bt-corn'. *American Journal of Agricultural Economics*, 86(2): 345-358.

James, C. 2009. 'Global Status of Commercialized Biotech/GM Crops: 2009'. ISAAA Brief No.41. Ithaca, NY: The International Service for the Acquisition of Agri-biotech Applications (ISAAA).

Johnson, R.B. and A.J. Onwuegbuzie. 2004. 'Mixed Methods Research: A Research Paradigm'. *Educational Researcher*, 33(7): 14–26.

Joseph, R.K. (2007). 'Is Genetically Modified Technology Desirable? The Law and Economics of Bt-cotton'. In K.D. Raju (ed.), *Genetically Modified Organisms: Emerging Law and Policy in India*, New Delhi: The Energy Research Institute, pp. 1–14.

Just, R.E. and R.D. Pope. 2003. 'Agricultural Risk Analysis: Adequacy of Models, Data, and Issues'. *American Journal of Agricultural Economics*, 85 (5): 1249–1256.

Just, R.E. and D. Zilberman. 1983. 'Stochastic Structure, Farm Size and Technology Adoption in Developing Agriculture'. *Oxford Economic Papers*, 35 (2): 307–328.

Karlhaloo, J.L. and P.A. Kumar. 2009. *Bt Cotton in India—A Status Report* (Second Edition). New Delhi: Asia-Pacific Consortium on Agricultural Biotechnology (APCoAB).

Kathage, J. and M. Qaim. 2012. 'Economic Impacts and Impact Dynamics of Bt (Bacillus thuringiensis) Cotton in India'. *PNAS (Proceedings of the National Academy of Sciences of the United States of America)*, 109 (29): 11652–11656.

Kranthi, K.R., S. Naidu, C.S. Dhawad, A. Tatwawadi, K. Mate, E. Patil, A.A. Bharose, G.T. Behere, R.M. Wadaskar and S. Kranthi. 2005. 'Temporal and Intra-plant Variability of Cry1Ac Expression in Bt Cotton and its Influence on the Survival of the Cotton Bollworm, Helicoverpa armigera (Hubner) (Noctuidae: Lepidoptera)'. *Current Science*, 89(2): 291-298. Available at http://www.iisc.ernet.in/currsci/jul252005/291.pdf (accessed on 14 March, 2011).

Kurungati, K. 2009. 'Bt Cotton and the Myth of Enhanced Yields'. *Economic and Political Weekly*, 44(2): 29-33.

Kurosaki, T. 1998. *Risk and Household Behaviour in Pakistan's Agriculture*. Tokyo: Institute of Development Economics.

Lalitha, N. 2004. 'Diffusion of Agricultural Biotechnology and Intellectual Property Rights: Emerging Issues in India'. *Ecological Economics*, 49 (2): 187-198.

Lalitha, N., B. Ramaswami and P.K. Viswanathan. 2009. 'India's Experience with Bt-cotton: Case Studies from Gujarat and Maharashtra'. In R. Tripp (ed.), *Biotechnology and Agricultural Development: Transgenic Cotton, Rural Institutions and Resource-poor Farmers*, London and New York: Routledge, pp. 135-167.

Lalitha, N. and P.K. Viswanathan. 2011. 'Pesticide Use in Cotton and the Need for Environmentally Sustainable Farm Management Standards in India'. Paper presented at the Sixth Biennial Conference of the Indian Society for Ecological Economics, Centre for Economic and Social Studies, Hyderabad, 20–22 October, 2011.

Liebowitz, S. and S.E. Margolis. 1990. 'The Fable of the Keys'. *Journal of Law and Economics*, 33 (1): 1-26.

Losey, J.E., L.S. Rayor and M.E. Carter. 1999. 'Transgenic Pollen Harms Monarch Larvae'. *Nature*, 399 (6733): 214.

Mishra, S. 2006a. *Suicide of Farmers in Maharashtra*. Report submitted to the Government of Maharashtra. Mumbai: Indira Gandhi Institute of Development Research.

Mishra, S. 2006b. 'Farmers' Suicides in Maharashtra'. *Economic and Political Weekly*, 41 (16): 1538–1545.

Mishra, S. 2008. 'Risks, Farmers' Suicides and Agrarian Crisis in India: Is There a Way Out?' *Indian Journal of Agricultural Economics*, 63 (1): 38-54.

Morse, S., R. Bennett and Y. Ismael. 2004. 'Why Bt-cotton Pays for Small-scale Producers in South Africa'. *Nature Biotechnology*, 22 (4): 379-380.

Murugkar, M., B. Ramaswami and M. Shelar. 2007. 'Competition and Monopoly in the Indian Cotton Seed Market'. *Economic and Political Weekly*, 42(37): 3781-3789.

Naik, G., M. Qaim, A. Subramanian and D. Zilberman. 2005. 'Bt Cotton Controversy: Some Paradoxes Explained'. *Economic and Political Weekly*, 40 (15): 1514-1517.

Newbery, D.M. and J.E. Stiglitz. 1981. *The Theory of Commodity Price Stabilization: A Study in the Economics of Risk*. Oxford: Clarendon Press.

Pannell, D. 1991. 'Pests and Pesticides, Risk and Risk Aversion'. *Agricultural Economics*, 5 (4): 361–83.

Parsuraman, S. and T. Rajaratnam. 2001. 'Agriculture, Food Security and Nutrition in Vidarbha: A Household Level Analysis'. *Economic and Political Weekly*, XLVI (19), 42-50.

Perrin, R. and D. Winkelman. 1976. 'Impediments to Technical Progress on Small Versus Large Farm'. *American Journal of Agricultural Economics*, 58(5): 888-894.

Pray, C.E., P. Bengali and B. Ramaswami. 2005. 'The Cost of Biosafety Regulations: The Indian Experience'. *Quarterly Journal of International Agriculture*, 44 (2005): 267–289.

Qaim, M. 2003. 'Bt Cotton in India: Field Trial Results and Economic Projections'. *World Development*, 31(12): 2115–2127.

Qaim, M. 2005. 'Agricultural Biotechnology Adoption in Developing Countries'. *American Journal of Agricultural Economics*, 87 (5): 1317–1324.

Qaim, M., A. Subramanian, G. Naik and D. Zilberman. 2006. 'Adoption of Bt Cotton and Impact Variability: Insights from India'. *Review of Agricultural Economics*, 28 (1): 48-58.

Qaim, M. and D. Zilberman. 2003. 'Yield Effects of Genetically Modified Crops in Developing Countries'. *Science*, 299 (5608): 900-902.

Qayum, A. and K. Sakkhari. 2005. *Bt-cotton in Andhra Pradesh—3-year assessment: The First Sustained Independent Scientific Study of Bt Cotton in India*. Hyderabad: Deccan Development Society.

Ramasundaram, P., S. Vennila and R.K. Ingle. 2007. 'Bt Cotton Performance and Constraints in Central India'. *Outlook on Agriculture*, 36 (3): 175-180.

Ramaswami, B. 1992. 'Production Risk and Optimal Input Decisions'. *American Journal of Agricultural Economics*, 74 (4): 860-869.

Ranganathan, T. and S. Gaurav. 2013. 'An Inquiry into the Composition of Farm Revenue Risk: The Case of Six Vidarbha Districts', *Economic and Political Weekly*, 48 (26/27): 125-128.

Rao, N.C. and S.M. Dev. 2009a. 'Biotechnology and Pro-poor Agricultural Development'. *Economic and Political Weekly*, XLIV (52): 56–64.

Rao, N.C. and S.M. Dev. 2009b. 'Impact of Bt Cotton on Farm Income and Employment in Andhra Pradesh'. *Agricultural Economics Research Review*, 22 (Conference Issue), 461-470.

Reddy, D.N. and S. Mishra (eds). 2009. *Agrarian Crisis in India*. New Delhi: Oxford University Press.

Roy, D. 2010. 'Of Choices and Dilemmas: Bt-cotton and Self-identified Organic Cotton Farmers in Gujarat'. *Asian Biotechnology and Development Review*, 12 (1): 51-79.

Roy, D. 2012. 'Cultivating Bt Cotton in Gujarat (India): Self-identified Organic Cotton Farmers Revisited'. *Asian Biotechnology and Development Review*, 14 (2): 67-92.

Roy, D., R.J. Herring and C.C. Geisler. 2007. 'Naturalizing Transgenics: Official Seeds, Loose Seeds and Risk in the Decision Matrix of Gujarati Cotton Farmers'. *Journal of Development Studies*, 43 (1): 158–176.

Sadashivappa, P. and M. Qaim. 2009. 'Bt Cotton in India: Development of Benefits and the Role of Government Seed Price Intervention'. *AgBioForum*, 12(2): 172-183.

Sahai, S. and S. Rahman. 2003. 'Performance of Bt-cotton: Data from First Commercial Crop'. *Economic and Political Weekly*, 30 (30): 3139-3141.

Scott, J.C. 1977. *The Moral Economy of the Peasant: Rebellion and Subsistence in Southeast Asia*. New Haven & London: Yale University Press.

Smale, M., P. Zambrano and M. Cartel. 2006. 'Bales and Balance. A Review of the Methods Used to Assess the Economic Impact of Bt-cotton on Farmers in Developing Economies'. *AgBioForum*, 9(3): 195-212.

Smale, M., P. Zambrano, G. Gruère, J. Falck-Zepeda, I. Matuschke, D. Horna, L. Nagarajan, I. Yerramareddy and H. Jones. 2009. 'Measuring the Economic Impacts of Transgenic Crops in Developing Agriculture during the First Decade: Approaches, Findings and Future Directions'. *IFPRI Food Policy Review*, 10.

Stanley-Horn, D.E., G.P. Dively, R.L. Hellmich, H.R. Mattila, M.K. Sears, R. Rose, L.C.H. Jesse, J.E. Losey, J.J. Obrycki and L. Lewis. 2001. 'Assessing the Impact of Cry1Ab-Expressing Corn Pollen on Monarch Butterfly Larvae in Field Studies'. *Proceedings of the National Academy of Sciences*, 98 (21): 11931-11936.

Stone, G.D. 2002. 'Both Sides Now: Fallacies in the Genetic-modification Wars, Implications for Developing Countries and Anthropological Perspectives'. *Current Anthropology*, 43 (4): 611–630.

Stone, G.D. 2011. 'Field versus Farm in Warangal: Bt-cotton, Higher Yields and Larger Questions'. *World Development*, 39 (3): 387–398.

Stone, G.D. 2012. 'Constructing Facts: Bt-cotton Narratives in India'. *Economic and Political Weekly*, 47 (38): 62–70.

Subramanian, A. and M. Qaim. 2009. 'Village-wide Effects of Agricultural Biotechnology: The Case of Bt-cotton in India'. *World Development*, 37(1): 256–267.

Subramanian, A. and M. Qaim. 2010. 'The Impact of Bt-cotton on Poor Households in Rural India'. *Journal of Development Studies*, 46 (2): 295–311.

Swaminathan, M. and V. Rawal. 2011. 'Are There Benefits from the Cultivation of Bt-cotton? A Comment Based on Data from a Vidarbha Village'. *Review of Agrarian Studies*, 1 (1): 101–124.

Tashakkori, A. and C. Teddlie. 1998. *Mixed Methodology: Combining Qualitative and Quantitative Approaches*. Thousand Oaks: Sage.

Viswanathan, P.K. and N. Lalitha. 2010. 'GM Technology and Sustainable Agriculture Future: Empirical Evidences from Bt-cotton Farmers in Maharashtra and Gujarat in India'. *Journal of Development and Agricultural Economics*, 2(1): 7-17.

Walker, T.S. and J.G. Ryan. 1990. *Village and Household Economies in India's Semi-arid Tropics*. Baltimore: The John Hopkins University Press.

# 6

# Impact of Price Controls on Diffusion of Bt Cotton and Profitability of Seed Providers in India

*Anchal Arora* and *Sangeeta Bansal*

## 1. Introduction

In 2002, Mahyco (Maharashtra Hybrid Seed Company) in collaboration with Monsanto introduced Bt cotton technology in India. Around 54,000 farmers in India grew officially approved Bt cotton hybrids on 50,000 hectares of land in 2002. Since then there has been a remarkable increase in the area cropped under Bt cotton in India—increasing to 10.6 million hectares by 2011.

Until 2005, Mahyco-Monsanto Biotech (MMB) dominated the market for cotton hybrids, either directly through selling hybrid seeds or indirectly through sub-licensing to private seed companies. India's regulatory system gave MMB a temporary monopoly on the Bt gene[1]. The domestic companies that licensed Bt trait from MMB were required to pay a one-time licence fee as well as a royalty for availing the gene. This led to a large price differences between Bt and non-Bt hybrids. In 2004, Bt hybrids cost $19 more per acre compared to non-Bt hybrids (Murugkar *et al.* 2007). The price for official Bt cotton seeds in India in 2006 was around Rs. 1600 per packet of 450 gram, which was around four times the price of non-Bt hybrid. Of the seed price of Rs. 1600, Rs. 1250 was charged by MMB as the trait value or royalty.

The large gap between the price of Bt cotton hybrid and non-Bt hybrid led to fears that monopolistic market structure was prevailing in the cotton seed market that had resulted in excessive seed prices. Concerns were raised that high seed prices may restrict access to technology for resource poor farmers (Lalitha 2004). The state of Andhra Pradesh imposed a ceiling of Rs. 750 (inclusive of technology fee) on Bt cotton seed price in Andhra Pradesh to make the technology affordable and accessible to small and marginal farmers in the state. The other states of India also imposed the same ceiling.

The government imposed price control has been one of the major challenges facing Indian firms involved in Bt cotton seed provision. The drastic reduction in seed prices led to the concerns that if seed prices and trait values are fixed at low arbitrary levels, farmers' benefits might increase in the short run but the incentive to invest in the development of new technologies might reduce due to shrinkage of company revenues. Research and development (R&D) expenditures on new and patentable genetic traits and seeds are an important part of the production cost of seeds. Over the last few decades, private sector R&D expenditures in agriculture have increased sharply as applications of new biotechnologies have become associated with exclusive property rights for genetic traits. This has contributed to an increase in seed prices (Krull *et al.* 1998). The price mark-up acts as an incentive for them to develop new technologies and, therefore, in the long run, price controls could have negative implications for product development. The literature suggests that profitability of the early years of Bt cotton sales would increase R&D investments until the price controls are imposed, after which R&D on cotton would decline (Pray and Nagarajan 2010). Price controls may delay new product launches, due to which farmers may lose in the long run as they will not get access to improved seeds.

In this context, the present chapter investigates the effect of seed price controls on the gross revenue and profitability of seed providers. We begin by discussing studies that have analysed implications of different pricing strategies of Bt cotton seeds for farmers and seed providing companies.

Analysing expected level of demand for Bt cotton seeds in Argentina under different pricing regimes, Qaim and de Janvry (2003) find that a high seed price is a barrier to adoption, especially for small holder farmers. They argue that reducing Bt cotton seed prices would not only increase farmers' profits, but would also be more profitable for the seed producing company. Reducing the seed prices could result in more than three times increase in the profit of the technology provider.

In contrast, analysing adoption of Bt cotton in India in the light of government seed price interventions, Sadashivappa and Qaim (2009) find a high willingness to pay (close to the official market price) for Bt cotton seeds. According to them, the take-off phase for Bt cotton had already begun before 2006 and thus the government seed price interventions had little impact on aggregate Bt cotton adoption. They apprehend that seed price controls might reduce the incentive of the company to innovate in the future.

Pray and Nagarajan (2010) reviewed the history of research and innovation in the seed/biotech industry in India to see if there is evidence that price controls are reducing research and development and innovation by this industry. Based on data on the total number of seed packets sold from industry sources, they have computed revenue realised by technology providers (MMB) and seed firms in India from 2002 to 2010 and postulated that the implementation of price controls in 2006 was followed by an immediate, large decline in the profits of seed and biotech firms. They argue that the seed price controls in case of Bt cotton in India would benefit farmers in the short run but in the long run, biotech companies may reduce their investments in research to develop or import new plant technology for India because of lower than expected or uncertain revenues from innovation. Farmers, thus, tend to lose in the long run as they would not get access to improved hybrids and genes developed by private firms. However, there isn't yet any quantitative evidence that firms have reduced their research or their innovations due to the lower returns to companies that provided new biotech in India.

The above discussion raises the following important questions for policy makers:

(a) What has been the impact of these price controls on the revenue and profitability of the seed providers in India?
(b) Are these price controls curbing the incentive of the company to innovate in future, which could reduce farmers' ability to get access to new technology, or are these beneficial for the farmers as well as seed providers in terms of increased adoption and profitability from technology?

If the prevailing market price for Bt cotton seeds is the profit maximising price, a reduction in price cannot increase firms' profitability. If the price charged, however, is sub-optimal, then price controls may result in an increase in profitability. This could be because of the firms' incomplete knowledge of the demand curve for Bt seeds. Qaim and Janvry (2003) find that Bt cotton seed prices in Argentina were sub-optimal and this was due to the firm's lack of assessment of the demand curve. They show that a reduction in prices could lead to a large increase in demand which in turn could increase the profits of the company. Another possible reason for sub-optimal prices could be that the seed prices are set to maximise global profits rather than local profits.

We quantify the impact of seed price interventions on gross revenue and profitability of seed providers, by simulating and comparing gross revenue and profit curves under two alternative scenarios: (a) no seed price intervention; and (b) seed price intervention using estimated coefficients from our model.

As noted, the Bt technology in India is developed by MMB, which has sub-licensed marketing rights to other companies. The benefits are distributed between MMB and the other seed companies according to the licence agreements between them. It is difficult to divide the benefits among firms; therefore, the seed providers in our analysis include both seed companies as well as technology providers. The study also computes seed price elasticity of demand and finds it to be highly elastic.

The commercial release of plant varieties produced through genetic engineering requires approval from the Genetic Engineering Appraisal Committee (GEAC)[2]. Initially the GEAC followed a 'case-by-case approval process', which mandates extensive testing of each hybrid under trial. In 2006, the approval process was changed to an 'event based approval system' from that of a case-by-case approval process. Accordingly, extensive bio-safety and agronomic testing is not necessary for an approved event. It was argued that this would speed up the introduction of new and diverse products for the Indian farmer, stimulate competition and offer a wider choice without compromising bio-safety and environmental safety (GOI 2006).

Indeed after the initial approval of three hybrids, many more Bt cotton hybrids have been approved for commercial cultivation in India. The initial hybrids contained the single Bt gene *cry1Ac* owned by Monsanto which licensed the gene to MMB. Subsequently, MMB has sub-licensed the gene to other firms in India. In 2005, 20 cotton hybrids using this gene construct were approved for cultivation by the GEAC, some of which were specifically suited for agro-ecological conditions of the northern zone. In May 2006, another Bt hybrid developed by MMB called Bollgard II[3] with stacked *cry 1 Ac* and *cry 2 Ab* genes was approved for commercial release in the central and northern zones. In the same year,

JK Agri Genetics Ltd. and Nath Seeds Ltd. also released their own approved events of Bt cotton. The number of approved hybrids increased from 62 in 2006 to 131 in 2007 and further to 780 in 2010 (James 2008). This rise in the approved varieties of Bt cotton hybrids resulted in an increase in the number of varieties available in the market, expanding the portfolio of choice for farmers, and helped them in finding the variety most suitable for their agro-climatic conditions. This effect of the number of Bt hybrids approved for commercial cultivation has been termed as varietal approval.

Studies that have attempted to estimate farmers' willingness to pay for Bt cotton technology in India have used contingent valuation methodology. Contingent Valuation (CV) techniques are often used to analyse individual preferences and elicit the WTP (willingness to pay) of consumers for goods that are non-marketable or not yet marketed. In case of technologies like Bt cotton, the WTP depends on a number of interrelated factors, including socio-economic and demographic characteristics, risk perceptions, awareness, and trust in the food safety and regulatory authorities. For the CV questions in a survey, a double bounded dichotomous choice approach is employed. In the questionnaire, respondents are asked whether they would be willing to purchase the specified good at a certain price bid. Price bids are randomly assigned to the questionnaires. Depending on the answer to the first bid, a second random bid is given, which is higher than the first bid for a 'Yes' response, and lower for a 'No' response. Based on the responses, a log likelihood function is maximised and finally, the mean willingness to pay is estimated.

Sadashivappa and Qaim (2009), using a contingent valuation (CV) choice approach, estimated the WTP for Bt cotton technology. They found that for 2002-03 growing season, mean WTP was Rs. 1633 per packet of Bt seeds, which was very close to the official market price. After the 2004-05 season, mean WTP even increased to Rs. 2595. This reflects the farmer's satisfaction towards the new technology despite Bt seed prices being relatively high. They also estimated the mean WTP for 2006 season, the year when government seed price interventions began (seed prices were reduced from Rs. 1600 to Rs. 750 for a packet of 450 gms of Bt cotton seeds). Interestingly, they found that the mean WTP estimated for 2006-07 was still much higher than the maximum retail price of Rs. 750.

Ramaswami *et al.* (2008) also estimated the mean WTP for both approved and unapproved Bt cotton seeds using the CV approach. Based on a survey of cotton growers in Gujarat in 2004, they found the mean

WTP for approved Bt seeds to be Rs. 778 and the median at Rs. 880. These estimates imply that approved seeds are overpriced and therefore have not been adopted widely. However, the mean WTP for unapproved seeds was found to be Rs. 3050.

Krishna and Qaim (2006) used a similar approach for *ex-ante* analysis of adoption of insect resistant Bt eggplant technology in India. They conducted interview based farm surveys in three major eggplant producing states, viz. Andhra Pradesh, Karnataka and West Bengal in 2005, and estimated the farmers' willingness to pay using the CV method. They found that the average WTP for Bt hybrids was more than four times the current price of conventional hybrid seeds.

In contrast to the above studies, we analyse diffusion of technology in terms of actual acreage adopted under the new technology, and examine economic factors affecting them.[4] To analyse diffusion of Bollgard I, we use a panel data study that covers nine major cotton growing states, viz. Punjab, Haryana, Rajasthan, Madhya Pradesh, Maharashtra, Gujarat, Andhra Pradesh, Karnataka and Tamil Nadu over 2002 to 2008 period. Until 2008, around 75 per cent of the total Bt cotton area was occupied by Bollgard I, thus this study takes diffusion of Bollgard I as a proxy for diffusion of Bt cotton hybrids in India from 2002 to 2008. In later years, however, there has been a shift towards Bollgard II. The percentage area under Bollgard II increased from 4 per cent of the total Bt cotton area in 2006 to over 75 per cent by 2011. Since state-wise data on adoption of Bollgard II is not available, we analyse the impact of economic factors on its diffusion using all-India level data over the period 2006–2011. Thus, this study also brings out the dynamism that came in the Bt cotton seed market with the introduction of Bollgard II.

The studies estimating diffusion path have showed that the cumulative adoption path often behaves like a logistic curve. It is based on the premise that diffusion occurs through interpersonal contacts among a group of homogenous adopters. When a new technology is introduced, adoption increases slowly at first because initially, only the most progressive and less risk averse adopt and then it increases more rapidly as information spreads to other producers and finally, it slows when nearly all producers who find the technology profitable have adopted, the process then reaching a stable ceiling. This produces the classic 'S-shaped' diffusion curve, first introduced by Griliches in 1957. The logistic function is often used to represent the S-shaped diffusion process for agricultural innovations.

The area under Bollgard II increased from 1.6 per cent of the total cotton area in 2006 to over 70 per cent by 2011. The corresponding area under single gene Bt hybrids (Bollgard I) declined from 40 per cent in 2006 to around 15 per cent in 2011. This suggests that Bollgard II, which was more expensive than Bollgard I, became popular among farmers and they seem to prefer it over single gene Bt cotton hybrids. This study also aims to bring out the dynamism that came in the Bt cotton seed market with the introduction of Bollgard II. It analyses the pattern of diffusion of double gene Bt cotton hybrids in India over the years 2006-11, and the impact of economic factors on it.

The rest of the chapter is organised into five sections. In section 2, we describe the dynamic logistic model used in the study and present the regression results. We then estimate percentage area under Bt cotton using the estimated coefficients. The estimated percentage area under Bt cotton from this model is then used to analyse the impact of price controls on gross revenue of seed providers. Section 3 compares the profitability of seed providers under alternative cost conditions, section 4 discusses the pattern of diffusion of Bollgard II in India and the factors responsible for it and section 5 provides conclusions and policy implications.

## 2. Dynamic Logistic Model

The traditional models of diffusion treat diffusion parameters as scalar constants and directly estimate them as functions of time-varying exogenous variables. These are termed as static diffusion models. The functional form used is as follows:

$$P = \frac{K}{1+e^{-(1+bt)}}$$

where $P$ measures the proportion of the innovation that is adopted. This can be expressed either as the percentage of producers adopting the innovation or as the percentage of acreage where the innovation is applied. $K$ is the adoption ceiling or upper limit on $P$, representing the maximum possible percentage area under Bt cotton in the long run equilibrium. The term $t$ is a time trend, $a$ represents origin parameter capturing the date of availability of Bt cotton while variable $b$ represents the slope coefficient or the relative speed of diffusion.

A limitation of static models is that they cannot be used to determine the impact of economic factors on the diffusion parameters, and thereby on diffusion paths. We, therefore, use a dynamic diffusion model that allows parameter of diffusion *b* to change overtime. It allows us to determine the impact of observed economic variables, such as prices, technology, etc., on the process of diffusion. In other words, the model assumes that the rate of adoption is a function of economic variables. Since the slope coefficient *b* is assumed to be a function of economic variables, the model is termed as variable slope dynamic logistic model (Frisvold 2004, Fernandez-Cornejo 2002 and Jarvis 1981).

The basic functional form of the variable slope dynamic logistic model is given by

$$P_{it} = \frac{K}{1+e^{-(a+b(z))t}} \qquad ...(1)$$

where *P* denotes percentage area under Bt cotton and is computed as (area under Bt cotton/area under cotton) multiplied by 100; *K* is the adoption ceiling or upper limit on *P* representing the maximum possible percentage area under Bt cotton in the long run equilibrium. The term *t* is a time trend and *a* represents origin parameter capturing the date of availability of Bt cotton. Variable *b* represents the slope coefficient or rate of diffusion of Bt cotton. The vector *z* denotes economic variables that are influencing the rate of diffusion, *b*.

Using the variable slope dynamic logistic model described above, we first analyse the impact of economic factors like seed prices, varietal approval and cotton prices on diffusion of Bt cotton in India (Arora and Bansal 2012). We find that seed price controls have significantly impacted diffusion of Bt cotton in India. We also find cotton prices and varietal approval (defined in terms of number of Bt hybrids approved for commercialization) to be important factors impacting the diffusion rates. We then examine the impact of price intervention in Bt cotton seeds on gross revenue and profitability of seed providers. In contrast to the study by Pray and Nagarajan (2010), which relied on data from industry sources, our analysis is based on the area under Bt cotton estimated from the above model.

Equation (1) can be transformed to

$$\log\left(\frac{P}{K-P}\right) = a + b(z)t \qquad ...(2)$$

where *b* is a linear function of seed prices, cotton prices and number of approved Bt hybrids, that is,

$$b_{it} = \beta_1 p_{s_{it}} + \beta_2 p_{c_{it}} + \beta_3 T_{it} \quad \ldots(3)$$

where $p_{sit}$, $p_{cit}$ and $T_{it}$ denote seed prices, cotton prices and number of approved Bt hybrids, respectively, in state *i* in year *t*. The adoption ceiling *K* is fixed at 100 and the coefficient of diffusion *b* is allowed to be a function of exogenous factors. The origin parameter *a* is assumed to be fixed. The study uses a panel data that covers nine major cotton growing states, viz. Maharashtra, Gujarat, Madhya Pradesh, Andhra Pradesh, Tamil Nadu, Karnataka, Punjab, Haryana and Rajasthan over the period 2002-08.

Plugging (3) in (2) and introducing year dummies, the dynamic logistic model takes the following form[5]:

$$\log\left(\frac{P_{it}}{K-P_{it}}\right) = a + \beta_1.p_{s_{it}}.t + \beta_2.p_{c_{it}}.t + \beta_3.T_{it}.t + \beta_4.T_{05}.t + \beta_5.D_{04}.t + \beta_6.D_{05}.t + \beta_7.D_{06}.t + \beta_8.D_{07}.t + \mu \quad \ldots(4)$$

where $D_{0j}$ is a dummy variable for *200j, j* = 4, 5, 6, 7, and $T_{05}$ is the interaction term ($T_{05} = d_{05} * T$) where $d_{05}$ is the dummy for the year 2005 and *T* denotes varietal approval.

The term $\frac{P_{it}}{K-P_{it}} = y$ is the ratio of actual to potential percentage area under Bt cotton and can be considered as a proxy for diffusion. The above model is estimated using generalised least squares (GLS).

Data on the area under Bt cotton was taken from ISAAA Brief 41 (James 2009). Data on aggregate area under cotton cultivation was obtained from the Cotton Corporation of India (CCI). The data on seed prices for a packet of 450 gm was taken from internal Mahyco-Monsanto estimates. For some of the years, the data represents an average rather than an exact price. For data on cotton prices, we have used the average price of *kapas* (raw cotton) for major varieties that include Bengal Desi, LRA, J-34, H-4, S-6 and DCH-32. These prices have been taken from the statistics published by Cotton Corporation of India Limited (Government of India). The variable $T_{it}$ denotes the number of cotton hybrids containing the cry 1 Ac gene approved for commercial cultivation in India. This data was taken from the Indian GMO Research Information System (year-wise list of commercially released varieties of Bt-cotton hybrids by the GEAC, IGMORIS[6]).

**Table 6.1:** Estimation Results of Dynamic Logistic Model

| *Dependent variable* | *Log(Pit / K – Pit )* |
|---|---|
| Independent variables | Model |
| Constant | –4.76895 |
| $p_{s_{it}} \cdot t$ | –.00038* |
| | (.00021) |
| $p_{c_{it}} \cdot t$ | .00039*** |
| | (.00008) |
| $T_{it} \cdot t$ | .00367** |
| | (.00193) |
| $D_{04}\,.t$ | .25570 |
| | (.21027) |
| $D_{05}\,.t$ | .00595 |
| | (.31659) |
| $M\,.t$ | .16428* |
| | (.10016) |
| n $D_{07}$ .t | .10896 |
| | (.06986) |
| $T_{05}$. T | .03952* |
| | (.02444) |
| R Squared | 0.8267 |
| Number of observations | 54 |

*Note:* The figures in brackets denote the standard error.
***, **, * denote significance at 1 per cent, 5 per cent, 10 per cent, respectively.
*Source*: The values for all the Figures and Tables were derived from the model used in this chapter.

The regression results are presented in Table 6.1. The results indicate that while seed prices significantly and negatively impact diffusion, cotton prices and varietal approval impact diffusion positively. The mean elasticity of diffusion with respect to seed prices is –1.8. We also find that the effect of varietal approval is stronger in 2005 than other years. This can be attributed to the introduction of around 20 new hybrids in 2005, some of which were specifically suited for the agro-ecological conditions of the northern zone. Further, the year 2006 dummy was significant, implying that apart from seed prices, varietal approval and cotton prices, some other factors impacted diffusion rates in 2006. Possible reasons include the approval of three other gene constructs approved for commercial release: Bollgard II, developed by MMB, and two others developed by JK Agri-Genetics Limited and Nath Seeds Limited. Another possible reason could be the change in the approval process of Bt hybrids from a case-by-case to an event based approval system. This may have stimulated the introduction of more hybrids, thereby resulting in wider

**Table 6.2:** Estimated Percentage Area under Bt-cotton from Dynamic Logistic Model

| Years | Central Zone | Southern Zone | Northern Zone |
|---|---|---|---|
| 2002 | 0.949 | 0.949 | – |
| 2003 | 1.498 | 1.498 | – |
| 2004 | 2.771 | 2.771 | – |
| 2005 | 17.671 | 11.042 | 4.337 |
| 2006 | 46.019 | 43.750 | 15.133 |
| 2007 | 56.506 | 52.687 | 36.517 |
| 2008 | 93.060 | 91.539 | 71.940 |

*Source*: Estimated from the dynamic logistic model used for this study.

choice for the farmers. Thus, although the government's seed price interventions were significant in impacting the increased percentage of area planted with Bt seeds, the dramatic increase in the percentage of area planted with Bt seeds in 2006 cannot be attributed entirely to the governmental seed price interventions.

Using the estimated coefficients from our regression, we compute log $y$, antilog of which gives us estimated $y$. Substituting $K = 100$ in the expression $y$, we obtain estimated percentage area under Bt cotton, $P_{it}$, for all the nine states under study. Table 6.2 presents the estimated percentage area under Bt cotton for three major cotton growing zones from 2002 to 2008. Using these estimates, we compute the seed price elasticity of demand in the next section.

## *2.1 Impact of Price Controls on Gross Revenue*

From economic theory we know that the price elasticity of demand is crucial in determining the effect of a price change on gross revenue. Specifically, the gross revenue of a firm increases with a fall in the product price if the demand is elastic. To examine the effect of seed price control on revenue of seed providers, we first compute elasticity of percentage area under Bt cotton with respect to seed prices for individual states.

Differentiating the left hand side of equation (1) and rearranging, we obtain the expression for elasticity of percentage area under Bt cotton with respect to seed prices,

$$\eta_{P_{P_s}} = \left(\frac{K-p}{K}\right)\beta_1 t p_s$$

where $\eta_{P p_s}$ denotes the elasticity of $P$ with respect to $p_s$, and $t$ is the time trend. This can be termed as the (seed) price elasticity of demand. The elasticity estimates are computed by plugging in estimated value of the coefficient of diffusion $\beta_1$, the estimated percentage area under Bt cotton (from Table 6.2), and $K = 100$ in the above expression. Table 6.3 presents the seed price elasticity of demand for three major cotton producing states, viz. Maharashtra, Gujarat and Andhra Pradesh. Note that the elasticity estimates are varying overtime.

**Table 6.3:** Elasticities of Percentage Area under Bt-cotton (P) w.r.t. Seed Prices ($p_s$) for Three Major Cotton Producing States

| *Year* | *Maharashtra* | *Gujarat* | *Andhra Pradesh* |
|---|---|---|---|
| 2002 | –0.60432 | –0.60603 | –0.60369 |
| 2003 | –1.20629 | –1.19286 | –1.20495 |
| 2004 | –1.70046 | –1.71219 | –1.71281 |
| 2005 | –2.01428 | –2.35189 | –1.86127 |
| 2006* | –0.58278 | –1.14808 | –0.20878 |
| 2007 | –0.16860 | –1.07202 | –0.07234 |
| 2008 | –0.00166 | –0.75831 | –0.03719 |

*Year when price controls began.

Our computations show that the seed price elasticity of demand for all three states for the year 2005 (the year preceding seed price interventions) is highly elastic—it is greater than 2 for Maharashtra and Gujarat and close to 2 for Andhra Pradesh. This suggests that a small fall in seed price from Rs. 1600 would result in an increase in the gross revenue of firms. However, since there was a drastic reduction in seed prices from Rs. 1600 to Rs. 750, we need to explicitly compute gross revenues with seed price intervention and compare them with gross revenues if such interventions had not taken place. To predict gross revenue curve in the absence of seed price intervention, we use the model developed above.

### *Scenario I: Seed Price Intervention*

Gross revenue is the product of price and quantity sold. Since the price of seed is for a packet of 450 gm of seeds, we need to express quantity in terms of number of packets of seed sold. We convert the estimated percentage area under Bt cotton into estimated area under Bt cotton in acres using the relation:

Estimated area under Bt cotton = {Estimated percentage area ($P_{it}$) *Total area under cotton}/100.

The estimated area under Bt cotton was then converted into estimated number of packets sold assuming that a 450 gm packet of Bt cotton seed is spread across one acre land (Murugkar *et al.* 2007). This was done for all the nine states and over the years from 2002-08. The estimated number of packets sold was multiplied with seed prices to arrive at gross revenue estimates. The state-wise estimates were then aggregated to compute the gross revenue of seed providers for India as a whole.

*Scenario II: No Seed Price Intervention*

Since the focus of this chapter is on determining the effect of seed price intervention on the gross revenue and profitability, the second scenario traces gross revenue if the government had not intervened in seed prices, and the seed prices had continued at the rate of Rs. 1600 per packet. To be able to do that, we computed estimated percentage area under Bt cotton for all nine states using the dynamic logistic model (equation 1) assuming seed prices to be Rs. 1600 from 2002-08 keeping all other variables intact. The estimated percentage area without intervention was converted into estimated number of packets sold and multiplied by seed prices (Rs. 1600) to arrive at state-wise gross revenue estimates without intervention. The state-wise estimates were then aggregated for India as a whole.

To compare the gross revenue curves under the two alternative scenarios we are assuming that except for the seed prices, the values of all other variables remain the same as in the base scenario. To be able to do so, the implicit assumption is that variables are not interrelated.

Table 6.4 presents the estimated gross revenue of seed providers with and without price interventions. These are illustrated in Figure 6.1. It can be seen from Figure 6.1 that gross revenue curve of seed providers with intervention lies above the gross revenue curve without interventions for 2006 and 2007. This shows that government seed price reductions resulted in increasing the revenue of seed providers in the short run, which is also consistent with our elasticity estimates.

Comparing increase in gross revenue of seed providers from 2005 to 2006 for different states, we can also examine regional variation in the impact of seed price control. Our results show that the largest increase in gross revenue in 2006 as compared to 2005 was obtained from the state

**Table 6.4:** Comparison of Gross Revenue of Seed Providers under Two Alternative Scenarios

| Years | Estimated Gross Revenue with Intervention (Rs. million) | Estimated Gross Revenue without Intervention (Rs. million) |
|---|---|---|
| 2002 | 263.20 | 263.20 |
| 2003 | 424.93 | 424.93 |
| 2004 | 782.88 | 782.88 |
| 2005 | 5536.50 | 5536.50 |
| 2006* | 8723.70 | 6380.70 |
| 2007 | 11051.34 | 7509.60 |
| 2008 | 26789.20 | 23029.90 |

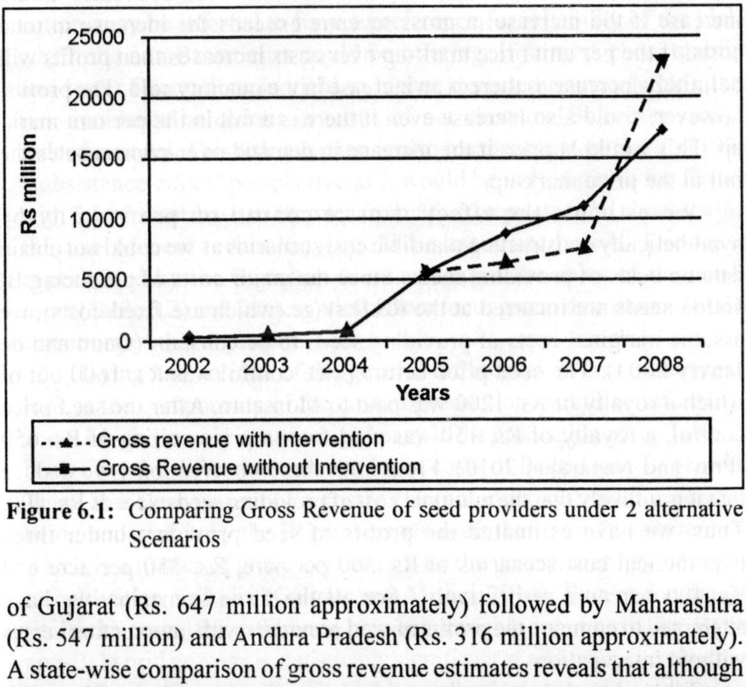


**Figure 6.1:** Comparing Gross Revenue of seed providers under 2 alternative Scenarios

of Gujarat (Rs. 647 million approximately) followed by Maharashtra (Rs. 547 million) and Andhra Pradesh (Rs. 316 million approximately). A state-wise comparison of gross revenue estimates reveals that although Maharashtra is the largest cotton producing state in India, Gujarat incurred the largest increase in revenues. This result is consistent with our elasticity estimates where we find that the seed price elasticity of demand for Gujarat in 2005 is highest among the three cotton producing states (Table 6.3).

The largest increase in gross revenue of seed providers from Bt cotton for the state of Gujarat as compared to other states for year 2006 could be attributed to a reduction in the spread of illegal Bt seeds in the state in the post intervention period. Illegal Bt seeds were priced between Rs. 800 and Rs. 1200 per packet of 450 gm as compared to a price of Rs. 1600 for official Bt cotton seeds (Murugkar *et al.* 2007). Thus government price controls could have probably contributed to a reduction in the illegal Bt area and an increase in the availability of legal Bt cotton hybrids in the state which in turn generated more revenues for the seed providers.

## 3. Impact of Seed Price Controls on the Profitability of Seed Providers

To ascertain the effect of price controls on the profitability of seed providers apart from the gross revenue, we also need to take into account the changes in total costs. The profitability of seed providers would increase if the increase in gross revenue exceeds the increase in total costs. If the per unit price mark-up over costs increases then profits will definitely increase as there is an increase in the quantity sold. The profits, however, could also increase even if there is a fall in the per unit mark-up. This would happen if the increase in demand over-compensates the fall in the price mark-up.

We estimate the effect of price control on profitability by hypothetically constructing plausible cost scenarios as we could not obtain data on costs of providing seeds. Since the major costs of producing Bt cotton seeds are incurred at the R&D stage (which are fixed costs), we assume marginal costs of providing seeds to be constant (Qaim and de Janvry 2003). The seed price before price control was Rs. 1600 out of which a royalty of Rs. 1200 was paid to Monsanto. After the seed price control, a royalty of Rs. 150 was paid from a price ceiling of Rs. 650 (Pray and Nagarajan 2010). From both these observations, we deduce that it is unlikely that the marginal cost of providing seed exceeds Rs. 400. Thus, we have estimated the profits of seed providers under three hypothetical cost scenarios of Rs. 300 per acre, Rs. 350 per acre and Rs. 400 per acre respectively.[7] For all the three scenarios, we have attempted to compare the profits of seed providers with intervention versus without intervention.

Tables 6.5–6.7 depict the estimated profits of seed providers with and without intervention under different cost conditions. Table 6.5 compares the estimated profits with intervention with that of profits without intervention assuming marginal costs to be Rs. 300 per acre. The figures clearly show that estimated profits have increased with intervention. We have depicted these profits in Figure 6.2. The gap between the two lines shows the change in the profitability due to price

**Table 6.5:** Estimated Profits with Intervention versus without Intervention, at Marginal Costs Equal to Rs. 300 per Acre Over the Years 2002-08

| *Years* | *Estimated Profits with Intervention keeping cost = Rs. 300 per acre (in Rs. million)* | *Estimated Profits without intervention keeping cost = Rs. 300 per acre (in Rs. million)* |
|---|---|---|
| 2002 | 190.740 | 190.740 |
| 2003 | 300.933 | 300.933 |
| 2004 | 636.299 | 636.299 |
| 2005 | 4545.050 | 4545.052 |
| 2006 | 5661.167 | 5119.286 |
| 2007 | 6630.837 | 6025.539 |
| 2008 | 9793.924 | 18509.130 |

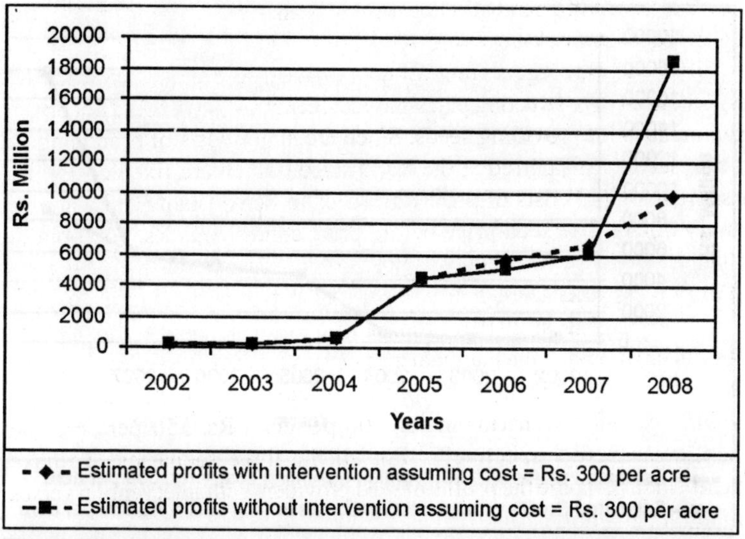


**Figure 6.2:** Profitability of seed providers (this includes seed firms and technology providers) with and without intervention assuming cost = Rs. 300 per acre

controls. It is evident that profits have increased with intervention for years 2006 and 2007. We have done the same exercise, assuming marginal cost of providing seeds to be Rs. 350 per acre. Table 6.6 and Figure 6.3

**Table 6.6:** Estimated Profits with Intervention Versus without Intervention with Cost = Rs. 350 per acre over the years 2002-08.

| *Years* | *Estimated Profits with Intervention keeping cost = Rs. 350 per acre (in Rs. million)* | *Estimated Profits without intervention keeping cost = Rs. 350 per acre (in Rs. million)* |
|---|---|---|
| 2002 | 183.22 | 183.22 |
| 2003 | 289.35 | 289.26 |
| 2004 | 611.73 | 611.73 |
| 2005 | 4378.59 | 4378.50 |
| 2006 | 5151.22 | 4908.50 |
| 2007 | 5886.94 | 5776.77 |
| 2008 | 8556.00 | 17751.00 |

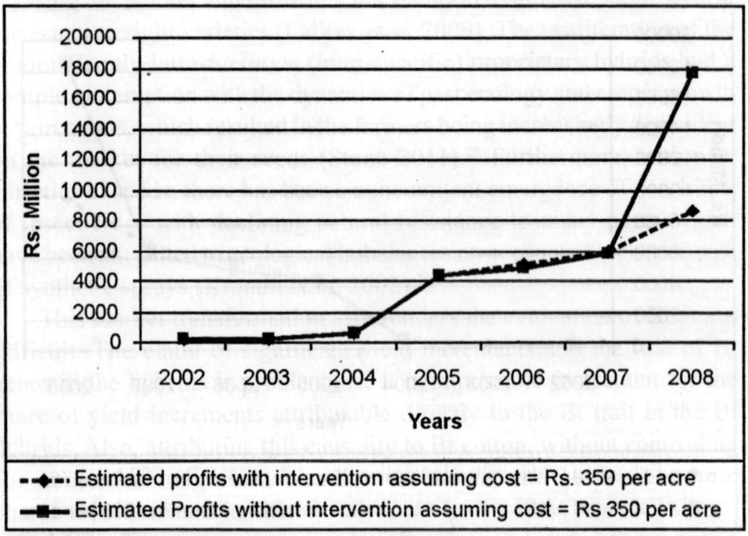


**Figure 6.3:** Profitability of seed providers with and without intervention assuming cost = Rs 350 per acre

depict the results of this exercise. Here also we find that profitability of seed providers have increased marginally with intervention for 2006 and 2007.

We, however, find that the profitability of seed providers is declining for both 2006 and 2007 if marginal cost of providing seeds is Rs. 400.

**Table 6.7:** Estimated Profits with Intervention versus without Intervention with Cost = Rs. 400 per acre over the years 2002-08.

| *Years* | *Estimated Profits with Intervention keeping cost = Rs. 400 per acre (in Rs. million)* | *Estimated Profits without intervention keeping cost = Rs. 400 per acre (in Rs. million)* |
|---|---|---|
| 2002 | 176.067 | 176.067 |
| 2003 | 277.784 | 277.784 |
| 2004 | 587.353 | 587.353 |
| 2005 | 4214.501 | 4214.502 |
| 2006 | 4640.274 | 4698.560 |
| 2007 | 5157.318 | 5530.488 |
| 2008 | 7300.548 | 17001.930 |

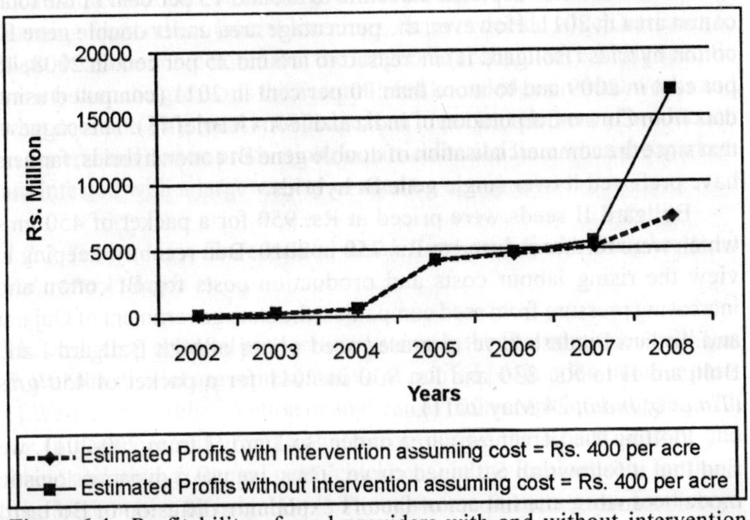


**Figure 6.4:** Profitability of seed providers with and without intervention assuming cost = Rs 400 per acre

This can be clearly seen in Table 6.7. Also, the profit curve without intervention in Figure 6.4 lies above the profit curve with intervention.

The above analysis suggests that the overall impact of price controls on profitability would depend on the cost conditions. If the marginal cost of providing seeds is less than Rs. 400 per acre, price controls would increase the profits in the short run whereas if cost Rs. 400 per acre or greater, firms' profits would decline with price reductions.

## 4. Diffusion of Double Gene Bt Cotton Hybrids in India

The GEAC approved the first two gene event MON 15985, commonly known as Bollgard II for commercialisation in May 2006. Bollgard II developed by Mahyco and sourced from Monsanto, has a double gene technology containing both *cry 1 Ac* as well as *cry 2 Ab* gene. The company claims that Bollgard II is a better resistance management tool than Bollgard I (single gene Bt cotton hybrids). In addition to bollworm, it also claims to offer protection against another pest, the spodoptera caterpillar, thereby ensuring higher yields.

In the first year of its commercialisation in 2006, Bollgard II occupied around 1.6 per cent of the total cotton area in India whereas the corresponding figure was around 40 per cent for Bollgard I. The percentage area under Bollgard I increased to around 60 per cent until 2008 and thereafter depicted a decline to around 15 per cent of the total cotton area in 2011. However, the percentage area under double gene Bt cotton hybrids (Bollgard II) increased to around 25 per cent in 2008, 46 per cent in 2009 and to more than 70 per cent in 2011 (computed using data from Cotton Corporation of India and ISAAA Brief 43). This suggests that since the commercialisation of double gene Bt cotton hybrids, farmers have preferred it over single gene Bt hybrids.

Bollgard II seeds were priced at Rs. 950 for a packet of 450 gms which were further reduced to Rs. 750 in 2010. But, recently, keeping in view the rising labour costs and production costs for Bt cotton and increasing pressure from seed companies, the state government of Gujarat and Andhra Pradesh have increased seed prices of both Bollgard I and Bollgard II to Rs. 830 and Rs. 930 in 2011 for a packet of 450 gms (*Times of India*, 24 May 2011).

Plotting the percentage area under Bollgard II from 200-2011, we find that it follows an S-shaped curve. Thus, we use a dynamic logistic model to analyse the impact of factors explaining diffusion of Bollgard II in India. The model used is as follows:

$$\log\left(\frac{P_t}{K - P_t}\right) = \beta_0 + \beta_1 . p_{st} . t + \beta_2 . p_{ct} . t + \beta_3 . T_t . t + \mu$$

Here $P_t$ refers to the percentage area under Bollgard II in India over the years 2006-11 (computed using data from Cotton Corporation of India and ISAAA Brief 43). The term $p_{st}$ denotes the seed prices of Bollgard II from 2006-11 on an all-India level which were taken from internal Mahyco

Monsanto estimates. The variable $p_{ct}$ refers to the average annual cotton prices which were taken from the *Handbook of Statistics*, Cotton Corporation of India, while the term $T_t$ refers to the number of double gene hybrids approved of Bt cotton in India.

The dynamic logistic model was estimated using generalised least square (GLS). The results of this model are discussed in Table 6.8.

The model has an adjusted $R^2$ of 0.96. All the coefficients are significant and have the expected signs. While cotton prices and varietal

**Table 6.8:** Estimation Results of Dynamic Logistic Model

| *Dependent variable:* | *Log( Pt / K – Pt )* |
|---|---|
| Independent variables | Model |
| Constant | .0293643 |
| $p_{s_{it}} \cdot t$ | –.0009799** |
| | (.0002024) |
| $p_{c_{it}} \cdot t$ | .0001893** |
| | (.0000235 ) |
| $T_{it} \cdot t$ | .000576** |
| | (.0001438) |
| R Squared | 0.96 |
| Number of observations | 6 |

*Note:* The figures in brackets denote the standard error.
***, **, * denote significance at 1 per cent, 5 per cent, 10 per cent, respectively.

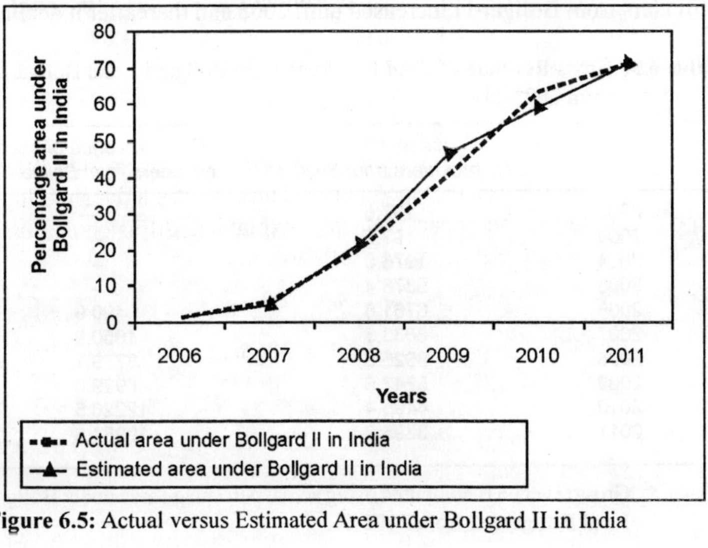


**Figure 6.5:** Actual versus Estimated Area under Bollgard II in India

approval are impacting diffusion ($z$) positively, seed prices are negatively impacting diffusion of Bollgard II in India. Using the estimated coefficients from the model, we computed the percentage area under Bollgard II in India ($P_t$). Figure 6.5 plots the actual versus estimated area under Bollgard II in India.

As discussed previously, the adoption of Bt cotton technology spread rapidly between 2002 and 2011. In 2006, when the price controls were implemented, adoption rates recorded a striking increase of 192 per cent over the previous year. The corresponding increase was 63 per cent in 2007, 23 per cent in 2008 and 12.7 per cent in 2011. Arora and Bansal (2012) discussed that the increase in the adoption of Bt cotton in 2006 cannot be attributed entirely to the seed price interventions of the government of India; the approval of Bollgard II and certain other specific events may also have contributed to an increase in the adoption. Thus, this analysis brings out the dynamism that came in the Bt cotton seed market with the introduction of Bollgard II and the factors responsible for it. Clearly, we find that since its introduction in 2006, farmers are displaying a preference for Bollgard II and despite it being more expensive, Bollgard II is substituting Bollgard I. This suggests that even without government intervention, adoption rates could increase because of the introduction of a superior quality product.

Also, comparing the gross revenues of seed providers[8] from Bollgard I and Bollgard II (Table 6.9), we find that the gross revenue of seed providers from Bollgard I increased until 2008 and thereafter it declined

**Table 6.9:** Gross Revenue of Seed Providers from Bollgard I and Bollgard II from 2002-11

| *Year* | *Gross revenue of seed providers from Bollgard* | *Gross revenue of seed providers from Bollgard II* |
|---|---|---|
| 2002 | 197.6 | – |
| 2003 | 197.6 | – |
| 2004 | 1976.0 | – |
| 2005 | 5378.4 | – |
| 2006 | 6761.6 | 490.9 |
| 2007 | 8633.5 | 1050.9 |
| 2008 | 8926.5 | 3779.1 |
| 2009 | 5747.6 | 8929.0 |
| 2010 | 4495.4 | 12226.5 |
| 2011 | 3895.2 | 19984.8 |

*Note:* * Gross revenue is computed using actual percentage area under Bollgard I and Bollgard II from 2002-11.

and in the later years 2009-11, seed providers extracted more revenues from Bollgard II. Thus, the government imposed price controls in case of Bollgard I increased the gross revenue of seed providers in the short run for two years 2006 and 2007. However, in the later years, seed providers captured the market share by practicing product differentiation and bringing a superior quality product.

## 5. Conclusion and Policy Implications

In 2006, the government of Andhra Pradesh imposed a ceiling of Rs. 750 on the price of a 450 gm packet of Bt cotton seeds. There is interest in the literature in analyzing the impact of price controls on the farmers as well as seed producing companies. We have analysed the impact of such price reductions on diffusion of technology in India, the gross revenue and profitability of seed providers. We have also analysed the dynamism that came in the Bt cotton seed market with the introduction of Bollgard II in 2006.

We find that price controls were significant in impacting diffusion of technology in India. We computed seed price elasticity of demand for three major cotton producing states and found it to be highly elastic in the year 2005. This suggests that a reduction in seed price would result in an increase in gross revenue for the seed providers. This indeed was the case. We found that the gross revenue of the seed providers increased for two consecutive years following the price controls, i.e. in 2006 and 2007. Comparing the gross revenue under two alternative scenarios—with and without seed price intervention, we find that seed price reductions have increased the gross revenue of seed providers, at least in the short run. However, comparing the gross revenues of seed providers in the long run using data on actual area under Bollgard I and Bollgard II, we find that gross revenue of seed providers from Bollgard I increased until 2008, thereafter, seed providers extracted greater revenues by introducing a superior quality product (Bollgard II) which is more preferred by farmers even though it is more expensive than Bollgard I. Thus, we find that the price controls increased the gross revenues of seed providers in the short run; however, the long run witnessed the substitution in the choices of farmers from Bollgard I to Bollgard II which ultimately caused an increase in the gross revenues of seed providers.

We have also attempted to study regional variation of the seed price control on gross revenue. We find that the largest increase in gross revenue in the post intervention period was incurred by Gujarat, although

Maharashtra is the largest cotton producing state. This is an interesting result and could be attributed to the reduction in the spread of illegal Bt seeds in Gujarat in the post intervention period.

Regarding the impact of seed price controls on profitability of seed providers, our results suggest that it would depend on the cost conditions. If marginal costs of providing seeds exceed Rs. 400 per acre then profits would decline with government price reductions. This result is close to Pray and Nagarajan (2010) who find that profits of both seed companies as well as technology providers have reduced in the post intervention period. This suggests that the price controls could hamper the incentive of the seed providers to innovate in future and thus could restrict the accessibility of Indian farmers to some important new technologies.

However, if cost of producing Bt seeds is less than Rs. 400 per acre then profits would increase with price controls. In that case, the government imposed price control might have improved the access of beneficial technologies to farmers without curbing the incentive of the company to innovate in the future. The results of this case are closer to Qaim and Janvry (2003) where they found that reducing the seed prices for Bt cotton in Argentina was not only beneficial for farmers but also beneficial for the seed companies. This could be possible if the seed prices in the pre intervention period were not profit maximising prices. Thus, the overall impact of price controls on the profitability of seed providers would depend on the cost of producing Bt seeds.

The excessive prices charged for the official seeds may strengthen the incentive to cheat. There was widespread adoption of illegal seeds in Gujarat, which were priced between Rs. 800 and Rs. 1200 per packet of 450 grams as compared to a price of Rs. 1600 for official Bt cotton seeds (Murugkar *et al.* 2007). Thus, the government imposed price control for legal Bt seeds could have probably contributed to a reduction in the illegal Bt seeds.

Apart from government interventions in the pricing of Bt cotton seeds, an alternative policy measure to increase the benefits for the farmers as well as seed providers would be to allow competition among alternative gene providers which could reduce the seed prices on its own (rather than the government doing it) and could ultimately increase the gross revenue as well as profitability of the seed providers as a whole. The government could also encourage competition in the seed market by supporting indigenous research by farmer scientists who develop their own hybrids.

## Notes

1 MMB derived a measure of protection for its gene through India's biosafety laws. As biosafety approvals are obtained for the composite of the gene and the germplasm, hybrids that incorporate MMB's gene but do not go through the biosafety process are considered illegal. Thus, most of the firms have chosen to license the Bt technology from MMB, as regulatory authorities are unlikely to approve a Bt hybrid that includes an unlicensed version of the MMB gene. However, in 2006, some non-MMB genes entered the market but their ability to compete in the technology market was handicapped by the first mover advantage of MMB.

2. The GEAC was formerly known as the Genetic Engineering Approval Committee.
3. We have named the single gene Bt hybrids as Bollgard I and double gene Bt hybrids as Bollgard II in this study.
4. The papers adopting such a methodology are Frisvold (2004), Fernandez-Cornejo (2002) and Jarvis (1981).
5. Refer to Arora and Bansal (2012) for a complete description and justification for the model used and results of variations of the above model.
6. Indian GMO Research Information System.
7. Pray and Nagarajan (2010) have done the analysis assuming that marginal cost of providing seeds went up by 35-40 per cent from the 2006-07 crop year due to increased cost of production. They also didn't have data on cost of seed production for every year so they applied the increases in the cost of seed production from 2006-07 year.
8. This gross revenue is computed using data on actual area under Bollgard I and Bollgard II which is taken from ISAAA, Brief 43 and multiplying it with the corresponding prices of Bollgard and Bollgard II.

## References

Arora, A. and S. Bansal. 2012. 'Diffusion of Bt Cotton in India: Impact of Seed Prices and Varietal Approval'. *Applied Economic Perspectives and Policy* 34(1): 102-118.

CCIL. Average Annual Prices of Kapas for Important Varieties, Statistics. Cotton Corporation of India Limited, Government of India. Available at www.cotcorp.gov.in.

Fernandez-Cornejo, J. and W.D. McBride. 2002. 'Adoption of Bioengineered Crops: Agricultural Economic Report No. 810'. United States Department of Agriculture, Washington, D.C.

Frisvold, G.B. 2004. 'Diffusion of Bt Cotton and Insecticide Use'. Paper prepared for presentation at the Western Agricultural Economics Association Annual Meeting, 30 June–2 July, 2004, Honolulu, Hawaii.

GOI. 2006. 'Bt Cotton to Cover 81 Lakh Acres in the Country by 2006'. Press Information Bureau: Government of India. Ministry of Environment and Forests, 3 July, 2006. Available at http://pib.nic.in/newsite/PrintRelease.aspx.

Griliches, Z. 1957. 'Hybrid Corn: An Exploration in the Economics of Technological Change'. *Econometrica*, 25(4): 501-22.

Griliches, Z. 1960. 'Hybrid Corn and the Economics of Innovation'. *Science*, 132 (3422): 275-80.

James, C. 2008. 'Global Status of Commercialised Biotech/GM Crops': (ISAAA Brief No.39). Ithaca NY: International Service for the Acquisition of Agri Biotech Applications.

James, C. 2009. 'Global Status of Commercialised Biotech/GM Crops': (ISAAA Brief No.41). Ithaca NY: International Service for the Acquisition of Agri Biotech Applications.

Jarvis, L.S. 1981. 'Predicting the Diffusion of Improved Pastures in Uruguay'. *American Journal of Agricultural Economics* 63(3): 495–502.

Krishna, V.V. and M. Qaim. 2006. 'Estimating the Adoption of Bt Eggplant in India: Who Benefits from Public Private Partnership'. *Food Policy*, 32: 523-543.

Krull, D.F., J.M. Prescott and C.W. Chum. 1998. *Seed Marketing and Distribution.* In *Maize Seed Industries in Developing Countries*. Boulder, Colorado: CIMMYT, pp. 125-141.

Lalitha, N. 2004. 'Diffusion of Agricultural Biotechnology and Intellectual Property Rights: Emerging Issues in India'. *Ecological Economics*, 49(2): 187–198.

Lalitha, N., C.E. Pray, and B. Ramaswami. 2008. 'The Limits of Intellectual Property Rights: Lessons from the Spread of Illegal Transgenic Cotton Seeds in India'. Gujarat Institute of Development Research Working Paper No.182, Ahmedabad, India.

Murugkar, M., B. Ramaswami, and M. Shelar. 2007. 'Competition and monopoly in the Indian Cotton Seed Market'. *Economic and Political Weekly*, 42(37): 3781–3789.

Pray, C. and L. Nagarajan. 2010. 'Price Controls and Biotechnology Innovation: Are State Government Policies Reducing Research and Innovation by the Ag Biotech Industry in India?'. *AgBioForum*, 13(4): 297-307.

Qaim, M., and A. de Janvry. 2003. 'Genetically Modified Crops, Corporate Pricing Strategies and Farmers Adoption: The Case of Bt Cotton in Argentina'. *American Journal of Agricultural Economics* 85(4): 814-828.

Sadashivappa, P. and M. Qaim. 2009. 'Bt Cotton in India: Development of Benefits and the Role of Government Seed Price Interventions'. *AgBioForum,* 12(2): 172-183.

IGMORIS. Year-wise List of Commercially Released Varieties of Bt Cotton Hybrids by the GEAC (Year 2002 upto May 2012), Indian GMO Research Information System. Available at www.igmoris.nic.in.

# 7

# Do Farmers Make an Informed Choice?

## An Analysis of the Selection of Seeds and Insecticides by Cotton Growers in Gujarat

*N. Lalitha* and *P.K. Viswanathan*

### 1. Introduction

Agricultural technology adoption process depends on a host of factors ranging from access to information about technology, credit, suitability to local conditions, easy adaption and cost of technology, to mention a few. Following the commercial release of Bt technology (cotton) in India in 2002, several studies have pointed out the faster adoption of the technology by the cotton growers particularly in Gujarat. The case of Gujarat is distinct as historically farmers in the state had adopted many of the cotton hybrids ever since they were introduced in the 1970s. One major difference in the composition of the cottonseed market, since the introduction of Bt cotton in Gujarat and elsewhere, is that, there are large number of private players and varieties now compared to a decade ago when the public sector had a higher share in the cotton seed market[1]. The seed market presents a complex scenario where the information flow of the product would depend on the reach of the company's marketing network and extension services. Hence, the decision to purchase a particular seed could depend on factors such as farmers' own experience with a seed variety, knowledge of the company that produces it, number of farmers (influential and otherwise) who have adopted it earlier or simply because the seed dealer had recommended the product, etc. In the context of agricultural technologies, 'farmers do not simply acquire information on a seed or other technology but rather learn how practices and technologies perform together under variable conditions' (Stone 2007: 211).

Terming this process as skilling, Stone (2007) describes the Bt seeds adoption in Warangal in Andhra Pradesh as agricultural 'deskilling', where farmers had adopted the Bt seeds more on the basis of 'imitation'. However, probing the Bt adoption in Gujarat in the early years where the 'unapproved' varieties were very popular, Stone (2007) argues that the 'tortured history of Bt technology in Gujarat has been instrumental in them (farmers) becoming re-involved in experimentation, assessment and even developing their own seeds' (Stone 2007: 228) which he calls as agricultural 're-skilling'. Applying this framework in the context of Bt adoption in Gujarat and Maharashtra, Lalitha *et al.* (2009: 135–167), found the 'imitation effect' to be much stronger in Maharashtra than in Gujarat in 2007. With nearly complete adoption of Bt cotton now, it is true that the farmers' reskilling chances are becoming less evident with more farmers becoming 'passive customers of seed firms', which is likely to be the case if farmers are using hybrid seeds of seed firms for their supply. Farmers' skills need to be oriented towards using the technology appropriately, so that the technology is useful for more number of years. In the context of Bt technology, we argue that farmers need to have appropriate information about the suitability of the seeds to their soil, and the pesticide use should be based on need assessment in the field; otherwise, farmers lose by inappropriate choice of seeds and excess or lower use of pesticides which may have consequences both on crop productivity as well as the environment.

In this context, the focus of this chapter is to determine whether: (1) there is any change in the Bt adoption intensity among different size class of farmers; and (2) there has been any change in the pesticide use among the different size classes after Bt adoption. The reason for the focus of analysis on seed and pesticide is that the Bt trait in the plant is supposed to provide the plant resistance against bollworm, which however will depend on the level of toxin available with the plant. Hybrids with higher toxin concentrations warrant greater reductions in the pesticide use compared to varieties with lesser level of toxin concentration (Tripp 2009). Hence, the variety chosen plays an important role in determining the pesticide use. But since seed and insecticides[2] may be categorised as products with 'credence quality' which are difficult to assess even after use, at least without some type of expensive testing (Tripp 2001: 249-264), and hence lack of information on the specific use of inputs could lead to heavy transaction costs for the farmers. Hence, it is of interest to analyse the information that the farmers have as well as the information available with the dealers who sell these two critical inputs. Such an analysis of the farmer choice of seeds and use of pesticide products may provide useful insights about the performance of the technology with respect to its effect on bollworm control (and use of toxic insecticides thereon) across size classes.

### *1.1 Data*

In order to analyse the farm level scenario in the selection and use of seed and pesticides, this chapter draws heavily from the information collected from three primary surveys conducted during the kharif seasons of 2007-08 and 2010-11. While two farmer surveys were conducted during both the years, the dealer survey was done during 2007-08.

The survey among farmers was conducted in five cotton growing districts of Gujarat, viz. Ahmedabad, Bhavnagar, Rajkot, Surendranagar and Vadodara. The sampling followed a random selection process. From each district, four talukas were selected randomly and from each taluka, two villages were selected randomly. From each village, five cotton growers were chosen randomly. Thus, in each district, we had selected 40 cotton growers, with a total coverage of 200 farmers from five districts. During the 2007–08 kharif season, we collected the required farm level information through three separate visits to each farmer. The first visit was just after sowing and much of the seed related information was collected during this visit. During the second visit, we collected information on pesticide use and other farm management practices up to the time of the visit. The last visit was after the harvest, during which, the information on the yield, and the pesticide use, if any, after our second visit, were recorded. The same villages were visited again in 2010-11 during the kharif season to gather information from the same set of farmers. But due to the non-availability of many of the farmers, in some villages we had to make fresh selection of a few/all the five farmers[3]. Information from these 200 farmers in both these years constitutes the first set of data. The sample districts—Ahmedabad, Bhavnagar, Rajkot, Surendranagar and Vadodara—accounted for 1.8, 15.7, 17.5, 9.1 and 7.3 per cent of the total cotton growing areas of the state in 2007–08 and together constitute almost 54 per cent of the total cotton area in the state. From each farmer, information regarding the total plots under cotton was obtained. From this, the larger three plots were chosen; all the information regarding seed selection, pesticide use, yield etc. was confined to these three plots of cotton land only. The third set of data is based on the information collected from a total of 82 seed dealers, randomly selected from the same talukas from where the villages were chosen for the survey of farmers.

This chapter is organised as follows. Before we analyse the information sourcing as well as seed and pesticide use practices of the farmers, it is necessary to understand the nature and functioning of the seed and insecticide industries that provides the two critical inputs in India which is provided in Section 2 following this introduction. Section 3 provides an analysis of the major sources and types of information that the farmers receive from the seed dealers regarding seeds and pesticides. Section 4 discusses the seed selection process, pattern in continuity of seed use over the two points of time with the farmers. Pesticide selection, spray decision and spray schedules are discussed in Section 5. The last Section 6 presents the conclusion.

## 2. Features of Seeds and Pesticides Industry in India

### *2.1 Status of Seeds Industry and Its Regulation*

The Indian seed industry comprises both organised and unorganised segments. The organised seed segment consisting of both public and private sectors account for 15-20 per cent of the total seed supplied in the country and the remaining comes from the unorganised sector (State of Indian Agriculture 2011–12: 51). This scenario has drastically changed over the years from a conservative structure dominated by public sector corporations to the present scenario where the total seeds sales turnover is shared in the ratio of 60:40 per cent[4] between the private and public sectors. The present Indian seed system fits perfectly in the four stage evolution of national seed systems which begins with farmer seed supply; progresses to the emergence of public plant breeding programmes; continues to the development of commercial seed enterprises often marketing public varieties; and culminates with private firms producing and marketing most varieties with some basic plant breeding research still managed by the public sector (Quoted in Tripp and Louwaars 1997: 433-446). The Seeds Act 1966, the Seeds Rules 1968 and the Seeds Control Order 1983 are the legal instruments which regulate the production, distribution and quality of the seeds produced in the country. The Plant Varieties Protection and Farmer's Rights Act (PVPFRA) that came into being in 2001, provides protection for the different varieties of the plants developed by breeders and farmers.

The Seed Act of India 1966 (amended in 1972), provides for regulating the quality of seeds for sale and thus mentions the role of the seed committee, central and state government seed laboratories, regulation of sales of seeds of notified varieties, role of certification agency, seed inspectors and seed analysts. Section 24 of the Seed Act (1966) provides that 'Nothing in this Act shall apply to any seed of any notified kind or variety grown by a person and sold or delivered by him on his own premises direct to another person for being used by that person for the purpose of sowing or planting' which encourages farmer to farmer exchange or sale of seeds. Seed certification in India is mandatory for public sector seed corporations and voluntary for the private sector (Tripp and Louwaars 1997) but 'truthful labelling', which mandates mention of genetic purity and germination rate of the seed including the 'truthfully labelled' seeds sold by the private sector[5].

The amendment made in 2006 to the Seed Control Order mentions that 'every dealer of seeds in notified kind or variety or other than notified kind or variety of seeds shall ensure that the standards of quality of seeds claimed by him shall conform to the standards prescribed for the notified kind or variety of seeds under Section 6 of the Seeds Act, 1966 (54 of 1966) and any other additional standards relating to size, colour and content of the label as may be specified'.[6] Cotton seed is also under the Essential Commodities Act, 1955 that has been amended in 1986 and in 2006, and thus can be brought under price control order of the government.

The formulation of the *sui generis* Plant Varieties Protection and Farmer's Rights Act (PVPFRA) that came into being in 2001 and provides ownership rights of plant varieties to the plant breeders and farmers. Such ownership rights are expected to promote the growth of the seed industry and ensure availability of high quality seeds to the farmers. The plant registration authority (PRA), set up in May 2007, has been accepting applications for registration since then. It provides the owners of such varieties the rights to produce, sell, market, distribute, import or export the variety registered under the PVPFRA. Under the PVPFRA, protection is available for four kinds of plant varieties, viz. essentially derived variety, farmers' variety, extant variety and new variety[7]. Protection is available for a period of 18 years for trees and vines, 15 years for other crops, and for extant varieties that are notified, protection is available for 15 years from the date of notification.

Varieties seeking registration will have to satisfy the guidelines set for distinctiveness, uniformity and stability known as the DUS criteria. Cotton varieties are tested for DUS characteristics in the Central Institute for Cotton Research (CICR) at Nagpur and Coimbatore, University of Agricultural Sciences, Dharwad, Haryana Agricultural University, Hisar and Punjab Agricultural University, Ludhiana.

Between 2007 and March 2013, the PRA received a total of 1115, 1876 and 1446 applications from the public sector, private sector and farmers respectively. The cotton applications constitute 9.2 and 39 per cent of the total applications in public and private sectors respectively. There were no applications from the farmers for any of these varieties which indicates that satisfying the 'DUS' characteristics may not be possible for small and medium farmers. The relatively large number of tetraploid cotton application can be linked to those private sector seed companies which have backcrossed their hybrid varieties with the Bt gene owned by Monsanto and other sources and have sought protection of their varieties. This has obviously increased the seed choices before the farmers since 2006.

### *2.2 Seed Distribution*

Companies generally appoint a distributor or stockist for a region which may comprise more than one state. Dealers are appointed at the district level. Depending on the companies' network, the dealers may be appointed at talukas who take care of the retail sales. Mostly dealers deal with more than one company's product and sell pesticides, fertilisers, and other agricultural implements. Normally 15–20 per cent commission is given to distributors who pass 12–17 per cent on to the dealers, while the retailers get 9–14 per cent gross discount on sales. If some portion of the seed remains unsold, companies, on request, help in revalidation of the seed by the certification agency at the dealers/distributor's cost. A few companies have the policy of getting back the unsold stocks so that they are not sold as fresh stocks in the subsequent season by the channel, which might damage the image of the company among the farmers (Singh and Asokan 1997: 421–440).

The distributor/dealer indent seed with the companies in December. Though, the seed multiplication agencies have to undertake the seed activity well in advance, no forecasting method is used to estimate the demand of the seeds (Alagh 2004: 124). Hence, the seed multiplication agencies both in the private sector and the public sector assess the demand based on the past sales, but the actual sales would depend on a host of other factors.

### *2.3 Pesticides Industry*

India is the largest producer of agrochemicals after the US and China (Dave 2012: 37), and has about 30 per cent of the total cultivated area under pesticide cover. The agrochemical market is highly fragmented. A number of pesticides formulations have become generic and about 600 generic companies are operating in this field in India. Hence, the same chemical would be sold by different trade names by different manufacturers. Patented new molecules are held mainly by the multinationals. The pesticide industry consists of both organised and unorganised manufacturers.

The use of insecticides is regulated under the Insecticides Act, 1968 which regulates the import, manufacture, sale, transportation, distribution and use of the pesticides with a view to prevent 'risks to human beings or animals and for matters connected therewith'. The Insecticides Rules, 1971 governs registration, license to manufacture, labelling of the product etc. As of November 2012, 241 insecticides have been registered for use in India and 32 products have been banned from manufacture, import and use in India[8].

Recent data reveals that pesticide consumption in India had reduced from 72.13 thousand tonnes in 1991–92 to 55.54 thousand tonnes in 2010–11 (GOI 2012: 60). In this regard, the important difference between India and developed countries is that India uses more insecticides (80%), while other countries depend on agronomic practices and rely on weedicides (Sreenivasa 1992: 79–88). Also, the small size of land holdings in India and the availability of farm labour explains to a certain extent the lower use of weedicide compared to the large holdings in the developed countries which rely on weedicides due to lack of availability of farm labour. But the fungicide and herbicide sectors are growing rapidly in India (Dave 2012). The relative share of cotton in total pesticide consumption which was estimated to be more than 50 per cent in India (Shetty 2004: 261-267), appears to be changing after the introduction of Bt cotton in the country (Table 7.1).

Over the years, there has been a shift in the kind of pesticides being used. Organochlorines are highly persistent both in the environment and biological systems, and are prone to development of insecticide resistance. Organophosphate group of pesticides are more toxic to parasitoids and predators than to insect, pests, and results in pest resurgence of secondary pests. Carbamates are highly toxic in nature to parasitoids and predators.

**Table 7.1:** Trends in Pesticide Usage in Cotton in India, 2000-10

(*Rs. Million*)

| *Year* | *Insecticides on cotton* | *Fungicides on cotton* | *Herbicides on cotton* | *Total insecticides in agriculture* | *Total pesticides in agriculture* | *Insecticides used on cotton as % of total insecticides* |
|---|---|---|---|---|---|---|
| 2000 | 8390 | 96 | 15 | 20520 | 29725 | 40.9 |
| 2001 | 10520 | 64 | 10 | 22680 | 32070 | 46.4 |
| 2002 | 5970 | 34 | 5 | 16830 | 26223 | 35.5 |
| 2003 | 9250 | 78 | 34 | 21460 | 31474 | 43.1 |
| 2004 | 10320 | 60 | 39 | 24550 | 35814 | 42.0 |
| 2005 | 6490 | 85 | 78 | 20860 | 24388 | 31.1 |
| 2006 | 5790 | 111 | 118 | 22230 | 22960 | 26.1 |
| 2007 | 7330 | 248 | 215 | 28800 | 46970 | 25.5 |
| 2008 | 7906 | 315 | 261 | 32820 | 52933 | 24.1 |
| 2009 | 8337 | 522 | 455 | 39090 | 69988 | 21.3 |
| 2010 | 8804 | 669 | 869 | 42830 | 76836 | 20.6 |

*Source*: Kranthi *et al.* 2011: 19.

Repeated application of synthetic pyrethroids has resulted in resurgence of mites, aphids, white fly etc. and also phytotoxicity. Neonicotinoids though do not have the limitations of the above mentioned nature yet are generally costly (Kumar *et al.* 2010). Industry sources point out that organochlorines, organophosphate and carbamates that were introduced during 1950-1970, and synthetic pyrethroids that were introduced during 1970-1990 and are out of intellectual property rights (IPR) protection. But most of the neonicotinoids are under intellectual protection and hence there are no generic formulators for them in India, which make this group of pesticide costly. Hence, comparison of costs could lead to more use of organochlorines or carbamates and recent studies on cotton have observed the mixed use of all these groups of pesticides in the context of India (Mancini *et al.* 2005: 221-232, Peshin *et al.* 2008: 73-80).

With this background information on the seeds and pesticides industries in India, in what follows, we discuss the seed and pesticide use scenarios as reported by the cotton growers in Gujarat.

## 3. Aspects of Information Sourcing of Farmers: An Analysis with respect to Seed and Insecticides

### *3.1 Profile of Seed Dealers*

The proliferation of cotton hybrids in the market necessitates that the companies set up proper dealer network to reach out to the farmers and the dealers are supposed to provide information required by the farmers. Dealers also act as an important link between the company and the farmers. In our survey, of the total sample of 82 dealers, 19.5 per cent each were located in Ahmedabad and Vadodara, 17 per cent in Bhavnagar, 20 per cent in Surendranagar and 23 per cent in Rajkot.

Selling cotton seed and pesticides is the main business activity for majority of the dealers while 77 per cent of the dealers also deal with other seeds. Only 7 per cent is engaged in cotton trading. About 77 per cent of the dealers have their shop in the talukas headquarters, perhaps to take advantage of the wider market potential that in talukas places offer. However, 16 per cent of the dealers are located in the villages. A small segment (7%) of the dealers operates in both village and talukas places. While 10 per cent of the dealers have been engaged in the same business even before 1980, 33 per cent entered the business between 1991 and 2001. Importantly, 45 per cent of the dealers have started their seed business after 2002 and as many as 14.6 per cent of the dealers came in to business in 2005 which essentially coincides with the government approval for more companies to sell Bt cotton hybrid seeds.

### *3.2 Seed Selling Practices*

Fifty per cent of the dealers said that they do recommend specific varieties to the farmers and 44 per cent also provide credit to purchase cotton seeds. Of those who extend credit, 33 per cent provided credit on all varieties and only 11 per cent of the dealers said that they offer credit on seeds of a particular company only. Though this is a small percentage, the availability of credit could induce the farmer to buy seeds of a particular company. Contrary to our expectations, we found that brand specific credit was extended not only for Bt hybrids but also for non-Bt hybrids and *desi* varieties. Interestingly, the company specific schemes to attract farmers pertain to providing inexpensive gifts such as purse, bag, calendar, key chain, diary and casserole. However, name of a particular company was mentioned by a dealer which announced gifts such as refrigerator, cellphone etc. for farmers with good production performance. Nearly 70 per cent of the dealers said that they do not enjoy any particular benefit extended by the companies, about 18 per cent said that they would prefer discounted prices at the time of booking and 12 per cent prefer commission on selling seeds.

While, 93 per cent of the dealers said that there is no compensation/

insurance offered by the company for seed failures, the rest of the dealers positively confirmed about specific company initiatives. Interestingly, though the dealers mentioned seven popular companies that provide compensation for seed germination failures, only three companies had actually paid compensation according to a particular dealer. Yet another company replaced the seeds free of cost for germination failure.

Regarding the price and size of the seed packet, 81 and 99 per cent of the dealers' response converged on the cost of Bt packet being sold at Rs. 750 (2007-08) and the size of the packet being 450 gram respectively. However, only 25 per cent of the dealers said that the Bt seed packet sold by them consisted of non-Bt seeds (refuge) as well.

### *3.3 Popular Varieties Sold by the Dealers*

Popularity of varieties sold in a region depicts the revealed preference of the farmers at different points of time and could differ across regions/ districts. The information on the popular seed variety sold by the dealers indicated that they sell majority of Bt hybrids developed by the private sector, besides which they mentioned to have sold six *desi* varieties and four non-Bt hybrids. Both these 10 varieties were developed by the public sector. The dealers were asked to name five top varieties sold by them in the past three years. Though for 2007, the dealers could collectively recall the names of 345 varieties, for 2006 and 2005, they could recall only 188 and 96 varieties. In Table 7.2, we have listed only those popular varieties which got at least five and more responses from the dealers.

Thus, we find 7 varieties to be most popular in 2005. In 2006 and 2007, 11 and 20 varieties were reported respectively. Interestingly, while RCH and Mallika have remained as the top two selling varieties, the ranking of other varieties have changed in 2007 compared to the earlier years. The variety ranking 3rd in 2005 has moved to 9th position in 2006 and was further pushed to 18th position in 2007. The variety ranking 4th and 5th in 2005 have also moved to 6th and 7th in 2006 and to positions 14th and 17th in 2007. Certainly, the introduction of more varieties has had a role to play. Vikram 5 and Ankur 651 have moved to the 4th and 5th position in 2007 while Ganga Kaveri which was in 8th position in 2006 has moved to 3rd position in 2007. Obviously, the sales reflect the revealed choice of the farmers. Given this, in what follows, we examine this important issue of the information on the varietal attributes being made available to the farmers by the seed dealers.

**Table 7.2:** Popular Seed Varieties Sold by the Seed Dealers 2005

| *2005 (96)* | | *2006 (188)* | | *2007 (345)* | |
|---|---|---|---|---|---|
| *Name of variety* | *Dealer responses %* | *Name of variety* | *Dealer responses %* | *Name of variety* | *Dealer responses %* |
| RCH Rasi 2 | 18.75 (1) | RCH Rasi 2 | 19.15 (1) | RCH Rasi 2 | 14.78 (1) |
| Mallika | 14.58 (2) | Mallika | 14.89 (2) | Mallika | 9.57 (2) |
| Bollgard 12 | 9.38 (3) | Vikram 5 | 6.91 (3) | Ganga Kaveri | 5.22 (3) |
| Bollgard 162 | 7.29 (4) | Ankur 651 | 4.26 (4) | (J K 205) | |
| Bt 6301 (Mahyco) | 6.25 (5) | Bunny | 3.72 (5) | Vikram 5 | 4.64 (4) |
| Vikram 5 | 5.21 (6) | Bollgard 162 | 3.19 (6) | Ankur 651 | 4.06 (5) |
| Bunny | 5.21 (6) | BT 6301 (Mahyco) | 3.19 (7) | Akka (Bollgard) | 3.77 (6) |
| | | Ganga Kaveri (J K 205) | 2.66 (8) | Parasbhrama | 3.48 (7) |
| | | Bollgard 12 | 2.66 (8) | Kanak (Bollgard) | 2.90 (8) |
| | | Vishvanath | 2.66 (8) | Alto (Mahyco) | 2.90 (8) |
| | | TV 333 TATA (11) | 2.66 (8) | Bombino | 2.90 (8) |
| | | | | Vishvanath | 2.90 (8) |
| Popular varieties sold (%) | 66.67 | | 65.95 | | 57.12 |
| Less popular varieties sold (%) | 33.33 | | 34.05 | | 42.88* |
| Total no. of dealer responses | 96 | | 188 | | 345 |

*Notes:* 1. Figures in brackets are respective rankings in total number of varieties being reported by the seed dealers surveyed. *- This also includes other less prominent varieties being sold by the dealers, viz. Bunny, Ankur 9, Bt 6301 (Mahyco), Tulsi 4, Vikram 9, Bollgard (Mahyco 12), JK Varun, Super Maruti (BG II).

2. Varieties which had at least 5 dealers' responses are considered as 'popular variety' here.

*Source*: Seed Dealer Survey in Gujarat covering 82 dealers, 2007-08.

### *3.4 Varietal Characteristics Known to the Dealers*

The dealers were asked to list the important characteristics of the varieties they sold to assess the information that they had acquired/provided by the company to be passed on to the farmers. Almost 100 per cent responses asserted on the time frame of the crop, number of bolls per shoot, size of the boll, expected production in irrigated conditions and watering required for that variety. Interestingly, 28 per cent of the dealer responses indicated that they were not aware of the likely performance of the variety in unirrigated conditions. About 13 per cent did not know the length quality of the fibre. While 82 per cent told that the varieties sold by them gave protection against bollworms, 86 per cent of them could not name the type of bollworm that the trait would address. However, dealers seem to know about the present trait of the Bt which do not provide protection against sucking pests. Hence, their responses on the technology protecting against sucking pests varied in that the majority of the dealers (49%)

responded by stating the effect to be bad (meaning no effect), followed by 33 per cent reporting medium effect and only 16 per cent reporting good effects. At the same time, as many as 95 per cent of the responses indicated that they were unaware of the names of the sucking pests that affect the plant.

### 3.5 Dealer Information on Pesticides

Except for one dealer, the rest of the dealers sold pesticides. The most popular pesticides sold by the dealers in 2007-08 were monocrotophos, acephate, imidochloprid and acetamiprid. The first two belong to organophosphate, while the other two belong to nitroguanidine and chloronicotinoids category. In terms of the hazardous nature of the pesticide, monocrotophos belong to the highly hazardous category and the other three belong to the slightly hazardous category. According to the dealers, more than 80 per cent of the pesticides sold by them were meant to be used against sucking pests. Interestingly, for the 13 per cent of the name of the pesticides reported, the dealers did not know the name of the company that is selling the product. Tata Rallis, Bayer and United Phosphorous Products were the three popular companies among the dealers.

Since all the dealers were selling Bt cotton seeds, it was of interest to know whether the sales of the pesticides have been affected after the introduction of the Bt technology. While 18 and 19 per cent of the dealers mentioned that there has been a reduction and no change in sales respectively, 60 per cent of the dealers reported an increase in the sale of pesticides. However, while 14 per cent reported a 50 per cent increase in sales, hardly 2 per cent mentioned a 100 per cent increase in pesticide sales after Bt introduction. The major reason attributed for the increase in pesticides sales was infestation by sucking pests, while the reduction was attributed to the effectiveness of Bt in the control of bollworms. Dealers who did not observe any adverse change in sales had attributed this to the incidence of sucking pests which demanded the farmers now to continue the sprays for control of these pests as compared to the earlier sprays meant for control of bollworms. It is quite interesting to observe that even dealers endorsed the point that farmers continued to employ equal number of sprays against sucking pests in the post-Bt scenario. About 73 per cent of the dealers also mentioned providing extension services to the farmers which, however, were confined to the use of pesticides and not on seed selection. On probing, we realised that the extension service was limited to advising the farmers about the use of pesticides and informing the company to arrange for visit to the farmers' fields.

While summing up this section, it may be observed that seed dealers seem to possess more information about the attributes of the seed varieties than about the pesticides. This could be because the lack of integration between the seed and pesticide markets in terms of information flows which poses an important issue as it is most likely that the farmers get unduly influenced by the dealers in choosing an insecticide product without knowing much about its impacts or usefulness on the pest complex.

## 4. Seed Choice by the Farmers

In this section, it would be apt to describe the seed market situation that was prevailing in Gujarat in 2007-08. As observed in a few writings, the presence of the unapproved Bt cotton fastened the process of wider adoption of the technology in Gujarat (Stone 2007, Herring 2007: 150-187, Ramaswami 2012: 177-188) compared to the other cotton growing states. A well 'organised' unorganised seed market was thriving in the 'farmer to farmer' sale facilitated by the Section 24 of the Seed Act, 1966 mentioned earlier, particularly before 2006. This was aided by clandestine plant breeding, and hybrid seed production and distribution. Seeds in this segment were guaranteed by the 'farmer seed seller' for their production and the farmers had more faith in them as these 'farmer seed sellers' are known to them. Also, as cited in Feder *et al.* (1985), from the experience of Tunisian farmers, domestically developed new varieties will be received more favourably by farmers than unfamiliar varieties. More importantly, the seed locally developed was more suitable for the agro ecological conditions, as these seeds were developed by the farmers based on long years of trials in own fields (Shah 2005: 4629-4639). The fact that there was no difference in the yield performance between the approved and unapproved seeds (Ramaswami 2012) and the early approved varieties that did not necessarily have all the agronomic qualities also facilitated the spread of the unapproved seeds. Before the price reduction of the approved Bt seeds by the government, the unapproved seeds were sold at relatively lower prices encouraging more adoption. After the price reduction of the approved seeds and particularly after the introduction of more Bt varieties, the unapproved seeds price was further reduced, encouraging the farmers to go for F1 seeds and reducing the need to use farm saved seeds totally[9]. Thus at the time of the

2007-08 survey, a complex seed market was there before the farmers which offered a wider choice of Bollgard I varieties, with equal choice of varieties from the unapproved segment as well and the newer Bollgard II variety that offers protection against bollworm and spodoptera. At the time of 2007-08 survey, though Bollgard II was already there in the market, farmers mostly preferred using Bollgard I as indicated by the sales record of a few of the companies in Gujarat exhibiting more sales of Bollgard I than Bollgard II.[10] However, there was a major turnaround later on, as the 2010-11 survey indicated the wider prevalence of the Bollgard II that offers protection against bollworm and spodoptera.

### 4.1 Brief Farmer Profile

Table 7.3 provides a comprehensive picture about the sample farmers. Of the total 200 farmers, in 2007-08, 53, 101 and 46 (constituting 26.5 per cent, 50.5 per cent and 23%) belong to the small, medium and large size holding categories respectively. In 2010-11, 60, 103 and 37 (constituting 30, 51.5 and 18.5%) belonged to the small, medium and large category. This trend in the number of farmers, indicating an increase in the number of small farmers, perhaps reflects on the 'scale neutrality' or the inclusiveness of the Bt technology, such that more small and marginal farmers started growing more of Bt cotton. Importantly, all the sample farmers had been engaged in cotton cultivation for several years. The average age of the farmers in both the survey periods was 46 years. In terms of educational status, it was found that during both the periods (2007-08 and 2010-11), a majority of the farmers reported having completed primary (43 and 38.5%) and secondary (36 and 43%) schooling, and a smaller proportion having college education (9.5 and 8%). Notably, the number of illiterate farmers was very few with 11.5 and 10.5 per cent respectively during 2007-08 and 2010-11 respectively. The proportion of land under Bt cotton to total cotton land had steadily increased from 54 per cent in 2003-04 to 83 per cent in 2010-11, which suggests that by and large farmers have tended to allocate more land for growing Bt cotton encouraged by the returns.

It is important to note that total cultivable land with the sample farmers had declined by about 6 per cent in 2010-11 as compared to 2007-08 and the overall land under irrigation has increased by 7 per cent thanks to the initiatives by the state government and private investment in tube wells (Viswanathan and Lalitha 2012). A comparison of land allocated for cotton in 2007-08 and 2010-11 indicate that overall, the percentage of land under

**Table 7.3:** Select Profile of the Sample Cotton Growers in Gujarat

| *No. of farmers* | *2007–08* | | | | *2010–11* | | | |
|---|---|---|---|---|---|---|---|---|
| | *Small* | *Medium* | *Large* | *Total* | *Small* | *Medium* | *Large* | *Total* |
| | 53 | 101 | 46 | 200 | 60 | 103 | 37 | 200 |
| Total cultivable land (ha) | 104.60 | 588.93 | 854.38 | 1547.91 | 103.43 | 600.26 | 745.53 | 1449.21 |
| Total irrigated land (ha) | 77.80 | 449.98 | 596.91 | 1124.69 | 99.48 | 533.09 | 570.83 | 1203.40 |
| Total land under cotton cultivation (ha) | 84.49 | 398.23 | 533.28 | 1016.01 | 93.59 | 422.73 | 336.90 | 853.22 |
| Total land under Bt cotton cultivation (ha) | 76.99 | 378.96 | 450.61 | 906.56 | 87.76 | 392.33 | 227.79 | 707.87 |
| Total land of the three chosen plots (ha) | 79.00** | 359.25 | 428.72 | 866.97 | 91.62 | 382.04 | 297.13** | 770.80** |
| Seed used for the chosen plots (Kg/per hectare) | 2.18 | 3.26 | 3.45 | 3.25 | 3.51 | 3.72 | 3.98 | 3.8 |
| Seed expenditure (Rs./ha) | 2301.2 | 2144.6 | 1067.5 | 1626.3 | 2592.1 | 2375.3 | 1962.2 | 2242.8 |
| Number of Bt varieties cultivated per farmer | 2.2 | 2.9 | 3.0 | 2.7 | 1.7 | 2.3 | 3.1 | 2.3 |
| Use of pesticides (Kg/ha) | 7.9 | 7.5 | 4.2 | 5.9 | 5.9 | 6.0 | 3.031 | 4.8 |
| Pesticide expenditure (Rs./ha) | 4117 | 4348 | 3229 | 3773.4 | 4194 | 4985 | 2428 | 3906 |
| Yield of cotton (Kg/ha) for the chosen plots | 2139.8 | 2175.7 | 1241.1 | 1710 | 2268.2 | 2248.3 | 2054.1 | 2176.7 |

** indicates the presence of non-Bt land under cultivation.

*Source*: Farm Household Survey, 2007-08 and 2010-11.

cotton has declined from 66 to 59 per cent. Further, while the small size holders have increased the area under cotton from 80 to 90 per cent, only a marginal increase is evident in the case of medium land holders. It emerges that the large farmers have actually reduced the land under cotton to 82 per cent in 2010-11 from 89 per cent in 2007-08. Thus, both the increase in the case of small and medium size holders and the reduction in the case of the large size holders are reflected in the total land of the chosen plots. The most used sources of irrigation for the farmers in both the years were well and bore well respectively which indicates the private investment in this sphere. A small percentage of farmers also reported using lakes, canals and river to irrigate their land.

### *4.2 Farmer Responses on Choice of Seeds and Input Market Dependence*

On an average, a farmer used about three varieties of Bt seeds during both the rounds of the survey (2.73 during 2007-08 and 2.3 during 2010-11), though the number indicated a decline by the last round, probably indicating the farmer's distinct choice for specific varieties by the second round. After considering the three big plots of each farmer (where more than 3 prevailed), we get a total of 417 plots and 400 plots respectively in 2007-08 and 2010-11. In what follows, we present the major aspects of information sourcing practices of the study farmers during the two rounds of the survey.

Farmers in both the years besides giving the name of the variety could also provide us the name of the seed company. While we got 100 per cent response regarding the company name (source of seed purchase) in 2007-08, the proportion was bit less in 2010-11, suggesting that about 95.5 per cent of the farmers could name the company whose seeds they had bought for the three major plots. Only seven could not name all the three companies and less than 1 per cent did not know the name of the company. Regarding the name of the variety, similar to the observation made by Tripp and Pal (1998) and Tripp (2001) about farmers' decision making on pearl millet seeds in Rajasthan, we also observed that though farmers were able to identify the company, their ability to name the precise variety was less evident. Farmers were also less aware of the agronomic characteristics of the seeds they bought. For instance, in our survey in 2010-11, 26 per cent of the plots were sown with Bt cotton seeds that were identified with the name of the company as Bollgard. Thus, when we say X seeds of Y Company is the most popular and adopted by the farmers, it could be any of the variety of the Y Company and it is common that most farmers use a generic name for all the varieties of the company. So if the same variety is known by a different name in another study village, it would be reckoned as a distinct variety of that village in the study.

Overall, the quantity of seeds used per hectare had marginally increased from 3.3 kg/ha in 2007-08 to 3.8 kg/ha in 2010-11 season. Perhaps this indicates the strategy of the farmers to increase the planting density as a precaution to prevent any type of plant loss. Seed quantity used by the small farmers has increased noticeably from 2.2 kg to 3.5 kg. Also, we find that in 2010-11, farmers' reliance on the organised sector[11] for their seed requirement has increased significantly compared to 2007-08 (Table 7.4). Further, the percentage of farmers who have used both the organised and unorganised sector seeds have also reduced to 1 per cent in 2010-11 indicating the formalisation of the seed market in Gujarat. One probable reason for the increased dependence on the unorganised sources in 2007-08, could have been due to the fact that 58 per cent of the farmers had to re-sow seeds again due to the heavy rains. Of this 58 per cent of the farmers, 85 per cent had re-sown once while 10 per cent had re-sown thrice. As mentioned elsewhere in the paper, the strategy of the unapproved seed producers to match the price reductions of the approved seeds had perhaps helped them to garner a larger share of the market in 2007-08. Also in 2007-08, except for closer to 10 per cent of the farmers who had adopted Bollgard II variety in 1 per cent of the total land, others had chosen the Bollgard I variety. In 2010-11, majority reported use of 'double Bt' and it was difficult to gauge the presence if any of the Bollgard I for the researcher.

**Table 7.4:** Source of Seed Purchase

| *Source of seed purchase* | *2007-08* | *2010-11* |
|---|---|---|
| Organised sector % | 35.13 | 90.92 |
| Unorganised sector % | 54.76 | 8.07 |
| Both organised + unorganised | 10.11 | 1.01 |
| Total quantity (kg) | 2825.6 | 2929.4 |
| Seed quantity per farmer (kg) | 14.1 | 14.6 |

*Source*: Farm Household Survey, 2007-08 and 2010-11.

The shift to organised seed sources also shows that farmers' adoption decisions get modified over time on the basis of new information (Feder *et al.* 1985). Also, the aggressive marketing network that companies have

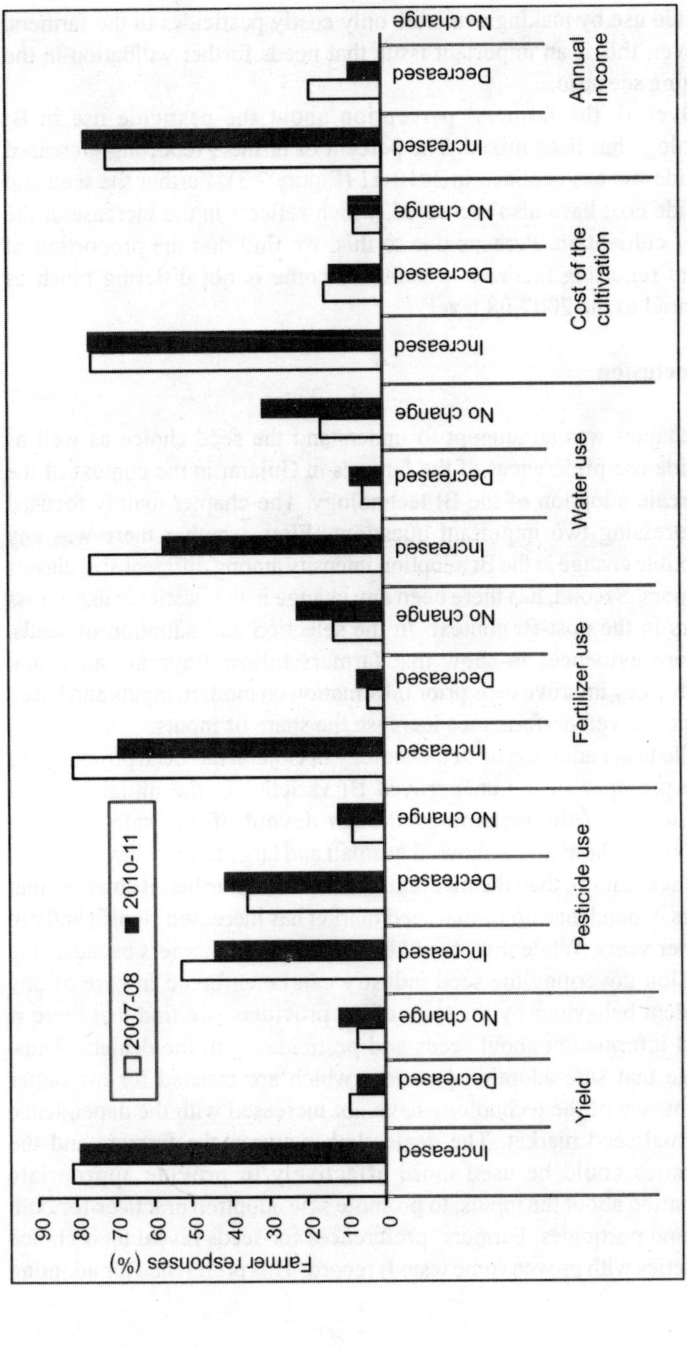


*Source:* Farm Household Survey, GIDR 2007-08 and 2010-11.

**Figure 7.3:** Farmer Responses about Effects of Bt Technology, 2007-08 and 2010-11

pesticide use by making available only costly pesticides to the farmers. However, this is an important issue that needs further validation in the emerging scenario.

Overall, the farmers' perception about the pesticide use in Bt technology has been mixed. The percent of farmers reporting increased pesticide use has declined in 2010-11 (Figure 7.3). Further the seed and pesticide cost have also increased, which reflects in the increase in the cost of cultivation. Perhaps due to this, we find that the proportion of farmers reporting increase in annual income is not differing much as compared to the 2007-08 level.

## 7. Conclusion

This chapter was an attempt to understand the seed choice as well as pesticide use preferences of the farmers in Gujarat in the context of the wide-scale adoption of the Bt technology. The chapter mainly focused on addressing two important questions. First, whether there was any perceptible change in the Bt adoption intensity among different size classes of farmers. Second, has there been any change in the pesticide use across farmers in the post-Bt context. In the selection and adoption of seeds, there are evidences to show that farmers follow Bayesian adoption, whereby they improve their prior information on modern inputs and based on the observed performance increase the share of inputs.

The faster adoption of Bt technology in Gujarat has been promulgated by the presence of the unapproved Bt varieties in the initial years of introduction of the technology though devoid of any safe adoption procedures. The analysis shows that small and large farmers have adopted newer seeds more than the medium land holders. Further, it emerges that farmers dependence on formal seed market has increased significantly in the latter years. While this should be helpful to the farmers because, the regulation governing the seed industry can be enforced in case of any fraudulent behaviour by the technology providers, we find that there is limited information about seeds and pesticides with the dealers. Thus, we note that safe adoption practices which are insisted for the better performance of the technology have not increased with the dependence on formal seed market. The dealer link between the farmers and the companies could be used more effectively to provide appropriate information about the inputs, to promote safe adoption practices for both seeds and pesticides. Farmers' preferences for seeds reveal their choice of varieties with proven (time tested) record. This preference for adopting 'time tested' varieties seems quite logical as the analysis demonstrate that newer varieties flooded in the market have not resulted in yield increase for majority of the small and medium farmers in Gujarat.

The analysis further shows that though there was a notable decline in the use of pesticides, the cost of pesticide incurred by the farmers continues to remain at high levels. This could be attributed to two or three major reasons. First, while there was a decline in the use of pesticides for control of bollworm owing to the Bt technology, the need to spray for sucking pests have nevertheless increased. Still, it is not clear whether the infestation of sucking pests is due to the environmental factors, mono-cropping or pests developing resistance to the pesticides being applied. Secondly, a few farmers indicated the use of newer pesticides which are least hazardous to humans and environment and these newer varieties are under intellectual property protection. Hence, the lack of generics, make the farmers dependent on these few newer pesticides. Third, since, the farmers most often mix more than two chemicals during the spray, which may totally change the composition of chemicals leading to resistance by the pests requiring more sprays. The lack of guidance on the timing of sprays and the absence of information on the probable combination of chemicals to be used, results in greater use of pesticides. This is a serious concern regarding the sustainability of cotton cultivation in India. Though, there is evidence from the analysis that farmers observe the pests pressure in their field and then resort to spray, information need to reach farmers regarding the appropriate use of chemicals. It has been evident in the case of cotton that several hybrids with better yield performance have had a natural death due to pests and faulty pest control practices. In the case of Bt cotton, already, farmers have shifted from Bollgard 1 to Bollgard II not only in Gujarat but also in other cotton growing regions. If the lack of safe adoption practices and wrong plant protection methods lead to pests developing resistance to Bollgard II, our apprehension is that, there is nothing in the shelf for the farmers till a final decision is taken on further adoption of GM technology per se in the country. These issues open new vistas of empirical scrutiny in future along with streamlining regulatory regimes as well as policy formulations for sustaining the initial dynamism cast by the Bt technology in the country.

## Notes

1. The Gujarat State Seed Corporation for instance reported a reduction in the sale of its hybrids since the introduction of Bt cotton in the state.
2. In this chapter, the terms insecticides and pesticides have been used interchangeably.
3. While 66 farmers remained common to both the surveys, the rest 134 (67%) farmers were freshly interviewed.
4. Available at http://seednet.gov.in/Material/Structure.pdf, accessed on 19 March, 2013.
5. A typical truthful label provides information on the following on its cover: crop, variety, stage, pure seed (minimum), inert matter (maximum), germination (minimum), genetic purity (minimum), other crops seeds (maximum), weed seeds (maximum), moisture, net weight, label number, year, lot number, date of test, date of packing, and the date of validity.
6. Available at http://seednet.gov.in/Material/Seed_Control_Order (Amendment)_2006.pdf; http://nsai.co.in/images/filepicker_users/6587628292-62/seed%20rules/Seed_Control_Order (Amendment)_2006.pdf
7. Essentially derived variety means a variety essentially derived from existing variety by (a) genetic engineering, (b) mutation, (c) tissue culture, (d) back cross derivative and any other ploidy change etc. Those varieties that are traditionally cultivated and evolved by the farmers in their fields or in a wild relative or land race or a variety about which the farmers possess common knowledge are called Farmers Variety. Farmers' variety is exempted from application, registration fees and these applications need not be accompanied with fees and affidavit for terminator technology. Existing varieties already in the market but not for more than a year can be applied for registration as a new variety. Older variety can be applied for registration as extant variety.
8. Available at www.cibrc.nic.in, accessed on 25 March, 2013.
9. Information obtained during the farmer interviews in 2007–08.
10. Discussion with officials of the Department of Agriculture, Government of Gujarat, in October 2012.
11. The organised sector here refers to the purchase from seed shop in a village, taluka or district, co-operatives, universities etc. while the unorganised sector refers to sources such as neighbour farmers, relatives, ginning factories.
12. Bracketed terms are figures for 2010-11.

## References

Alagh, Y. K. 2004. *An Overview* (*Vol.* 1), *State of the Indian Farmer, A Millennium Study*. Academic Foundation, New Delhi.

Dave, Pradip. 2012. 'Three Keys for India's Agrochemical Industry'. *CropCare*, Vol. 38, No.4: 37, October-December.

Feder, G., R. Just and D. Zilberman. 1985. 'Adoption of Agricultural Innovations in Developing Countries: A Survey'. *Economic Development and Cultural Change*, Vol. 33, No. 2: 255-298.

Government of Gujarat. 2010. *Gujarat Agriculture Statistics at a Glance 2009-10*. Directorate of Agriculture, Gandhinagar, Gujarat.

GOI. 2012. *Status of Indian Agriculture 2011–12*. Government of India, p. 60.

Herring, R.J. 2007. 'Stealth Seeds, Bioproperty, Biosafety and Biopolitics'. *Journal of Development Studies*, 43 (1):150-187.

Kranthi, K.R. 2011. 'Technology Needs Stewardship'. Interview in *Down to Earth*, 31 July 2011. Also available at http://www.downtoearth.org.in/content/technology-needs-stewardship.

Kranthi, K.R., M.V. Venugopalan, M. Sabesh, M.S. Yadav. 2011. *CICR–Vision 2030*. Central Institute of Cotton Research, Nagpur, p.16.

Kumar V., I.M. Maisuria, C.J. Patel, H.R. Desai, D.B. Sheth and Seema Bhadauriya. 2010. *Ecofriendly Management of Mealy Bug in Cotton*. Main Cotton Research Station, Navsari Agricultural University Surat.

Lalitha, N. and P.K. Viswanathan. 2010. 'Pesticide Applications in Bt cotton Farms: Issues Relating to Environment and Non Tariff Barriers'. *Asian Biotechnology Development Review*, Vol. 12 (2): 11-35.

Lalitha, N. and B. Ramaswami. 2007. 'Pesticides Use Pattern Among Cotton Cultivators in Gujarat'. In Ravindra Dholakia (ed.), *Frontiers of Agricultural Development in Gujarat.* Centre for Management in Agriculture, Indian Institute of Management, Ahmedabad, 77-98.

Lalitha, N., Bharat Ramaswami and P.K. Viswanathan. 2009. 'India's Experience with Bt Cotton: Case Studies from Gujarat and Maharashtra'. In Robert Tripp (ed.), *Biotechnology and Agricultural Development: Transgenic Cotton, Rural Institutions and Resource-poor Farmers*, UK: Routledge, pp.135-167.

Mancini, F., A.H.C. Van Bruggen, J.L.S. Jiggins, A.C. Ambatipudi, H. Murphy. 2005. 'Acute Pesticide Poisoning among Female and Male Cotton Growers in India'. *International Journal of Occupational and Environmental Health* 11(3): 221-232.

Meena, R.A., M.N. Mishra and R.G. Dani. 1999. 'Effect of Storage Containers on Seed Longevity in Cotton (Gossypium Hirsutum)'. *Indian Journal of Agricultural Sciences*, 69 (10): 740-2.

Peshin, R., A.K. Dhawan, K. Vatta, K. Singh. 2008. 'Attributes and Socio-Economic Dynamics of Adopting Bt Cotton'. *Economic and Political Weekly* 42 (52): 73-80.

*Biospectrum*. 2012. 'Profiles of Top 20 Biotech Companies'. *Biospectrum*, Vol.10, Issue 6.

Qaim, M. 2005. 'Agricultural Biotechnology Adoption in Developing Countries'. *American Journal of Agricultural Economics*, Vol. 87 (5): 1317-1324.

Ramaswami, Bharat, Carl Pray and N. Lalitha. 2012. 'The Spread of Illegal Transgenic Cotton Varieties in India'. *Biosafety Regulation, Monopoly and Enforcement. World Development*, Vol. 40 (1): 177-188, January.

Ryan, B. and N. Gross. 1943. 'The Diffusion of Hybrid Seed Corn in Two Iowa Communities'. *Rural Sociology*, Vol. 8: 15-24.

Shah, Esha. 2005. 'Local and Global Elites Join Hands: Development and Diffusion of Bt Cotton Technology in Gujarat'. *Economic and Political Weekly*, Vol. XL, No. 23: 4629-4639.

Shetty, P.K. 2004. 'Socio Ecological Implications of Pesticide Use in India'. *Economic and Political Weekly*, Vol. 39 (39): 261-67.

Singh, Gurdev and S.R. Asokan. 1997. 'Seed Industry: Problems and Prospects. In Bhupat Desai (ed.), *Agricultural Development Paradigm for the Ninth Plan under New Economic Environment*. New Delhi: Oxford & IBH Publishing Co. Pvt. Ltd., pp. 421-440.

Sreenivasa, S.R. 1992. 'Profile of Pesticide Industry and Import Policy on Technical Grade Pesticide'. In S.P. Seetharaman (ed.), *Agricultural Input Marketing*, New Delhi: Oxford IBH, pp.79-88.

Stone, G. 2007. 'The Birth and Death of Traditional Knowledge: Paradoxical Effects of Biotechnology in India'. In Charles McManis (ed.), *Biodiversity and the Law: Intellectual Property, Biotechnology and Traditional Knowledge*. Earthscan, pp. 207-238.

Tripp, R. 2001. 'Can Biotechnology Reach the Poor? The Adequacy of Information and Seed Delivery', *Food Policy*, 26 (2001): 249-264.

Tripp, R. 2009. *Biotechnology and Agricultural Development, Transgenic cotton, Rural Institutions and Resource Poor Farmers*. Oxon/New York: Routledge.

Tripp, R. and N. Louwaars. 1997. 'Seed Regulation: Choices on the Road to Reform'. *Food Policy*, Vol. 22, No. 5: 433-446.

Tripp, R. and S. Pal. 1998. 'Information Exchange in Commercial Seed Markets in Rajasthan'. *Agricultural Research and Extension Network (AGREN)*, Network Paper No. 83.

Viswanathan, P.K. and N. Lalitha. 2012. 'Economics, Environmental Compatibility and Social Responsibility of GM technology: Looking Beyond the Conventional Perspectives'. Paper presented in the International Seminar on Biotechnology in Indian Agriculture: Performance, Potential and Concerns, Centre for Economic and Social Studies, Hyderabad, January 18-19.

# 8

# Emerging Issues of Labour Use in Cotton Cultivation in Punjab

*Kamal Vatta* and *R.S. Sidhu*

## 1. Introduction

The success of green revolution brought significant changes in the cropping pattern, technology and policy in Punjab. Rising incomes and declining prices of capital, relative to that of labour, encouraged large scale mechanisation and brought structural changes in the crop sector employment. A decline of more than 25 per cent was observed in the use of human labour on Punjab farms during 1981–82 to 2005–06 (Sidhu *et al.* 2011). The decline was accompanied by increasing casualisation of labour and it adversely affected the livelihoods of the rural landless. Annually, a fall of 116 million man days of employment was observed in Punjab agriculture (*Ibid.*). Due to declining demand and a huge supply of labour, the real wages in agriculture remained stagnant for about a decade after the mid-1990s (Himanshu 2005). Distress prevailed amongst the rural poor comprising the old/disabled and those in need of productive employment at minimum wages to escape poverty (Hirway 2004). This resulted in the passing of the National Rural Employment Guarantee Act (NREGA) in 2005, providing guaranteed wage employment of 100 days in a financial year to every willing rural household. The implementation of NREGA has resulted in a significant reduction in the labour supply for important agricultural operations throughout India. For some operations, such as weeding and sowing, the reduction in supply has been to the extent of 30–53 per cent (Harish *et al.* 2011). There is a reduction in the area under labour-intensive crops due to shrinking labour supply (*Ibid.*).

Despite a severe decline in demand for labour in rice and wheat crops, there was no such decline in labour use in case of cotton in Punjab. This is due to the reason that while most of the other labour-intensive operations in rice and wheat (except the transplantation of rice) have been mechanised, the two major operations in cotton cultivation, i.e. cotton picking and hoeing, are still done manually. Punjab is not only a major food grain producing state with rice and wheat accounting for more than 80 per cent of the gross cropped area, but cotton crop is also grown on more than 6 per cent of the area and turns out to be an important crop of Punjab after wheat and rice in 2010-11 (GOP 2012). Hence, demand-supply dynamics in labour have significant implications for employment, wages and crop profitability.

The area under cotton, which peaked at 0.78 million ha in 1997-98, touched bottom at 0.45 million ha in 2003-04 with productivity levels of less than 450 kg/ha, owing to severe problem of cotton bollworm. A significant increase in the MSP (minimum support price) for rice also weaned away some area from cotton due to improvement in comparative profitability of rice (Figure 8.1). Due to the introduction of Bt cotton and its subsequent success, the area increased to 0.53 million ha with the total production of more than 2.29 million bales in 2008-09. The area under cotton has not been able to reach the earlier levels of about 0.78 million ha during 1997-98, though the production has increased by more than 75 per cent from 1.30 million bales to the current levels (GOP 2012).

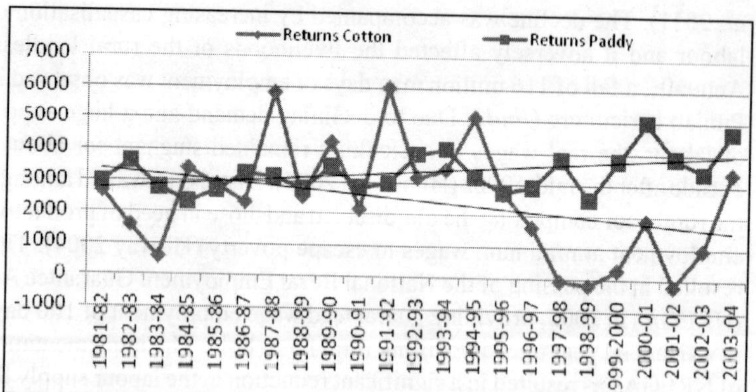


*Source:* Data taken from the published reports of Commission for Agricultural Costs and Prices.

**Figure 8.1:** Net Returns from Cotton and Paddy in Punjab (Rs./ha at 1981-82 prices)

It reflects that the gains in production of cotton over this period have come from increases in productivity which rose by more than 60 per cent (compared to 437 kg/ha in 2000-01) and reached to more than 743 kg/ha in 2008-09 as well as from the shift in area under *desi* cotton (to the extent of about 1 lakh ha) to Bt cotton during this period (*Ibid.*).

The agricultural wages in Punjab have grown comparatively faster over time (Sidhu *et al.* 2011) with the recent increase between January 2008 and December 2010 being more than 80 per cent (Labour Bureau, Shimla). The rise in wages coupled with decreasing supply of wage labour has serious implications for cotton cultivation in Punjab. Owing to the problem of over-exploitation of ground water resources, depleting soil health, increasing incidences of insect-pest attacks and falling profitability of the rice-wheat rotation, the state intends to diversify its cropping pattern and an increase in the area under cotton emerges as one of the potent options for such diversification (Johl 2002). However, the ever-increasing cost of labour use and subsequent squeezes in its supply may act as a deterrent to the future cotton production in the state. This chapter, thus, intends to examine the issue of labour use in cotton cultivation in the light of the rising costs and declining supply and highlights its implications for Punjab agriculture. The paper has been organised in eight sections. After introduction, the next section discusses the database and methodology used for this study. The third section highlights recent changes in the cost of cultivation while the fourth section provides an in-depth analysis of the labour use in cotton. The fifth and sixth sections provide the projections on the demand for labour use and farmers' perceptions on labour shortages, respectively. The seventh section highlights the shift of agricultural wage workers to the non-agricultural sector. The last section concludes with the policy implications.

## 2. Database and Methodology

This chapter is based on the primary data collected from 1000 rural households spread over four districts of Punjab, viz. Bathinda, Ferozepur, Sri Muktsar Sahib and Mansa, which account for almost 90 per cent of the area under cotton in Punjab. The survey was conducted during the year 2010-11 with the reference period of 2009-10. A total of 50 villages were selected from these four districts based on the proportion of cotton area under each district and 20 households were selected from each village for data collection. All the households in each selected village were classified into the landless, small (below 2 ha), medium (2 to 6 ha) and large (6 ha and above) cotton farming households based on the size of their operational area. The number of households to be selected from each farming category was based on the probability proportional to the size criterion, while the total number of landless and farming households in each village was selected in the ratio of 1:3, to give fair representation to each category of household. The survey covered 1000 households comprising 246 landless households having agricultural labour as major source of their livelihood, 135 small farmers (below 2 ha of operational area), 361 medium farmers (2-6 ha of operational area) and 258 large farmers (above 6 ha of operational area). Cotton appeared to be an important crop in the kharif season in the study area with 51.2 per cent of the net area sown under cotton crop. The data was collected by recall method for two points of time, i.e. 2004-05 and 2009-10.

The projections of demand for labour in cotton were made by multiplying the current labour use in cotton (per ha) with the area under cotton cultivation during 2009-10. The projections by operations were also made in the same manner. Also, the current distribution of operational holdings was taken into consideration for making the projections.

### 3. Recent Changes in the Cost of Cultivation of Cotton

At current prices, the operational cost of cultivation of cotton crop was Rs. 14,394 per acre in 2009-10 (Table 8.1), while it was about Rs. 8807 per acre in 2004-05 showing an increase of about 63 per cent. The cost of labour use more than doubled, while the increase in all the other items such as seed, machine use, manures and fertilisers, chemicals and irrigation showed less than overall average increase during this period. The proportion of labour use, in the total operational cost, jumped by almost 10 percentage points from 38.53 per cent to 47.72 per cent, during this period, while the proportion of all other items followed a decline. There is an increased proportion of labour cost even in real terms as has been pointed out earlier. It highlights that labour use is gaining significance in cotton cultivation in recent times. The rising proportion might be either due to increase in the labour use or rise in agricultural wages in relation to the prices of other inputs. While rise in productivity after the introduction of Bt cotton has increased the labour use in cotton, the wages have also increased significantly during this period giving a double push to the cost of labour use in cotton.

It comes out clearly that the operational costs (at constant 1980-81 prices) have increased stupendously by more than 33 per cent (Table 8.2).

**Table 8.1:** Recent Trends in the Cost of Cultivation of Cotton Crop in Punjab

*(Rs./ha)*

| *Particular* | *2004–05* | *2009–10* |
|---|---|---|
| 1. Seed | 1109.78 (13.60) | 1722.38 (11.97) |
| 2. Machine labour | 1073.86 (12.19) | 1564.77 (10.87) |
| 3. Manures and Fertilisers | 646.26 (7.34) | 1279.14 (8.89) |
| 4. Chemicals | 1890.05 (21.46) | 2080.10 (14.45) |
| 5. Irrigation | 398.88 (4.53) | 422.84 (2.94) |
| 6. Human labour | 3393.57 (38.53) | 6868.50 (47.72) |
| 7. Ot..ers | 79.48 (0.91) | 105.30 (0.72) |
| 8. Interest on working capital | 214.80 (2.44) | 351.08 (2.44) |
| Total | 8806.68 (100.00) | 14394.11 (100.00) |

*Note:* Figures in parentheses are percentages of the respective totals.
*Source:* Primary Survey, 2010-11.

**Table 8.2:** Changes in Labour Cost and Other Components of Cost of Cultivation in Recent Five Years in Punjab (at 1980-81 prices)

*(Rs./ha)*

| *Cost component* | *2004–05* | *2009–10* | *% change* |
|---|---|---|---|
| Labour cost | 888.15 | 1463.01 | 64.73 |
| Other costs | 1027.88 | 1085.78 | 5.63 |
| Total operational cost | 1916.03 | 2548.79 | 33.02 |

*Source:* Primary Survey, 2010-11.

Contrary to the previous trends, while the cost of labour increased by more than 65 per cent, the other components of cost of cultivation (taken together) grew only by 5.63 per cent during the last five years.

As there is no evidence of any significant reduction in labour use in cotton during the last five years as evidenced from the discussion with the sample farmers, sharp increases in wages are believed to have caused an abrupt rise in the labour cost. Such rise in real as well as monetary wages may be attributed largely to a sharp upward revision of wages under

NREGA activities in Punjab from Rs. 101 per day during 2005-06 to the current level of Rs. 166 (www.nrega.nic.in). The rise in wage rates was also facilitated by a sharp increase in the minimum support price (MSP) for wheat from Rs. 750/qtl to Rs. 1285/qtl and that of rice from Rs. 610/qtl to Rs. 1110/qtl during 2006-07 to 2010-11 and the market prices for cotton almost doubling during this period (GOI 2011). With the increase in the output prices, wage rate also tended to go up as the economic gains of agricultural labour are largely determined by the economic gains realised in the crop sector. There are some serious implications for such a trend, especially for cotton. The MSPs for rice and wheat as well as the agricultural wages are unlikely to fall down in the future keeping in mind the historical trends in output prices and regular upward revisions in the wages under NREGA. However, the cotton prices in all probability will not be able to retain such high peaks and will come down in the long-run, while nominal wages are not expected to fall in the future. It means, while gross returns from cotton cultivation may fall in the future, the labour costs may remain high in absolute as well as in proportionate terms, resulting in a decline in the net returns from the crop. In addition, this will alter the relative costs of different inputs and may lead to technological changes in order to change the input mix in a significant manner.

## 4. Labour Use in Cotton Cultivation

The study reveals that the overall labour use in cotton during the year 2009-10 was 127.80 labour days per acre (Table 8.3). The respective use on small, medium and large farms was 119.70 labour days, 127.50 labour days and 131.70 labour days, respectively. Almost 54-55 per cent of the total labour was used for the picking of cotton followed by manual weeding, which accounted for about 27 per cent of the total labour use. These two operations together accounted for more than 82 per cent of the total labour use. While activities such as land preparation, sowing and chemical sprays accounted for about 1-3 per cent of the labour use each, the share of irrigation and fertilisers hovered around 5-6 per cent. The pattern of labour use reveals that the possible demand-supply gap for labour use will arise only for the picking and weeding operations. Picking seems to be even more significant as compared to weeding as weeding operations are spread over a longer duration and require comparatively lesser labour than picking.

Further, component-wise examination of the labour use in cotton reveals that family labour accounted for about 31 per cent of the overall

**Table 8.3:** Pattern of Labour Use in Cotton across Different Farm Size Categories

*(Labour days/ha)*

| *Labour type* | *Small farms* | *Medium farms* | *Large farms* | *Overall farms* |
|---|---|---|---|---|
| 1. Land preparation | 2.43 (2.03) | 2.58 (2.01) | 3.10 (2.35) | 2.63 (2.05) |
| 2. Sowing | 0.75 (0.63) | 1.52 (1.19) | 1.20 (0.91) | 1.32 (1.04) |
| 3. Irrigation | 7.60 (6.35) | 8.40 (6.57) | 8.53 (6.48) | 8.33 (6.51) |
| 4. Chemical sprays | 3.97 (3.32) | 3.53 (2.75) | 4.00 (3.03) | 3.72 (2.91) |
| 5. Manures and fertilisers | 6.45 (5.39) | 6.17 (4.83) | 5.87 (4.47) | 6.13 (4.79) |
| 6. Manual weeding | 32.95 (27.52) | 35.00 (27.35) | 36.20 (27.48) | 35.07 (27.45) |
| 7. Picking | | | | |
| Male | 10.15 (8.48) | 7.50 (5.86) | 9.85 (7.48) | 8.72 (6.83) |
| Female | 55.40 (46.28) | 63.25 (49.44) | 62.95 (47.80) | 61.88 (48.42) |
| Total labour days (per acre) | 119.70 (100.00) | 127.95 (100.00) | 131.70 (100.00) | 127.80 (100.00) |

*Note:* Figures in parentheses are percentages of the respective totals.
*Source:* Primary Survey, 2010-11.

labour use, permanent labour accounted for about six per cent (hired usually for a season and paid monthly wages, mostly in cash), local wage labour for about 55 per cent while migrant labour accounted for about 7 per cent of the total labour use (Table 8.4). The dependence on family labour declined considerably with an increase in the farm size as the proportion of family labour in the total labour use was observed to be 45.88 per cent, 34.53 per cent and 20.44 per cent of the total labour use on the small, medium and large farms, respectively. This was so because the availability (supply) of family labour per unit area declined as farm size increased. The respective use of permanent labour was 1.10, 4.35 and 16.22 labour days accounting for 0.92 per cent, 4.03 per cent and 12.32 per cent of the total labour use across small, medium and large farms, respectively. The large farms showed comparatively higher dependence on both the types of casual labour, i.e. local labour and migrant labour from Uttar Pradesh, Bihar and Rajasthan for cotton cultivation. The proportion of casual labour use on the small, medium and large farms

**Table 8.4:** Different Components of Labour Use in Cotton

*(Labour days/ha)*

| *Labour type* | *Small farms* | *Medium farms* | *Large farms* | *Overall farms* |
|---|---|---|---|---|
| Family labour | 54.93 (45.88) | 44.17 (34.53) | 26.93 (20.44) | 40.03 (31.32) |
| Permanent labour | 1.10 (0.92) | 4.35 (3.40) | 16.22 (12.32) | 7.87 (6.16) |
| Local wage labour | 60.60 (50.63) | 71.43 (55.82) | 75.23 (57.12) | 70.90 (55.48) |
| Migrant wage labour | 3.07 (2.57) | 8.00 (6.25) | 13.32 (10.12) | 9.00 (7.04) |
| Total | 119.70 (100.00) | 127.95 (100.00) | 131.70 (100.00) | 127.90 (100.00) |

*Note:* Figures in parentheses are percentages of the respective totals.
*Source:* Primary Survey, 2010-11.

**Table 8.5:** Extent of Male and Female Labour Use in Cotton Picking

*(Labour days/ha)*

| *Labour Use* | *Male* | *Female* | *Total* |
|---|---|---|---|
| Small Farms | 9.92 (15.15) | 55.63 (84.85) | 65.55 (100.00) |
| Medium Farms | 7.50 (10.60) | 63.25 (89.40) | 70.75 (100.00) |
| Large Farms | 9.85 (13.54) | 62.95 (86.46) | 72.80 (100.00) |
| Overall | 8.67 (12.32) | 61.93 (87.68) | 70.60 (100.00) |

*Note:* Figures in parentheses are percentages of the respective totals.
*Source:* Primary Survey, 2010-11.

was more than 53 per cent, 62 per cent and 67 per cent, respectively. The respective proportions of local casual labour were 50.63 per cent, 55.82 per cent and 55.48 per cent and that of migrant labour were 2.57 per cent, 6.25 per cent and 10.12 per cent of the total labour use on these farms.

Further, female labour is used only during the picking of cotton and accounts for almost 88 per cent of the total labour days during the picking operation (Table 8.5). NREGA is expected to squeeze the supply of labour force (both male and female) to agriculture and thus there is a need to keep such aspects in mind while generating future demand and supply scenarios for agricultural labour in cotton to understand the labour dynamics properly. As NREGA is devised to provide employment of the last resort, this is expected to benefit the female labour force more and will surely reduce the female labour supply for agriculture in the long-run.

## 5. Projections for the Demand for Labour in Cotton Cultivation

The analysis of the pattern of labour use in cotton across different farm size categories reveals that large farms are the most vulnerable farming category followed by the medium farms with respect to the widening demand-supply gap for labour. The estimates of the labour use in cotton were prepared by multiplying area under cotton cultivation in 2009-10 with the labour use in cotton (per ha). The total area under cotton (5.11 lakh ha) was further classified into different size categories based on the proportion of different holdings in the study sample. Based on the information collected through the survey and the pattern of distribution of operational holdings in the study area, the total demand for labour for cotton cultivation in Punjab during 2009-10 was 66.4 million labour days. Looking into the pattern of land distribution across various size categories, the large farms accounted for about 66 per cent of the total labour use in cotton followed by about 28.5 per cent share of the medium farms and just 5.5 per cent share of the small farms.

Table 8.6 provides the split of labour use in different activities in cotton cultivation in Punjab as explained above. Large farms account for 66 per cent of the labour use during manual weeding and their share in male and female labour during cotton picking was 69.72 per cent and 65.40 per cent, respectively. Put together, medium and large farms account for about 95 per cent of the total labour use in most of the activities of cotton cultivation.

Labour use across different kinds of labour (family, permanent and casual) was also explored and is presented in Table 8.7. While 67 per cent of the local wage labour in cotton is used by the large farms, the source of labour for a vast majority of farms (78%) is migrant labour. This analysis reveals that the medium and large farms are the major labour users in cotton and are most likely to be affected if the demand-supply scenario for labour in cotton worsens further. Any squeeze in the labour supply or an increase in the labour demand due to rise in area under cotton (owing to the attractive price rise in cotton) will adversely affect the returns of the large landholders and the wage rates are likely to rise further in near future.

**Table 8.6:** Pattern of Labour Use in Cotton across Different Farm Size Categories

*(Labour days in thousands)*

| Labour type | Small farms | Medium farms | Large farms | Overall farms |
|---|---|---|---|---|
| Land preparation | 74.14 | 381.24 | 1028.30 | 1483.68 |
| | (5.00) | (25.69) | (69.31) | (100.00) |
| Sowing | 23.01 | 225.71 | 398.19 | 6469.11 |
| | (3.56) | (34.89) | (61.55) | (100.00) |
| Irrigation | 231.91 | 1246.15 | 2835.48 | 4313.54 |
| | (5.38) | (28.89) | (65.73) | (100.00) |
| Chemical sprays | 121.25 | 521.60 | 1325.85 | 1968.70 |
| | (6.16) | (26.49) | (67.35) | (100.00) |
| Manures and fertilisers | 196.85 | 916.12 | 1955.95 | 3068.92 |
| | (6.41) | (29.85) | (63.74) | (100.00) |
| Manual weeding | 1005.04 | 5187.56 | 12024.51 | 18217.11 |
| | (5.52) | (28.48) | (66.00) | (100.00) |
| Picking | | | | |
| Male | 309.68 | 1111.48 | 3273.05 | 4694.21 |
| | (6.60) | (23.68) | (69.72) | (100.00) |
| Female | 1690.17 | 9377.45 | 20916.00 | 31983.62 |
| | (5.28) | (29.32) | (65.40) | (100.00) |
| Total labour days | 3652.05 | 18967.31 | 43757.33 | 66376.69 |
| | (5.50) | (28.58) | (65.92) | (100.00) |

*Note:* Figures in parentheses are percentages of the respective totals in rows.
*Source:* Primary Survey, 2010-11.

**Table 8.7:** Different Components of Labour Use in Cotton in Punjab on Different Size Groups of Sample Farms

*(Labour days in thousands)*

| Labour type | Small farms | Medium farms | Large farms | Overall farms |
|---|---|---|---|---|
| Family labour | 1675.56 | 6549.41 | 8944.00 | 17168.97 |
| | (9.76) | (38.15) | (52.09) | (100.00) |
| Permanent labour | 33.60 | 644.89 | 5390.90 | 6069.39 |
| | (0.55) | (10.63) | (88.82) | (100 00) |
| Local wage labour | 1049.03 | 10587.55 | 24994.18 | 37430.76 |
| | (4.94) | (28.29) | (66.77) | (100.00) |
| Migrant wage labour | 93.86 | 1185.46 | 4428.25 | 5707.57 |
| | (1.64) | (20.77) | (77.59) | (100.00) |
| Total | 3652.05 | 18967.31 | 43757.33 | 66376.69 |
| | (5.50) | (28.58) | (65.92) | (100.00) |

*Note:* Figures in parentheses are percentages of the respective totals.
*Source:* Primary Survey, 2010-11.

adopted also promotes formalisation of the market. For instance, Nuziveedu Seeds Private Limited (with popular brands like Bunny and Mallika) which ranks 3rd among the top 20 biotech companies and first among agribiotech companies in India, has got approval for 164 cotton hybrids, a marketing network spread across 17 states, involving 2500 distributors and 59,000 dealers. It has expanded the marketing network by co-partnering with KRIBHCO (Krishak Bharati Co-operative Limited) and Tata Chemicals. Through this network, they reach out to a total of 4 million farmers in India (*Biospectrum* 2012).

It was evident from the surveys that almost 73 and 78 per cent of the seeds have been brought by paying cash in both the years. Interestingly, though in 2011, 12 per cent of the farmers said that they were persuaded by the dealers to buy a particular seed only 3 per cent of them actually bought the same seed, implying that farmers make an informed choice while choosing the Bt seed.

To register any complaint against the seed company in case of seed (germination) failure, farmers require a proper bill from the seed sellers to confirm their purchase. In 2007, because of the presence of the huge unorganised seed segment, 60 per cent of the farmers said that they did not get proper bill on their purchase. In 2010-11, however, despite 90 per cent of the seed purchases have been sourced from the organised sector, only 63 per cent reported that they received proper bill proving their purchase. The farmers also do not get any concessions while buying the seeds.

### *4.3 Seed Selection*

As presented in Table 7.3, there was an increase in seed quantity and cost of seeds across size classes during 2010-11 as compared to 2007-08. In 2007, while 46 and 27 per cent of the farmers said to have adopted the variety after knowing about it from their neighbours and seed dealers, in 2011, varieties have been chosen based on a combination of factors, like the high yielding nature of the seed (28.7%), own experience (20.7%) and neighbour farmers' experience (13%).

Tripp (2001) gives examples from Africa and Pakistan, where even if farmers have been incurring crop loss due to crop diseases and pests, they may not be appreciating the differences in varietal susceptibility. This is evident from our study as well, when farmers listed the characteristics of the variety that they were aware of when they first bought that variety, the yield performance of the seed was mentioned by many (45 and 40% of the responses) while the resistance to pests which is the trait of the Bt was mentioned only by 21 and 5 per cent of the farmers in both the years. Hence, it may be surmised that while selecting the variety for the first time, farmers appear to give maximum weightage to the yield characteristics. For instance, in 2007, 73 per cent of the farmers' response indicated that they chose the variety because of their high yielding nature. The resistance to bollworm offered by the Bt technology, got only 4.5 per cent of the response. It should be, however, noted here that, the insect resistance nature of Bt seed also helps in improving the yield performance. More interestingly, having seen the variety in someone else's field got less than 1 per cent response, though a few of them got to know about the seed from their neighbours. Hence, it could be said that the yield performance factor was given larger weightage by the farmers.

Further, it appears as farmers are exposed to more varieties, more factors have been considered to choose a variety due mainly to the increased farming risks as time progressed. This is very similar to the experience of the Iowa farmers with corn hybrids, who even if the experience of the 'community laboratory' was available to them, would insist on their personal experimentation (Ryan and Gross 1943: 15-24). In this case, we find that farmers weigh several factors that are important to them and then adopt a variety. Thus in 2011, factors like high yielding nature of the variety (29%), own experience with the variety (21%), having seen the variety in fellow farmers' field (13%), suitability of the variety to the land (13.6%) and bigger bolls (7%) have been considered by the farmer in selecting the variety. It emerges that the sample farmers seed decisions fits in the 'Bayesian model whereby producers improve their prior beliefs on the basis of observed performance and thus are inclined to increase the share of the modern technology over time' (Feder *et al.* 1985: 264).

### 4.4 Bt Adoption: Experiment or Imitation?

Agricultural adoption frameworks discuss the influence of experimentation and imitation type of factors behind (Ryan and Gross 1943, Stone 2007) new agricultural technologies, especially, seed varieties. As there were hardly few farmers with less than five years of experience with cotton generally and had also adopted Bt cotton on a wider scale, we wanted to ascertain whether farmers have continued to cultivate Bt or not as time pass by. There were few farmers who had not cultivated Bt cotton continuously. However, this break was not due to Bt technology related issues, as farmers reported that they discontinued Bt cultivation for a particular season alone due to the non-availability of seed, higher cost of seeds and fertilizer and lack of irrigation. On the other hand, in 2007, out of the 200 farmers, only 3.5 per cent of the farmers said that they had a break in Bt cultivation in favour of other crops. These farmers discontinued Bt cultivation due to the non-availability of seed, higher cost of seeds and fertiliser and lack of irrigation. In 2010-11, 8.5 per cent of the farmers who had reduced the area under Bt cotton, cited reasons such as crop rotation, mealy bug attack and reduced cotton yield for discontinuing cotton. Further, in 2007-08 while majority had cultivated Bollgard I variety, around 20 farmers had begun Bollgard II cultivation and in 2010-11, almost all the farmers have adopted Bollgard II. Hence, in order to understand this transformation in the two time points, we looked at the land allocation for different varieties over a period of time.

If farmers continuously use the same seed varieties for more than one year, for how many seasons this trend is continued and how do they try to evaluate the pay offs? We found some evidence to this effect in our data (Table 7.5).

Of the 136 farmers who have continued one or more varieties for at least two years, we find that 50 per cent of them have tried one variety for two years and 34.6 per cent have tried two varieties for two years. As the number of varieties increase year after year, the percentage of farmers who would continue the variety decreases. Only a few farmers have continued with 1 variety for 7or 8 years, which indicates that the farmers have fixed a particular variety after initial years of trials. Thus, this data clearly demonstrates that most Bt farmers take decision to continue or discontinue a particular variety after trying it for a minimum of two or three years. This in other words suggests that as the number of varieties increases, the chance of them getting continuously adopted for a longer time also decreases. Perhaps, longer trials with a variety are done with maximum one variety which has gone up to 8 years in our study, and only 6 farmers in the entire sample have done so. It is also evident that larger numbers of farmers use a variety for two or three years and then discontinue it, which appears to be a normal phenomenon with the farmers.

As observed by Qaim (2005: 1317-1324) that 'disadoption' and

**Table 7.5:** Pattern in Continued Adoption of Specific Varieties among Farmers (%)

| *Seed variety* | *Continuity of the seed over the 8 years of Bt adoption* | | | | | | | |
|---|---|---|---|---|---|---|---|---|
| | *2 yr* | *3 yr* | *4 yr* | *5 yr* | *6 yr* | *7 yr* | *8 yr* | *(%) of farmers reporting discontinuous use* |
| 1 Particular seed | 50.0 | 62.4 | 90.4 | 88.9 | 100.0 | 100.0 | 100.0 | 82.7 |
| 2 Particular seeds | 34.6 | 30.7 | 5.8 | 11.1 | Tried other varieties | | | 12.3 |
| 3 Particular seeds | 7.4 | 4.0 | 1.9 | Tried other varieties | | | | 4.9 |
| 4 & above particular seeds | 8.1 | 3.0 | 1.9 | Tried other varieties | | | | 0.0 |
| Total Farmers (#) | 136 | 101 | 52 | 36 | 9 | 5 | 6 | 81 |

*Note:* The blank cells indicate that farmers have chosen newer varieties.
*Source*: Farm Household Survey, 2007-08 and 2010-11.

'readoption' decision of the seed technology are not irreversible and are part of the normal learning process for the farmers, we also observed that 67 farmers have disadopted and readopted a single variety over the course of time. Further, the readoption trend is also not uniform with farmers who have tried more than three varieties.

We do not find a consistent trend with increase or decrease of land under that particular variety and so does the yield. Except for one farmer who had just only one plot of land all the other five farmers had more than one plot of land under cotton at some/most point of the study period.

### *4.5 Varietal Continuity: Trend Among Size Groups*

The probe into the varietal continuity trend among different size groups of the farmers provides interesting insights on the adoption behaviour of the farmers. It points out that when farmers are adopting newer varieties, they undertake a calculated risk. This is evident from the fact that a certain percent of total land is maintained under familiar varieties to ensure, continuous returns to farming. Overall, the varieties that were common to both the years gained a total of 26 hectares in 2010-11. Both small and large farmers have reduced the land under these varieties in 2010-11 and the decline is sharper for the large farmers. This decline also means that both small and large farmers experiment more with newer varieties and are willing to take more risk with newer varieties. In contrast, the medium farmers have continued with about 40 per cent of the varieties in 2010-11 and also increased the land under new varieties.

**Table 7.6:** Varietal Continuity among Size Groups

| *Size class* | *% land in common variety to total land* | | *% of the number of common varieties to total number of varieties* | |
|---|---|---|---|---|
| | *2007-08* | *2010-11* | *2007-08* | *2010-11* |
| Small | 27 | 22 | 13 | 12 |
| Medium | 40 | 53 | 42 | 43 |
| Large | 24 | 22 | 17 | 17 |
| All | 47.5 | 58 | 50 | 52 |

*Source*: Farm Household Survey, 2007-08 and 2010-11.

What is the reason for this behaviour among the farmers? The evidence lies partially with the difference in the expected and realized yield in 2007-08, where farmers reported realising only 60 per cent of the expected yield. Among the size groups, while the small and medium size groups realized a yield of just above 60 per cent in 2010-11, the farmers from large size groups obtained only 57 per cent of the expected yield. This gap between the actual and realised yield was also reported across all size classes. While environmental factors could also have a role in this, the main reason attributed for the reduction in the yield is due to the infestation of mealy bugs (32%) and white fly (20%). About 19 per cent of the responses also indicated falling of the bolls due to rain.

Given the relatively lower performance of the yield of the variety chosen, the farmers were asked in (2007-08 survey) to report their possibility of continuing the same variety in the next season. While 45 per cent said they would continue with the same variety, 36 per cent said no (discontinue) and 18.5 per cent reported that they were not decided. About 74.5 per cent of the farmers also reported that they would allocate the same amount of land for cotton in the next season as well. Farmers cited reasons such as high cost of production, diseases, mealy bug infestation and higher cost of fertiliser and pesticides that they had incurred for the chosen variety for switching over to a newer variety. Interestingly, of these set of farmers who reported that they would switch-over to new varieties (in 2007-08 survey), 23 per cent observed that they would go for a legal variety, while 7.5 per cent said they would adopt illegal Bt and a minor segment of 3 per cent aspired to grow *desi* variety.

So, if such considerations have influenced the varietal choice in 2010-11 which had led to growing newer varieties in 42 per cent of the land with almost 78 per cent of land allocated for the same by the large and small farmers, it may be worth exploring 'if these decisions had yielded significant outcomes in the yield for the different size classes?'

Overall, yield has increased by 27 per cent and we find that the large farmers have benefited the most by yield increase compared to the small and medium farmers (Table 7.7). The probable reasons for this yield difference could be that, the newer varieties chosen by the large farmers are better suited for their land as compared to the small farmers' agro ecological conditions, relative endowment status as well as differences in the use of inputs like irrigation and fertilisers. In fact, Kranthi (2011) argues that the proliferation of varieties have lead to decline in the yield in the recent years as farmers use them without knowing the suitability of the seed to their land. This emphasizes the need to look for ways to give

**Table 7.7:** Yield Differences in the Size Class

| *Yield (Kg/ha)* | *Small* | *Medium* | *Large* | *Total* |
|---|---|---|---|---|
| 2007-08 | 2139.8 | 2175.7 | 1241.1 | 1710.0 |
| 2010-11 | 2268.2 | 2248.3 | 2054.1 | 2176.7 |
| % increase | 6 | 3 | 65 | 27 |

*Source*: Farm Household Survey, GIDR 2007–08 and 2010–11.

more information to the farmers about the critical inputs as well as the agro-ecological suitability of the varieties. One probable reason for the relatively lower yield could be the presence of the seeds from the unorganised sector. As the certified seeds are appropriately stored and packed, there would be less chance for quality deterioration. Seed quality deteriorates if stored in very high temperature or at very high moisture conditions (Meena *et al.* 1999: 740-42), hence it may be inferred that seeds that are bought from fellow farmers, and that most often do not stick to the standard packing, could yield less. That said, the reporting errors on yield from the field also cannot be ruled out in leading to such outcomes.

## 5. Pesticide Use: Information Seeking and Farmer Practices

Cotton cultivation is susceptible to a number of pests which attack the plant throughout the plant life. Since the mid 1990s, pests in cotton were not controlled even after 30 sprays (Shetty 2004: 5261-67). Hence, pesticide consumption in cotton in India had been traditionally higher, which suggests that, higher or less intensive use of modern inputs depends on whether the modern inputs are risk reducing or risk increasing and whether relative risk aversion is increasing or decreasing (cited in Feder *et al.* 1985: 261) with the increased adoption of the new technology. In this regard, Bt technology, which offers partial solution to the pests issue, has been credited with reducing the pesticide consumption (Table 7.1) which has also been confirmed in a few empirical studies mentioned earlier. As pesticides create as many problems as they solve (Stone 2007: 213), the focus of this analysis is to understand the factors considered by farmers before they spray and whether such considerations actually demonstrate a decline in the pesticides used.

Almost 98 per cent of the farmers in both 2007 and 2011 mentioned the names of the pesticides that they had used in their Bt cotton farms.

Most often these names were the names of the brand or local (as it is known to the farmers) or actual ingredient. But, unlike the seed purchases, here, while 20 per cent of the farmers mentioned the names of the pesticide companies of all the pesticides that they had used in 2007, 37 per cent could mention some names and 43 per cent were not aware of any name of the pesticides. In 2011, while 69.5 per cent could mention the names of the companies of all the pesticides that they had used, 24 per cent could mention some and only 6.5 per cent could not name one.

As per the Insecticides Rules, 1971, insecticides have been categorised into extremely toxic, highly toxic, moderately toxic and slightly toxic categories. These are indicated by bright red, yellow, blue and green colours. Each packet will have to bear a triangle filled with the appropriate colour indicating toxicity. To understand if the farmers had any knowledge in this regard, we had verified the farmer awareness about the hazard indications in colours, as printed on the pesticide packets. In 2011, 37 per cent (53% in 2007) of the farmers mentioned that they had seen the colour indicator on the pesticide cover. Of the farmers who had noticed the colour in 2007 and 2011, 15 and 1.3 per cent did not know what the colour meant. However, it appears that compared to 2007, where farmers have used red (36%) and yellow (5.5%) indicated pesticides, in 2011, 9 and 16 per cent of the farmers have used such pesticides respectively. Compared to the just 1.5 per cent who have used the moderately hazardous blue category pesticides in 2007, 10 per cent farmers have used pesticides with blue indicator in 2011. Hardly 2 and 1 per cent farmers have bought slightly toxic insecticides bearing the green colour in 2007 and 2011 respectively.

It is a common practice among the farmers to use some of the insecticides as a nutrient. A totally different scenario emerges regarding the micro nutrients used by the farmers. In 2011, as many as 88 per cent did not know the name of the nutrient that they had used which is much higher than the 52.5 per cent of the farmers who did not know the names of nutrients that they had used in 2007.

Interestingly, the reason for the selection of a particular chemical when they bought for the first time (in 2007-08) had been based on the recommendation of the dealers (57%), followed by the experience of the neighbour farmers (29%). Now, farmers go to the shop to buy the product that they had already decided (94%). Also, except for 11 per cent of the farmers who said they purchased from that particular shop for the first time, a majority reported having had a continued association with the same shop with 52.5 per cent buying pesticides for the past 2-5 years, followed by 6-10 years as reported by 21.5 per cent and more than 10 years in case of 13.5 per cent of the farmers who have been procuring their agri inputs from the same shop for 2-5 years, 6-10 years and more than 10 years respectively.

Unlike seeds, where nearly 75 per cent is bought on cash payment, in the case of pesticides, while 50 per cent paid cash and purchased the product, 26 per cent bought the product on credit and 9 per cent used both cash and credit to buy the product. About 31 per cent of farmers who have purchased their inputs on credit also mentioned of paying interest on credit, while 60 per cent of the farmers who availed credit did not pay interest. Hence, the decision to charge interest is perhaps based on the number of years the farmer has been purchasing the inputs from the shop, the past credit history and the number of products bought on credit.

### *5.1 Intensity of Pest Pressure Before and After Spraying*

Farmers reported about the targeted pest against which they had sprayed during both the periods. In 2007, 66 and 30 per cent of the sprays were targeted for various sucking pests and bollworms and 3 per cent were for different diseases of the plants. But from the farmers' response, it was clear that the infestation of bollworms have become almost nil in 2011, as hardly 1 per cent of the spray was aimed at control of bollworms, while 71 per cent of the sprays were targeted at sucking pests. Interestingly, 18 of the sprays in 2011 were also used as a growth regulator, nutrient and to improve boll formation. A small number of chemicals used in the sprays were also used as weedicide, fungicide and to restrict the disease condition of the plant.

As majority of the farmers do not follow any integrated pest management (IPM) practices, how does one decide to spray insecticides? It was evident from the survey that farmers assess the pest scenario before spraying as well as after spraying to see the effectiveness of pesticides spraying. Almost all the farmers had visited the field in both the years (94.5 and 96.5%) and ascertained the need to spray pesticides before actually doing so (Table 7.8), whereby it is clear that the need to spray either for sucking pests or bollworms was known to the farmers.

**Table 7.8:** Intensity of the Pests Before and After Spray (%)

| *Intensity of pests* | *Before spray* | | *After spray* | |
|---|---|---|---|---|
| | *2007–08* | *2010–11* | *2007–08* | *2010–11* |
| High | 61.7 | 45.7 | 15.38 | 1.1 |
| Medium | 23.4 | 35.7 | 13.09 | 33.3 |
| Low | 13.6 | 17.5 | 43.75 | 62.2 |
| Don't know | 0.8 | 1.1 | 1.80 | 0.7 |
| No pests | 0.6 | — | 1.12 | — |
| Total | 100.0 | 100.0 | 75.13 | 97.2 |

*Source*: Farm Household Survey, 2007-08 and 2010-11.

Farmers' response indicates that before spraying the intensity of the pests have been high in 2007-08 (62%) and relatively less in 2010-11 (46%). Responses indicating medium and low intensity of the pests have been higher in 2010-11, indicating that perhaps there has been an overall reduction in the pest intensity in the year 2010-11. Interestingly all the sprays were reported to be targeted at variety of the sucking pests including the rising menace of mealy bugs. Though farmers spray the pesticides to prevent boll shedding, the primary indicators that farmers consider while deciding on spraying is the curling up of the leaves (78 and 80%), followed by damages to bolls and shoots that lead to boll shedding.

Once the spraying is done, farmers also assess the effectiveness of the pesticides by the visible freshness of the plant (72 and 57%). Interestingly while 18 per cent of the responses indicated the limited impact of the pesticides in 2007-08, 38.4 per cent of the responses in 2010-11 indicated visible reduction in the pests. Only 4 per cent of the responses in 2007-08 mentioned that the pesticides were ineffective, which in 2010-11 became as low as 0.2 per cent.

Have farmers been experiencing the emergence of any newer pests affecting cotton? To this question 60 per cent of the farmers' responses during 2007-08 indicated that they had seen some new pests which they had not seen earlier; surprisingly, this proportion went up to 85 per cent during 2010-11. Almost 77 per cent of the farmers in 2007 pointed out that they had not seen these newer pests before five years which indicates possible emergence of secondary pests. Interestingly, the proportion of farmer responses indicating mealybugs as the newer pest in 2010-11 was 84 per cent which was only 47 during 2007-08. This, in a way, could explain the increasing agonies as shared by the farmers leading to an increase in the use of pesticides in the post-Bt scenario as also discussed in Chapter 9 of this volume.

Does that mean that 'Bt technology while controlling the bollworm increases the secondary pests infestation?' Scientists do think so. The probable causes for the infestation of sucking pests could be: monocropping of cotton over wide area, indiscriminate use of pesticides, climatic variability, increase in soil moisture in irrigated tract, reduction in the number of sprays because of Bt technology during 45-120 days which may aid mealy bug and lack of information on new and minor pests (Kumar *et al.* 2010).

### *5.2 Pesticide Spraying Practices*

We observed that use of hired labour to spray pesticides was quite common in both the years (46 and 51%), followed by spraying by the self (40 and 37%). Use of family labour (other than the respondent) was reported by very few farmers.

Enquiries regarding the safe spraying practices adopted if any by the farmers indicated that such instances are very hard to come by. That is, 44 and 29 per cent of the farmers in 2007-08 and 2010-11 reported that they do not follow any precautions while spraying pesticides.

Of the remaining who took precautions, the most common safety measure was covering the nose and mouth with a piece of cloth (45 and 42% of the responses). The traditional wisdom of spraying in the opposite wind direction however is not practised by many. Similarly, using appliances with long nozzle also targets the spray on the specific area which again is reported by very few. Further, a simple precaution of washing the hands with soap after spraying has been reported to have been practiced by very small number of farmers (4 and 10%), all of which indicates the poor awareness regarding the harmfulness of the pesticides among the farmers. This is a serious problem, especially when we consider the fact a large number of the insecticide products being used by the farmers are highly or moderately hazardous and their exposure poses several health related issues.

Interestingly, preference to buy original products was reported by 2.5 per cent of the responses in the year 2007–08 alone, which perhaps come from the experience of using bogus products. General resistance of the pests to some of the chemicals, however, cannot be ruled out here. Importantly, though the farmers have reported this to the dealers, they were not aware of any action taken against their complaint.

Does that reduce the faith of the farmers on the companies that produce pesticides? Not really so, as 62 per cent of the farmers said there are trusted companies, in which companies like Bayer, Tata and Excel figured prominently.

One way of preventing harmful exposure of pesticides to humans is to adopt safe re-entry periods in the fields where pesticides have been sprayed. However, we found that majority (44%) of the farmers re-entered the field after one day, while 39 per cent re-entered fields after two days and 14 per cent re-entered the field after three days in 2011, which signals probable health risks for the farmers in the future.

Interestingly, despite the large ignorance of safety principles followed by the farmers, only 8 and 4 per cent of the farmers reported experiencing health problems like nausea, eye irritation and skin problems after spraying pesticides. This may be interpreted in such a way that most farmers are unable to relate the health related problems with the intensity of pesticide use practices that they follow. However, validation of this would need further empirical investigations.

### *5.3 Refuge Practices*

In order to derive the best out of the Bt technology and to delay the resistance to the pests, scientists recommend adoption of refuge practices in the Bt cotton plot, as detailed in Chapter 1. But in practice, refuge adoption has been low and the farmer surveys reveal that the refuge practice remains unknown to the farmers, even after a decade of introduction of the Bt technology (Table 7.9).

**Table 7.9:** Educational Level of Farmers with Refuge Practices (%)

| | *Farmer's education 2007–08 (n=36)* | | | | *Farmer's education 2010–11 (n=34)* | | | |
|---|---|---|---|---|---|---|---|---|
| | *Primary* | *Upto secondary* | *College* | *Illiterate* | *Primary* | *Upto secondary* | *College* | *Illiterate* |
| Small | 25.0 | 50.0 | 0.0 | 25.0 | 50.0 | 50.0 | 0.0 | 0.0 |
| Medium | 18.2 | 59.1 | 13.6 | 9.1 | 15.8 | 63.2 | 15.8 | 5.3 |
| Large | 16.7 | 83.3 | 0.0 | 0.0 | 9.1 | 72.7 | 18.2 | 0.0 |
| Total | 19.4 | 61.1 | 8.3 | 11.1 | 17.6 | 64.7 | 14.7 | 2.9 |

*Source*: Farm Household Survey, 2007-08 and 2010-11.

Table 7.9 shows that the extent of refuge adoption has been very low at 18 (36 of 200 farmers) and 17 per cent (34 of 200 farmers) respectively during 2007-08 and 2010-11. We note that the medium size class has

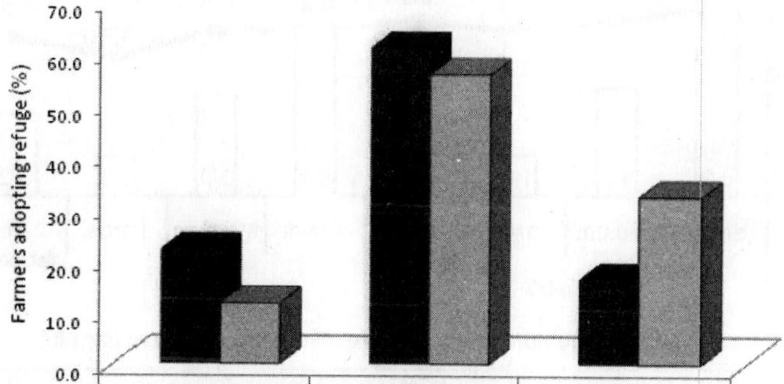


| | Small | Medium | Large |
|---|---|---|---|
| ■ 2007-08 | 22.2 | 61.1 | 16.7 |
| ■ 2010-11 | 11.8 | 55.9 | 32.4 |

*Source*: Farm Household Survey, GIDR 2007-8 and 2010-11.

**Figure 7.1:** Adoption of Refuge Practices by Size-class, 2007-08 and 2010-11

adopted refuge practice more than other farmers in both the years and the proportion of large farmers adopting refuge has doubled between the two years (Figure 7.1). Citing the research done in the context of Thailand farmers, Feder *et al.* (1985: 277), report that education affects the probability of adoption positively only above a threshold level of four years. We also observed in our data that refuge adoption was more among the farmers with education upto secondary level. Also, we note that unlike the lack of compliance with refuge practices among the illiterate and primary educated farmers in 2011, farmers with secondary education seem to follow refuge practices and a few more college educated farmers have certainly adopted refuge practice.

In 2007-08, (and 2010-11) small farmers accounted for 21 per cent (25) of the plots, 9 per cent (11.9) of the land and 23 per cent (24.9) of the sprays. Medium farmers accounted for 53 per cent (53) of plots and (56) sprays and 41per cent (50) of land. Large farmers had a share of 26 per cent (22) for the plot, 49 per cent (39) of land and 24 per cent (19) of the spray.[12]

Comparison of the spray schedule in 2007-08 and 2010-11 reveal that the number of sprays has declined for all size classes in 2010-11. It is evident that majority of the sample farmers have sprayed pesticides and only less than 1 per cent of the total responses indicated no spraying in both the years. While for the entire sample, there is no difference in

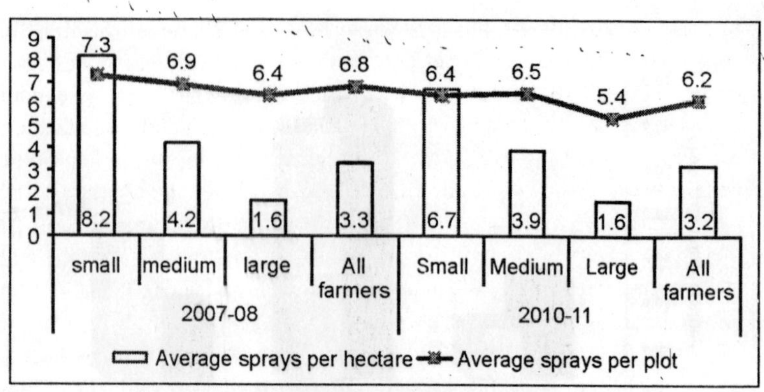


*Source*: Farm Household Survey, GIDR 2007-08 and 2010-11.

**Figure 7.2:** Average Number of Sprays per Plot and Hectare

**Table 7.10:** Spray Pattern among the Farmers

| *Sprays (no.)* | *Small* | | *Medium* | | *Large* | | *All farmers* | |
|---|---|---|---|---|---|---|---|---|
| | *2007-08* | *2010-11* | *2007-08* | *2010-11* | *2007-08* | *2010-11* | *2007-08* | *2010-11* |
| 1 | 13.1 | 16.4 | 14.2 | 15.0 | 14.5 | 17.6 | 14.0 | 15.9 |
| 2 | 26.2 | 32.0 | 28.2 | 29.8 | 28.4 | 35.0 | 27.8 | 31.4 |
| 3 | 39.3 | 46.2 | 41.8 | 44.4 | 42.3 | 51.9 | 41.3 | 46.4 |
| 4 | 52.2 | 59.7 | 54.2 | 58.5 | 54.6 | 67.8 | 53.8 | 60.7 |
| 5 | 64.1 | 72.1 | 65.0 | 71.3 | 66.0 | 80.8 | 65.0 | 73.4 |
| 6 | 74.9 | 81.7 | 73.9 | 81.2 | 76.4 | 90.4 | 74.7 | 83.2 |
| 7 | 83.5 | 89.0 | 81.5 | 89.2 | 84.2 | 96.0 | 82.6 | 90.6 |
| 8 | 90.3 | 94.2 | 87.4 | 94.9 | 90.3 | 99.6 | 88.7 | 95.8 |
| 9 | 94.8 | 96.5 | 91.0 | 96.6 | 93.9 | 99.8 | 92.5 | 97.4 |
| 10 | 96.6 | 97.3 | 93.0 | 97.5 | 95.9 | | 94.5 | 98.1 |
| Maximum sprays reported (no.) | 20 | 15 | 14 | 20 | 13 | 13 | 9 | 13 |
| Total no. of sprays | 649 | 1512 | 690 | 2851 | 614 | 1374 | 478 | 2466 |
| % to total number of sprays | 22.8 | 53.0 | 24.2 | 100 | 24.9 | 55.7 | 19.4 | 100 |
| Average sprays per plot | 7.3 | 6.9 | 6.4 | 6.8 | 6.4 | 6.5 | 5.4 | 6.2 |
| Average sprays per hectare | 8.2 | 4.2 | 1.6 | 3.3 | 6.7 | 3.9 | 1.6 | 3.2 |

*Note:* Figures pertaining to sprays 1-10 are cumulative percentages.
*Source*: Farm Household Survey, GIDR 2007-08 and 2010-11.

the average sprays per hectare (3.3 and 3.2) in both the years, sprays per plot have only declined slightly from 6.8 to 6.2 in 2010-11 (Figure 7.2). Though for all size classes, the total sprays have declined in 2010-11, only for large size class farmers, there is a reduction by nearly 5 per cent points. Also in 2010-11, for all size groups, we find that though the number of sprays has declined, the spraying has been intensive as the first five sprays cover more than 70 per cent of the total sprays, which was about 60 per cent in 2010-11 (Table 7.10). In 2007-08, though the spray schedule has extended to 20, 15 and 14 for the small, medium and large farmers, we find that the first five spray schedules account for more than 60 per cent of the total sprays for all the size classes. While the average sprays per hectare have reduced in 2010-11 for small and medium size holders, there is no change in the case of large farmers and they sprayed the least at 1.6 sprays per hectare. What are the reasons for this? One of the reasons is that large farmers had 16 per cent of their land under *desi* cotton in 2007-08 and sprayed totally 24 times. In 2010-11, *desi* accounted for 31 per cent of their land and sprayed just six times. It is known that desi varieties are resistant to many pests and require least number of sprays. Also, the relatively larger number of sprays per hectare (and also per plot) in the small size group require attention. Perhaps it follows the established fact that small farmers farm more intensively and use more resources. We also looked for clues for this in the pesticide spray behaviour in the varieties that were common in both years. Overall, the common varieties in 2007-08 and 2010–11 accounted for 55 and 68 per cent of the total sprays. For the small, medium, and large farmers the percentage of sprays on common varieties to total sprays on all varieties in 2007-08 and 2010-11 have been 22 per cent (both years); 42 and 58 per cent, and 29 and 28 per cent respectively. Thus, for the small and large farmers, the newer varieties account for more sprays, which again emphasises that information regarding the timing of the spray, could lead to reduction in the sprays.

We also looked at the spray pattern through the plant life of the different size groups to understand if there is higher or lower intensity of the pesticides sprays by any size groups (Table 7.11). A comparison of the spray pattern by days after sowing in both the years suggest that the intensity of the sprays had been by and large the same with 85 per cent of the sprays occurring between 31-120 days with relatively lower number of sprays after 91-120 days in all the size groups. Reductions have happened in sprays after 120 days in 2010-11 compared to 2007-08. In 2007-08, a fewer number of farmers have sprayed beyond 13 sprays while in 2010-11 farmers have sprayed maximum 13 sprays, of course, with differing intensities across different groups.

The explanation by Kumar *et al.* (2010) that reduction in the sprays between 45-120 days as one of the causes for increase in the mealy bug is perhaps refuted as evidenced by the farmers spraying behaviour. This

**Table 7.11:** Pattern of Pesticide Sprays Adopted by the Farmers, 2007-08 and 2010-11

| *Size class/ year* | *Spray pattern (days after sowing)* | | | | | | |
|---|---|---|---|---|---|---|---|
| | *1–30* | *31–60* | *61–90* | *91–120* | *121–150* | *151–180* | *181 above* |
| Small | | | | | | | |
| 2007–08 | 6.20 | 27.75 | 35.97 | 24.19 | 5.89 | 0 | 0.00 |
| 2010–11 | 9.14 | 36.22 | 33.61 | 15.66 | 3.10 | 0.82 | 1.47 |
| Medium | | | | | | | |
| 2007–08 | 5.11 | 30.41 | 33.07 | 20.12 | 6.97 | 2.86 | 1.46 |
| 2010–11 | 9.36 | 34.94 | 32.53 | 17.84 | 4.97 | 0.37 | 0.00 |
| Large | | | | | | | |
| 2007–08 | 4.69 | 26.98 | 33.87 | 23.46 | 8.80 | 2.20 | 0.00 |
| 2010–11 | 8.39 | 34.59 | 34.17 | 17.82 | 4.40 | 0.63 | 0.00 |
| All size classes | | | | | | | |
| 2007–08 | 5.3 | 29.0 | 33.9 | 21.8 | 7.2 | 2.0 | 0.8 |
| 2010–11 | 9.1 | 35.2 | 33.1 | 17.3 | 4.4 | 0.5 | 0.4 |

*Source*: Farm Household Survey, GIDR 2007-08 and 2010-11.

would be a concern as the spraying trend as observed above depicts the farmer spraying behaviour with Bollgard 1 variety (Lalitha and Ramaswami 2007: 77-98). Also, the practice of mixing more than one chemical for spraying which is often resisted by entomologists has been commonly observed with the Gujarat farmers (Lalitha and Ramaswami 2007; Lalitha and Viswanathan 2010: 11-35). For whatever reasons, there has been a drastic change in the adoption from Bollgard 1 to Bollgard II in Gujarat as well. An important issue here is that whether this spraying behaviour as reported by the farmers will reduce the life of the Bollgard II variety? If it happens so, what is next choice for the farmers?'

## 6. Bt Technology: Overall Performance

Has the lower use of pesticide impacted the pesticide cost per hectare? Not really so as presented in Table 7.3. This is because, it has been observed that since the introduction of Bt cotton, while there has been a reduction in the volume of pesticide used, the price of pesticide products has been increasing (Kranthi 2011). This trend suggests that farmers are perhaps shifting to costlier pesticides and this switch over to costlier pesticides could also be due to the fact that they are under IPR protection. In all probability, this would mean that seed cum pesticide dealers try to compensate for the losses that they incur in the event of the decline in

It needs to be mentioned that before the American Bollworm problem in cotton became uncontrolled, the area under cotton in Punjab reached the highest level of 0.78 million ha during 1997-98 and then reached its all time low of 0.45 million ha during 2002-03. After the introduction of Bt cotton, the area showed a slight increase to 0.53 million ha in 2008-09. We can think of the scenarios of the area reaching (i) halfway between the current area 0.51 million ha and the highest level of 0.78 million ha and (ii) the highest level of 0.78 million ha achieved so far. If any of these scenarios is realised, it will happen by replacing the area under rice crop. Labour use in rice is just about half that of cotton and thus we can say that there will be an additional demand for about 8 million and 16 million labour days in the first and second scenarios, respectively.

## 6. Farmers' Perceptions on Labour Shortages in Cotton Cultivation

It was important to explore farmers' viewpoint about the impending labour shortage, its impact and mitigation strategies. Table 8.8 reveals the

**Table 8.8:** Proportion of Farmers Expressing Existence of Labour Shortage during Cotton Cultivation in Punjab

*(Per cent)*

| Operation | Small Farms | Medium Farms | Large Farms | Overall |
|---|---|---|---|---|
| Land preparation | – | – | 0.40<br>(1) | 0.13<br>(1) |
| Sowing | – | – | – | – |
| Irrigation | – | – | – | – |
| Chemical Sprays | – | – | 0.40<br>(1) | 0.13<br>(1) |
| Manures and Fertilizers | 3.70<br>(5) | 8.31<br>(30) | 8.53<br>(22) | 7.56<br>(57) |
| Manual Weeding | 17.04<br>(23) | 24.65<br>(89) | 39.92<br>(103) | 28.51<br>(215) |
| Cotton Picking | 25.93<br>(35) | 35.46<br>(128) | 55.04<br>(142) | 40.45<br>(305) |

*Note:* The proportions are calculated on the basis of the response of all the 754 farmers belonging to different size categories. Figures in parentheses are the number of farmers reporting the shortage.

*Source:* Primary Survey, 2010-11.

proportion of farmers of different categories expressing the existence of labour shortage for various operations during the cultivation of cotton. Manual weeding and picking of cotton are the two most labour intensive operations where the farmers expressed the fear of major labour shortages. Overall, more than 40 per cent of the farmers experienced labour shortage during picking of cotton and about 29 per cent farmers expressed labour shortage during manual weeding in cotton. Medium and large farmers experienced greater labour shortages during picking and manual weeding, which strengthens the earlier argument of greater vulnerability of such farm types to the labour shortages.

The extent of labour shortage during these operations is presented in Table 8.9. The estimates of the shortages are based on the responses of the farmers from the primary survey. Overall, the farmers face labour shortages to the extent of about 34 per cent during picking and about 26 per cent during manual weeding, followed by about 12 per cent for the application of manures and fertilisers during cotton cultivation. The extent of labour shortage increased with an increase in the farm size. Once again, it reflects relatively more vulnerability of the medium and large farmers with respect to labour shortages.

**Table 8.9:** Average Extent of Labour Shortage during Cotton Cultivation in Punjab

*(% of total labour required)*

| *Operation* | *Small Farms* | *Medium Farms* | *Large Farms* | *Overall* |
|---|---|---|---|---|
| Manures and Fertilizers | 12.27 | 13.23 | 13.76 | 12.63 |
| Manual Weeding | 20.00 | 26.59 | 27.17 | 26.23 |
| Cotton Picking | 30.65 | 32.20 | 35.15 | 33.88 |

*Note:* Those farmers who responded positively to the labour shortages (their numbers are given in Table 8.8), were asked further questions on the extent of labour shortages—they reported the shortages in the intervals of below 20%, 20-40%, 40-60%, 60-80% and above 80% of the required labour. The estimates of the table are the weighted averages of those responses. The estimates for other operations are not given as almost no labour shortage was experienced.

*Source:* Primary Survey, 2010-11.

Multiple reasons were cited for the shortage of labour in cotton. Almost 95 per cent of the farmers facing labour shortage cited that workers were not ready to work at the current wages and demanded higher than the current wages, causing labour shortage (Table 8.10). More than 86 per cent of them pointed out the shift of workforce from farm to non-farm works as an important reason for labour shortage. The introduction of NREGA in the country has also been cited by 68 per cent of the farmers for reduction in the supply of wage labour for agriculture. Though there does not seem to be a clear significant impact of NREGA on causing labour shortages, the programme has actually helped in raising the agricultural wages as NREGA wages are being used as benchmarks in deciding the wage rate in agriculture in recent times in Punjab. Other issues such as shortage of migrant labour or appropriate chemicals of good quality appeared less important to the farmers as they were relying heavily on local labour for cotton cultivation. Most of the farmers attributed recent sharp increases in the cost of cultivation of cotton to the sharp increase in wage rates during this time.

**Table 8.10:** Farmers' Perceptions about Important Reasons for Labour Shortages in Cotton

*(% multiple response)*

| *Reasons for labour shortage* | *% response* |
|---|---|
| Shift of the workforce from farm to non-farm work | 86 |
| Unwillingness of the workforce to work at current wage rates | 95 |
| Shortage of migrant labour | 16 |
| Non-availability of appropriate and good quality chemicals | 6 |
| Labour shortage due to NREGA | 68 |

*Note:* The percentages have been calculated from only those farmers who reported labour shortages (the numbers are given in Table 8.8). Some of the farmers expressed that availability of appropriate chemicals/weedicide which can effectively control the weeds in cotton at later stages of growth can help in addressing the issue of labour shortage during weeding operations.

*Source:* Primary Survey, 2010-11.

In order to overcome the labour shortage, more than 45 per cent of the farmers paid higher wages, about 13 per cent farmers tended to use more of their family labour and about 35 per cent farmers sourced labour from outside the village (Table 8.11). Hiring labour from outside is also a costly proposition as the farmer has to arrange for the transportation of the workers to and from the farm. About six per cent of the farmers reported delay in their farm operations in order to overcome the labour shortages. Other measures such as use of machines, labour exchange between the

**Table 8.11:** Important Adaptation Strategies for Labour Shortage across Various Farm Size Categories in Punjab

*(% response)*

| *Mitigation Strategy* | *Small Farms* | *Medium Farms* | *Large Farms* | *Overall* |
|---|---|---|---|---|
| Provide higher wages | 34.55 | 42.90 | 54.45 | 45.46 |
| More use of family labour | 17.23 | 14.21 | 7.92 | 12.56 |
| Bring labour from outside | 31.03 | 36.73 | 32.87 | 34.47 |
| Use of machines | – | 0.80 | 0.79 | 0.66 |
| Delaying of the farm operations | 13.75 | 4.56 | 3.97 | 5.88 |
| Labour exchanges with other farmers | 3.44 | – | – | 0.57 |
| Higher use of chemicals | – | 0.80 | – | 0.40 |

*Note:* The percentages have been calculated from only those farmers who reported labour shortages.

*Source:* Primary Survey, 2010-11.

farm families and higher use of chemicals do not seem to be the preferred options amongst the cotton cultivators. It is due to the non-availability of viable mechanisation options for the operations and lack of any chemical options for controlling weeds at later growth stages. Also, due to majority of the cotton cultivators belonging to the medium and large farm holdings, there is little surplus family labour during peak periods which can be considered for the exchange option. Farmers were observed to have followed more than one strategy against the labour shortage, though we have reported only the most prominent strategies. Comparatively higher proportion of large and medium farmers were resorting to payment of higher wages and bringing labour from outside the village as compared to their smaller counterparts. As paying higher wages was more expensive, the small farmers were resorting relatively more to the greater use of family labour, delaying of the farm operations and labour exchanges, etc. These strategies seemed less feasible on the medium and large farms where the total labour requirements were comparatively much higher.

## 7. Shift of Agricultural Workers towards the Non-agricultural Sector

The landless labour households reported a sizable shift of the workforce from agriculture to non-agriculture sector during 2004-05 to 2009-10 (Table 8.12). Most of this shift is partial in nature as the workers have started shifting between wage work in the non-farm sector as well as in the farm sector depending on the availability of work. Currently, almost 85 per cent of the workers amongst the sample wage working households

**Table 8.12:** Perceptions on Increase in Non-agricultural Employment and Its Reasons

| *Reason* | *% response* |
|---|---|
| Shift of the workers towards non-agricultural employment in the last 5 years (full and partial) | 78.4 |
| **Major reasons for shift to non-agricultural work*** | |
| Declining employment opportunities in agriculture | 72.0 |
| Comparatively higher wages in non-agricultural work | 70.9 |
| Opportunities for regular employment in the non-farm sector | 56.1 |
| More employment opportunities in the non-farm sector | 56.1 |
| Impact of NREGA | 39.7 |
| Lack of interest to work in agricultural sector | 34.4 |
| Impact of government policies | 4.2 |
| Delay in wage payment in agriculture | 1.1 |

*Note:** The proportions for different reasons are multiple and have been estimated from only those who observed the shift to non-agricultural sector.

*Source:* Primary Survey, 2010-11.

were engaged in agriculture with their proportion declining by 6 percentage points since 2004-05. However, it should be kept in mind that almost all the wage workers reported their partial shift to the non-farm wage activities owing to comparatively lesser employment opportunities in agriculture. There was a variety of reasons for such shift from agriculture to the non-agriculture sector. The two most important reasons are declining employment opportunities in agriculture and comparatively higher wages being paid in the non-agricultural work as reported by 72 per cent and 71 per cent respectively of the landless labour households. About 56 per cent of these households find availability of more employment opportunities and more regular employment in the non-farm sector for such a shift. Almost 40 per cent of the households believe that NREGA has facilitated the shift of agricultural workforce towards the non-farm sector, while 34.4 per cent households cite lack of interest in agriculture of the youth as an important reason for the shift.

In nutshell, decline in the employment opportunities in agriculture, comparatively more employment opportunities in the non-farm sector with higher wages coupled with some impact of NREGA and the preferences of youth for non-farm jobs are some of the important reasons for causing a shift in the workforce from the farm to the non-farm sector. However, such a shift is not permanent for many workers, as non-agricultural sector is also largely distress driven. Currently, the landless labourers do not find any issues in labour supply to the agricultural sector (Table 8.13). Only about 13-14 per cent of the households recognize some labour shortage for manual weeding and picking of the cotton. They opined that labour shortages appear only when higher wages are sought by labour (in comparison to NREGA wages), which the cultivators are not willing to pay. It is, however, important to note that these households perceive such shortage to increase in the future. As preferences of the young workers lie mostly in the non-agricultural work, more and more of such shifts in the labour force is expected in the near future. Such a shift is likely to create shortage of casual labour in agriculture which will push the wages up. It should be kept in mind that the agricultural wages have almost doubled (from Rs. 70 to Rs. 150 per day) during the last five years. The shortage of labour force in the long run will have major implications for cotton cultivation and the resulting rise in wages may lead to mechanisation of cotton picking operations for which practically no mechanical options are available. The same is true for options of chemical weeding, when cotton plants attain three feet height.

**Table 8.13:** Proportion of Agricultural Workers Witnessing Any Shortage of Labour in Cotton Cultivation

| *Operation* | *% witnessing shortage of labour* | *Major reasons for shortage* |
|---|---|---|
| Land preparation | 4.2 | Higher wages |
| Sowing | 3.7 | Higher wages |
| Irrigation | 3.7 | Higher wages |
| Chemical sprays | 5.8 | Higher wages |
| Manures and fertilisers | 5.0 | Higher wages |
| Manual weeding | 12.9 | Higher wages (71%), shift to non-farm work (16%), impact of NREGA (13%), lack of effective weedicides (10%) |
| Picking of cotton | 14.1 | Higher wages (67%), shift to non-farm work (18%), impact of NREGA (12%) |

*Note:* The proportions in the second column have been estimated using the responses from 246 landless wage working households. Figures in parentheses are percentages of the workers expressing the shortage.

*Source:* Primary Survey, 2010-11.

The local landless labour households perceive some decline in the use of migrant labour for different operations of cotton due to the little decline in the inflow of migrant labour owing to the implementation of

NREGA in other states of India as well. About 50 per cent of the households perceived a decline in the use of migrant labour for chemical sprays, manual weeding, manures and fertiliser application and other operations in cotton, while only about 21 per cent of the local labour households agreed to the decline in migrant labour use for picking of cotton which may be due to the reason that picking is largely carried out by local female wage workers, though there is a decline in the inflow of workers from the adjoining villages and the neighbouring state of Rajasthan (Table 8.14). Picking of cotton is generally done by the local labour and that too by the females. Despite some decline in the migrant labour use, there does not seem to be a corresponding equivalent increase in the local labour. It may be another reason for a rise in agricultural wages as 80 to 99 per cent of wage working households cited labour shortage as the major reason for increase in wages.

**Table 8.14:** Perceptions of Local Agricultural Labourers on Labour Use and Wage Rates

| *Activity* | *% response (N=246)* | | |
|---|---|---|---|
| | *Incidence of decline in the use of migrant labour in cotton* | *Increase in local labour use in the event of decline in migrant labour* | *Incidence of increase in the daily wages due to labour shortage* |
| Chemical spray | 54.9 | 20.0 | 98.8 |
| Manual weeding | 48.8 | 23.4 | 79.1 |
| Manures and fertilisers | 53.7 | 16.7 | 88.6 |
| Picking of cotton | 20.9 | 31.3 | 86.7 |
| Others | 54.1 | 13.8 | 96.4 |

*Source:* Primary Survey, 2010-11.

## 8. Conclusion

Labour is an important component of the cost of cultivation of cotton with its share in the total cost of cultivation being almost 48 per cent. The share has followed an increase over time. While the other components of cost of cultivation registered a small increase during 2004-05 to 2009-10, the cost of labour use increased by about 65 per cent. Such an increase has happened largely due to a sharp increase in the wage rates during this period. On an average, the labour use in cotton amounted to 127.80 labour days per ha. Cotton picking and manual weeding emerged as two major components of labour use in cotton with the respective shares of 55 per cent and 27 per cent. An increased dependence on causal labour was witnessed with an increase in the farm size due to decline in the use of family labour.

In 2009-10, there was a demand for 66.4 million labour days in cotton and medium and large farms accounted for almost 95 per cent of this demand. It is believed that the medium and large farms are more vulnerable to the rising cost of labour use in cotton. It is also evident that these farms experience more labour shortages during the peak labour demand periods. The extent of such shortages varies between 20 and 35 per cent for manual weeding and cotton picking. The farmers perceived the shift of workforce to non-farm work, unwillingness of the labour to work at current prices and successful implementation of NREGA in the country as three important factors for labour shortages in cotton. Such shortages compel the farmers to pay higher wages.

Finally, the current situation of labour use and supply in cotton points towards the possibility of further rise in wage rates in the future. As the medium and large farmers are more vulnerable to labour shortages, the rise in wages will adversely affect their cost of cultivation. Such a trend may adversely impact the net returns from cotton cultivation as rise in cotton prices may fail to match the increase in wage rates. It may result in a significant long-term change in the pattern of labour use in cotton and may also encourage the mechanisation of important labour-intensive operations such as cotton picking and weeding.

In a scenario, where the supply enhancement of wage labour can be achieved only by increasing the wage rate with a consequent increase in the cost of cultivation of cotton, there is a potential to look for mechanisation options for picking of the cotton. As the medium and large farmers are the most affected segment of cotton farmers by the demand-supply mismatch of cotton, the probability of acceptance of such options will be quite high. Further, research into some chemicals, which can control the weeds at the later stage of growth of cotton and bring down the demand for manual labour for inter-culture operations, is another option. However, the implications of such mechanisation and chemical use will have severe issues of labour displacement from agriculture and will adversely affect the livelihoods of the poor landless households which still derive a significant proportion of their income from agricultural wages.

## References

GOI. 2011, *Agricultural Statistics At A Glance,* Directorate of Economics and Statistics. Ministry of Agriculture, Government of India, New Delhi.

GOP. 2012. *Statistical Abstract of Punjab,* Economic Advisor, Government of Punjab, Chandigarh.

Harish, B.G., N. Nagaraj, M.G. Chandrakanth, P.S. Srikantha Murthy, P.G. Chengappa and G. Basavaraj. 2011. 'Impacts and Implications of MGNREGA on Labour Supply and Income Generation for Agriculture in Central Dry Zone of Karnataka'. *Agricultural Economics Research Review,* Conference Number, Vol. 24: 485-94.

Himanshu. 2005. 'Wages in Rural India: Sources, Trends and Comparability. *Indian Journal of Labour Economics*, Vol. 48, No. 2: 375-406.

Hirway, Indira. 2004. 'Providing Employment Guarantee in India: Some Critical Issues'. *Economic and Political Weekly,* Vol. 39 (48): 5117-24.

Johl, S.S. 2002. *Agricultural Production Pattern Adjustment Programme in Punjab for Productivity and Growth,* Report of the Chief Minister's Advisory Committee on Agriculture Policy and Restructuring Adjustments, Government of Punjab.

Sidhu, R.S., Kamal Vatta and Jasdev Singh. 2011. 'Demand for Labour and Wages in Punjab Agriculture'. *Agricultural Economics Research Review,* Vol. 24: 549 (summary), Conference Number.

# 9

# Wheeling the Bt Technology on the IPM Cart

## A Case for Environmentally Sustainable Crop Protection Practices for Cotton in India

*P.K. Viswanathan* and *N. Lalitha*

### 1. Introduction

There is no denying of the fact that adoption of Bt technology has been one of the remarkable developments that India had witnessed in its history of agrarian transformation over the past 5-6 decades. Though mainly confined to the single crop of cotton, the diffusion and adoption of Bt technology in India has been so rapid and widespread that the extent of its adoption has almost reached peak levels in most of the traditional cotton growing regions/states in less than a decade's time[1]. As the voluminous literature as described in the foregoing chapters suggest, the rapid adoption of Bt technology in India and elsewhere gets legitimised and authenticated by virtue of its economic performance measured in terms of savings in yield arising from the control of crop (plant) damages caused mainly by the major pest, i.e. American Bollworm[2]. Since control of crop damage caused mainly by bollworm has been considered as the primary attribute of the technology, the larger claims of the 'incremental yield effect' could be considered as the 'incidental benefit' or the secondary attribute of the technology. Nevertheless, one could argue that the cotton sector in India has been able to overcome the historically persistent problem of 'low yield trap' (100-200 kg/ha) thanks to the commercialisation of Bt technology.

Given that the wider adoption of Bt technology had primarily resulted in a significant reduction (to the extent of 40-70% as reported by several studies) in the pesticide use across cotton growing countries[3], including India, it may be logically right to argue that the technology has immense potential in terms of providing sustainable environmental management outcomes. This argument holds good on the ground that historically, cotton cultivation has been subjected to maximum exposure to hazardous pesticides than any other crop across cotton growing countries of the world, especially, India[4]. Hence, the benefit of reduction in pesticide use as provided by the Bt technology alone renders it the status of 'environment or eco-friendly technology', with huge potential and scope for improving the state of agro-ecological environments where cotton is grown, as well as protecting the health of the farm workers and farmers engaged in the application of pesticides.

While a large number of empirical studies bring about the economic benefits realised from the adoption of Bt technology in India and elsewhere in terms of: (a) reduction in pesticide use[5]; (b) the resultant savings in crop output (yield gain); (c) income gains and the subsequent poverty reduction and welfare gains, there are hardly a few studies (for instance, Qaim and Janvry 2005; Subramanian and Qaim 2010; Baffes, 2011; Kouser and Qaim 2012; Ashok *et al.* 2012; Ashok *et al.* (Chapter 4, this volume) that deal with the environmental as well as health related positive outcomes of the technology in case of cotton. In a more recent study in Pakistan, examining the environmental and the health related benefits, Kouser and Qaim (2012) have derived economic values of the positive externalities arising from cultivating Bt cotton to the tune of US$ 79 per acre. While half of the total benefits (US$ 79) were attributed to health benefits the remaining was reckoned as benefits from environmental improvements due to a reduction in the pesticide use.

Nevertheless, many of these studies examining the impact of Bt technology are severely constrained by several problems, like, selection bias, limited coverage of sample farms, absence of long-term assessment/ monitoring of the social, environmental as well as health benefits, etc. More importantly, the long term impacts of the technology as well as its sustainable environmental management outcomes are contingent upon the agro-management practices being adopted by the farmers under diverse socio-economic and ecological as well as environmental settings. At the same time, it is also important to consider that it is too early to arrive at any conclusive evidences about the social and environmental impact of Bt technology which has an adoption history of hardly 10 years by now.

Incidentally, an important aspect that received less attention in the entire debate of Bt adoption is the sustainable environmental management outcomes of the technology as it has been deemed essential that Bt adoption should be integrated with the scientific farm management practices governed by Integrated Pest Management (IPM) and Insecticide Resistance Management (IRM) principles. Apparently, adoption of Bt with proper integration of the IPM/IRM practices will provide sustainable environmental management outcomes in multiple ways in terms of: (a) identifying appropriate insecticides so as to delay resistance; (b) effective control of target pests while allowing the beneficial pests to remain intact; and (c) conserving naturally occurring biological control for enhanced sustainability of ecosystems and environment. Moreover, it has also been reported that with the recent introduction of Bt cotton, novel eco-friendly pesticides and IRM strategies coupled with the trends in technology dissemination through area-wide farmer participatory approaches and farmer field schools, IPM programmes have become successful and sustainable in several country contexts (Kranthi 2007).

In this regard, the Monsanto Company (the monopoly provider of the technology in India) considers the integration of IPM/IRM practices as an important strategy to be followed by the farmers adopting Bt cotton and Bt corn[6]. It has been reported that growers of Monsanto Bt corn and cotton have found them to be highly compatible with their goals of IPM and sustainable agriculture with considerable environmental benefits arising from reduced insecticide use in particular. Growers could enhance the sustainability of their cotton and corn agriculture systems by following recommended IPM practices, including cultural and biological control measures, and appropriate use of pest thresholds and sampling. These measures are not only considered as important for non-Bt refuge acres, but are equally important for detecting and controlling non-target pests that exceed established thresholds on Bt crops. Given the socio-economic and environmental benefits due to integrating Bt adoption with IPM practices, the IPM checklist recommended by the Monsanto also stipulate: (a) employing appropriate scouting techniques and treatment decisions; (b) selecting insecticide treatments that have minimal negative impact on beneficial insects; (c) selecting cultivars well-adapted to the farm setting, giving appropriate attention to impact of crop maturity and timing of harvest on pest severity; (d) using recommended cultural methods to reduce pest over-wintering; and (e) destroy crop promptly after harvest and use other soil management practices to reduce overwintering insects (Monsanto 2013; Manjunath 2012).

There are also several evidences of increased socio-economic and environmental benefits (Atreya *et al.* 2011; Peshin 2012; van den Berg and Jiggins 2007) following the adoption of IPM as an environmentally sustainable crop protection strategy in case of cotton as well as other crops. The IPM approach has been promoted as an eco-friendly strategy of pest containment by exploiting the role of natural agents/forces in harmony with other pest management tactics and with the sole aim to effect minimum disturbance to the environment.

However, though the criticality of integration of IPM/IRM practices along with Bt adoption has been widely recognised and even stipulated as the essential pre-condition for adoption of the Bt technology, it emerges that the farmer adoption practices reveals interesting contradictions that need some reflections in the Indian context based on empirical data. An important question that needs probing in the Indian context in the event of the widespread adoption of the Bt technology in the cotton crop is that 'whether the farmers have been able to make a proper use of the technology integrating the IPM/IRM practices with potential implications with respect to sustainable environmental management outcomes?'

In this backdrop, this chapter critically reviews the dynamics of development and diffusion of Bt technology in India in the broader theoretical perspective of technology induced sustainable environmental management outcomes as arising from the proper integration of IPM/IRM practices while growing Bt cotton. In doing so, the chapter draws insights from the empirical studies on the comparative performance of farms adopting Bt cotton along with IPM/IRM practices *vis-à-vis* farms not adopting the same due to a host of constraints. To illustrate the lack of farm level compliance to the scientific farm management practices (IPM/IRM), the chapter also uses empirical data describing the farm level dynamics from the states of Gujarat and Maharashtra. While attempting this analysis, we expect that the adoption of Bt technology along with IPM/IRM practices by farmers in India might result in better payoffs with significant improvements (through learning from experience) in farming practices leading to sustainable environmental management outcomes. Based on the empirical analysis of the emerging farm management practices that are environmentally less sustainable, we bring out a strong case for a radical transformation in the existing cotton farming system in India in the advent of the Bt technology with its potential environmental outcomes when followed in compliance with the IPM/IRM practices. For conceptual clarity, we consider sustainable environmental management outcomes as specifically related to the judicious use of the technology with respect to sustainable crop protection and farm management practices such as use of critical inputs, including pesticides as well as the refugia practices along with integration of IPM/IRM practices.

The chapter is organised into four sections. Section 2 makes a critical review on the relevance of the use of the Bt technology with proper integration of the IPM/IRM principles among the major cotton growing states, viz. Andhra Pradesh, Punjab and Maharashtra in the post-Bt scenario. The discussion in the chapter assumes relevance in the context of the apparent contradictions in the adoption of Bt technology in India *vis-à-vis* other major cotton growing countries, such as the US, where adoption of Bt technology has been strictly regulated with respect to compliance of IPM/IRM as well as refuge management practices. Section 3 examines the dynamics of farmer adoption of Bt technology in the context of Gujarat as revealed by the empirical surveys. Section 4 concludes the chapter with some policy prescriptions for achieving sustainable environmental management outcomes in order to sustain the dynamism cast by the wide-scale adoption of the Bt technology among the Indian states.

To further the discussions, this chapter follows a conceptual framework (modified after Belis-Bergouignan *et al.* 2004) of 'technology induced sustainable environmental management outcomes' wherein Bt technology is expected to offer as a potential instrument for ameliorating the environmental as well as ecosystem damages supposed to have been caused by the conventional cotton farming practices (T1) involving the intensive use of agro-chemicals, especially, hazardous pesticides in containing the insects/pests as well as numerous plant diseases caused (Figure 9.1).

To elaborate the framework (Figure 9.1), we consider that the introduction of Bt technology leads to a new paradigm (T2) of farm resource use in which the environmental benefits are much higher in view of the reduction in pesticide use. The reduced pesticide use while enabling the farms to attain productive efficiency (PE1) also leads to better environmental performance (EP1). As per this, the conventional cotton farming is being considered as the old paradigm (T1) where, cotton production is characterised by the use of heavy dosage of agro-chemicals, use of conventional and probably low yielding seed varieties, indiscriminate use of other farm inputs, all of which result in poor environmental performance as well as low productive efficiencies at the farm level.

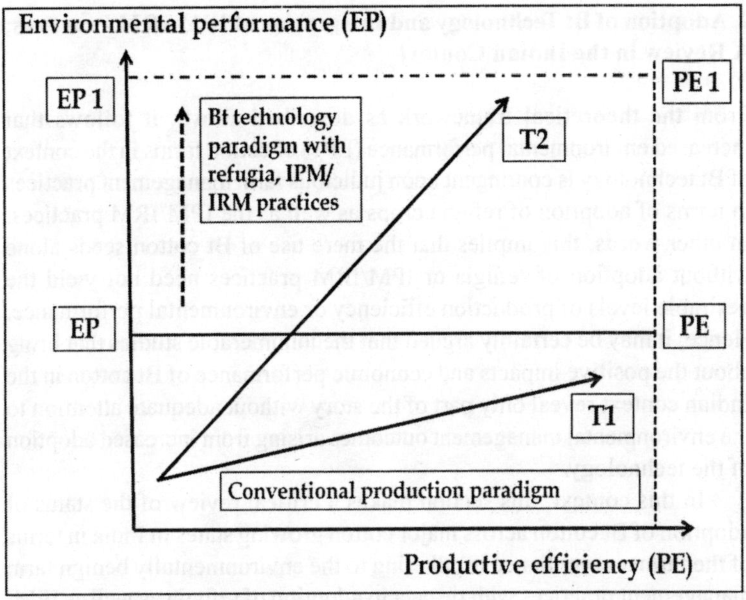


*Source*: Modified after Belis-Bergouignan *et al.* 2004.

**Figure 9.1:** Bt Technology and Environmental Performance: A Conceptual Framework

The frontier (horizontal line) of this first paradigm (T1) corresponds to the maximum level of environmental performance (EP) that can be reached with the conventional cotton farming system due to several inherent farm as well as technology related constraints. The horizontal line further signifies that, in order to exceed the level of environmental performance under the conventional farming regime (T1), farms have to adopt the new regime (T2) of Bt technology along with much improved farm management practices, including judicious use of agro-chemicals, adoption of refugia as well as IPM/IRM practices to achieve optimum levels of productive efficiency and environmental performance. In short, using this conceptual framework, we argue that theoretically, a shift from conventional cotton production system (T1) to the new paradigm of Bt technology adoption (T2) might turn out to be the best way of increasing farm production efficiency (PE1) along with high environmental performance (EP1) achieved through (a) reduced use of agro-chemicals (insecticides), (b) adoption of proper system of refugia, and (c) IPM/IRM practices in the post-Bt scenario.

## 2. Adoption of Bt Technology and Integration of IPM/IRM Practices: A Review in the Indian Context

From the theoretical framework as described above, it follows that increased environmental performance (EP1) of cotton farms in the context of Bt technology is contingent upon judicious farm management practices in terms of adoption of refugia crops as well as the IPM/IRM practices. In other words, this implies that the mere use of Bt cotton seeds alone without adoption of refugia or IPM/IRM practices need not yield the desirable levels of production efficiency or environmental performance. Hence, it may be certainly argued that the innumerable studies that bring about the positive impacts and economic performance of Bt cotton in the Indian context reveal only part of the story without adequate attention to the environmental management outcomes arising from increased adoption of the technology.

In this context, this section makes a critical review of the status of adoption of Bt cotton across major cotton growing states in India in terms of the farmer compliance in adhering to the environmentally benign farm management practices with respect to adoption of refugia as well as IPM/IRM practices. This section relies on the empirical literature across the major cotton growing states of Maharashtra, Andhra Pradesh, Gujarat and Punjab. Before examining the status of adoption of the refuge as well as IPM/IRM practices by Bt cotton growers in India, it may be quite contextual to explain how critical are these practices in achieving sustainable environmental management outcomes in the context of the wide-scale adoption of Bt technology in India and elsewhere.

As observed above, Monsanto's GM promotion activities on corn and cotton also propagate the proper integration of IRM along with use of Bt technology as the best way to preserve the benefits of the technology and the key component of the IRM plan is planting the refuge. In fact, in the US, the continued availability of Bt technologies (as well as the use of it by the farmers) is contingent upon the strict adherence to the refuge practices and the growers' compliance with the registration conditions of the Environmental Protection Agency (EPA). It is being observed that stringent compliance of refugia can address the problem of Bt resistance development. Though the percentage of refuge acres varies based on specific geographical conditions, historically the required refuge for Bt cotton has been 50 per cent in the U.S. Cotton Belt[7]. It is further reported that compliance with the refuge management in the US was higher than 88 per cent in five of six years from 1998 to 2003 (Carriere *et al.*

2005). In addition to mandating non-Bt crop refuges, the US Environmental Protection Agency requires monitoring for field resistance to provide early warning of resistance development (USEPA 2007). Like in the USA and Argentina, refuge compliance of 20 per cent is mandated in Australia through an agreement with the farmers. Grower mixing of non-Bt seeds with Bt technologies is not permitted. As a condition of registration of Bt products by the EPA, seed companies are required to conduct IRM compliance assessments during the growing season to ensure grower compliance. Failure to follow IRM guidelines and properly plant a refuge may result in the loss of access to Bt technologies.

It has been reported that being the dominant producer of cotton with long history of adoption of Bt and with the track record of strong compliance with the IPM practices, the US has been able to achieve significant gains in production efficiency in cotton by reducing insecticide and thereby preserving the cotton growing habitats[8]. The indirect benefits included: (a) significant improvement in populations of beneficial insects and wildlife in cotton fields; (b) reduced pesticides runoff; and (c) improved farm worker and neighbour safety as well as soil-related environmental improvements through changed tillage practices with herbicide tolerant varieties (ICAC 2004).

Having discussed the imperatives of synchronising the refuge as well as IPM/IRM practices in the process of Bt adoption, now we turn to examine the empirical scenario of Bt adoption along with the refugia management as well as IPM/IRM practices across the major cotton growing states as reported by some of the recent studies on the theme.

In fact, the IPM practices in the world were evolved during the 1960s through 1980s as measures to regulate the excessive use of pesticides among crops, including cotton. The IPM practices were developed to address the technological failure of chemical pesticides (insecticide resistance, secondary pest outbreak and pest resurgence) and the indiscriminate ways of application and their adverse effects on human health and environment. In the early stages of its development, IPM was a technical approach designed to reduce the number of pesticide applications. Subsequently, it developed into a methodology in which farmers were encouraged to develop IPM interventions themselves through a better understanding of the agro-ecosystems (Sexton *et al.* 2007; Pouchepparadjou *et al.* 2004). While the IPM had some successes in the industrialised world, the interventions were more focused towards the developing world. The developing countries, in turn, have adopted the IPM technology with the help of extension efforts by OECD countries, who, by then, had a very elabourate system of testing and analysis prior to the introduction of new pesticides, continuous re-evaluation and re-registration of existing materials and regulation of application procedures and residues (Pretty 2005).

However, despite five decades of integrated pest control strategies and four decades of IPM interventions in USA, Asia, Latin America, Australia, the widespread use of IPM practices has not been widely adopted in view of the complexities in the IPM, policy restrictions, and counteracting forces of the pesticide industry (Peshin 2012). Even then, some successful IPM outcomes in terms of effective functioning of the farmer field schools (FFS), community IPM, etc. have been reported from countries like Bangladesh, Cambodia, Nepal, China, India, Indonesia, Pakistan, Sri Lanka, Thailand, Vietnam (see van den Berg and Jiggins 2007), though the adoption and coverage is not sufficient to meet the global objectives.

In India, the IPM programme being launched since 1985 is an eco-friendly strategy of pest control by exploiting the role of natural agents/forces in harmony with other pest management tactics and with the sole aim of reducing the disturbance to environment. Cultural control, use of natural enemies and plant resistants are basically compatible and supportive tactics in the IPM strategy. The IPM/IRM (a component of IPM) strategies are designed to reduce the dependence on insecticides by resorting to the use of a rational and sensible sequence of insecticides that are effective on the target species, cause least disturbance to beneficial fauna and eases the process of selection of insecticides. The strategies include cultivation of sucking pest tolerant genotypes and chemical seed treatment to help delaying the first spray, thereby conserving the initial build-up of natural enemies (Kairon and Kranthi 1998; Abhilash and Singh, 2009).

The IPM/ IRM[9] programmes have been introduced especially in the high productivity zone of the North-West and coastal regions covering 103 districts growing rice and cotton, which accounted for 67 per cent of total pesticide use. The Central Institute for Cotton Research (CICR), Nagpur, implemented an IRM based IPM programme in 10 cotton-growing states in India. These 10 states accounted for 80 per cent of insecticide use in cotton (Russell 2004). Crop yield increased from 6.72-40.14 per cent in rice and 22.7-26.63 per cent in cotton fields with IPM compared to non-IPM fields. From a macro perspective, some of the visible impacts of the IPM interventions in India are: (a) a reduction in chemical pesticide sprays to the extent of 50-100 per cent in case of rice and 30 to 51 per cent in case of cotton; (b) more than 10 times increase in use of biopesticides/neem based pesticides from 123 MT during 1994-95 to 1262 MT during 2009-10; and (c) almost 45 per cent decline in the overall consumption of chemical pesticides in the country from 75033 MT (technical grade) to 41822 MT between 1990-91 and 2009-10 (MoA 2011).

After the introduction of Bt cotton, the IPM/IRM strategies were modified to further minimise insecticide sprays so as to move towards eco-friendly systems of pest management. Between 2002 and 2007, the IRM-IPM project was implemented over 196,000 ha across 1820 villages in 28 districts of 10 cotton-growing states of India with a funding of US$ 2.58 million from the Ministry of Agriculture, Government of India under the Technology Mission on Cotton (TMC) Mini-Mission (MM-II) programme. Based on the annual project reports for 5 years, the overall impact of the project between 2002 and 2006 in terms of the net financial gains to farmers, was estimated to be US$ 39.5 million due to US$ 23.0 million from yield increases and US$ 16.5 million from savings on pesticides (Anonymous 2007, as cited in Peshin *et al.* 2009). Following the implementation of IPM/IRM programmes for major crops including cotton and rice in major states, a detailed survey was carried out all through 2002–2007 to evaluate the overall impact of the project in terms of the net financial gains gained by farmers. By and large, the total economic benefit was estimated to be Rs. 120 crores, due to Rs. 78 crores from yield increase and Rs. 42 crores from savings on pesticides.

In this context, it may be contextually relevant to provide an analysis of the impact of IPM/IRM practices following the commercialisation of Bt cotton as it emerges from some of the recent studies undertaken in the major cotton growing states, especially, Andhra Pradesh and Punjab. It is being observed that the IPM/IRM strategies (Appendix Box 9.1 and Appendix Box 9.2) though adopted on a very limited scale by few of the states, such as Andhra Pradesh, Punjab, Maharashtra and Gujarat, yet have strengthened pest management systems by identifying appropriate insecticides so as to delay resistance, ensure effective control of target pests, and conserve naturally occurring biological control for enhanced sustainability of ecosystems. The remainder of the section mainly draws from the most recent studies undertaken about the impact of IPM/IRM practices in Andhra Pradesh (Singh *et al.* 2011), Punjab (Peshin 2005; Peshin *et al.* 2009; Peshin 2012; Vatta *et al.* 2009) and Maharashtra (Sharma *et al.* 2004) in the post-Bt adoption context. It also uses the data on IPM/IRM as gathered by the authors from Maharashtra (CICR, Nagpur) and Gujarat (Department of Agriculture, Government of Gujarat).

In Punjab, the IRM programme was implemented since 2002 by the Punjab Agricultural University (PAU), Ludhiana. It was implemented in three districts, viz. Bathinda, Ferozepur, and Mansa and these districts account for 70 per cent (356,000 out of 509,000 ha) of the total area under cotton cultivation in Punjab. Under the programme, farmers are provided training in IPM. From each district, a sample of five villages was selected randomly; thus, a total number of 15 villages were selected for the study. In each village, ten farmers under the IRM-IPM programme were considered thus, a total sample of 150 farmers (experimental group) selected for the study (Peshin 2012).

In Andhra Pradesh, the IRM strategies on cotton were implemented in three districts, viz. Guntur, Khammam and Kurnool. To disseminate IRM strategies, a total of 165 villages were adopted in these districts from 2008 to 2011 along with 75 villages that were selected as non-IRM villages for comparing the impact of IRM strategies. The strategies blend all crop production practices to incorporate proper and low use of insecticides so that natural enemy populations are least disturbed and, different groups of chemicals are judiciously used (Singh *et al.* 2011).

In Maharashtra, the dissemination of the IRM technology was carried out as a farmers participatory programme in 130 villages of Wardha and Yavatmal districts in collaboration with and technical guidance of the CICR Nagpur covering 12,000 farmers in about 6000 ha area. The programme has been reviewed for the last 10 years especially after the introduction of Bt in Maharashtra. The farmers were assisted in cotton pest management throughout the cropping season by conducting regular meetings, imparting knowledge on IRM strategies and insecticide use (CICR, Unpublished IRM Report, Maharashtra).

In Gujarat, while IPM/IRM programmes are yet to receive adequate attention among the farmers, the Junagadh Agricultural University has been undertaking monitoring experiments with respect to the spread and intensity of sucking pests and bollworms and damage caused by bollworms to fruiting bodies in different cotton hybrids, viz. conventional (G. Cot Hy-10; Vikram 5-NBt) and Bt hybrids (MECH-12 Bt; RCH-2 Bt; Vikram 5-Bt; KDCHH-441 Bt). Under the evaluation scheme, farm level data have been generated as regards the intensity as well as plant damages caused by sucking pests, bollworms, other cotton diseases, etc. Besides, an in-depth survey of mealybugs has also been conducted in about 56 villages of the 7 major cotton growing districts in the Saurashtra region, viz. Amreli, Junagadh, Bhavnagar, Porbandar, Jamnagar, Surendranagar and Rajkot.

A summary of the major impacts of the IPM/IRM interventions as revealed by the major studies in Punjab and Andhra Pradesh as well as the unpublished empirical data from Maharashtra and Gujarat is presented in Table 9.1.

**Table 9.1:** Major Impacts of IPM/IRM Interventions across Select States, Post-Bt Cotton Scenario

| *Major impacts* | *Andhra Pradesh* | *Punjab* | *Maharashtra* | *Gujarat* |
|---|---|---|---|---|
| 1. Average number of sprays/ha | 3.74 (7–8) | 7.50 (7–15) | 2.35 (5–7) | 5.60 (7–12) |
| 2. Cost of plant protection (Rs./ha) | 2200 | 2470 | 1072 | 2245 |
| 3. Average lint cotton yield (Kg/ha) | 2667 | 2300 | 856 | 2785 |
| 4. Savings in number of sprays (%) | 30.87 | –2.59 | 77.05 | 47.65 |
| 5. Savings in cost of sprays (%) | 30.53 | 23.43 | 68.19 | 10.85 |
| 6. Increase in yield (%) | 18.75 | –3.96 | 13.40 | 22.60 |
| 7. Gain in income (net profit) | 19.37 | –5.23 | 15.60 | 51.21 |

*Note:* Figures in parentheses indicate the range of number of sprays reported by non-IPM/IRM farmers.

*Source*: Authors' compilation based on: Peshin 2005; Vatta *et al.* 2009; Singh *et al.*; CICR (unpublished); Personal communication, Department of Agriculture, Government of Gujarat (IPM/IRM data, Junagadh Agricultural University up to 2008-09).

As evident from Table 9.1, the four states show distinct patterns with respect to the major IPM/IRM interventions in the post-Bt context. These indicators representing the major impacts are aggregations made over few years of implementation of the IPM/IRM practices and the figures are to be considered in relation to the farm holdings that do not follow IPM/IRM practices. The average number of sprays are the highest in Punjab (7.5/ha) and the lowest in Maharashtra (2.35/ha). Punjab also reports the maximum number of sprays (7-15/ha) undertaken in non-IPM/IRM fields also. Cost of plant protection is reported to be lower in Maharashtra in comparison to rest of the states, which may be attributed to the fact that a major share of Bt cotton is grown under irrigated conditions in all the three states *vis-à-vis* sheer lack of irrigation facilities accessible to a vast majority of cotton farmers in Maharashtra. This is also an important reason for the lowest level of yield reported from Maharashtra compared to other states.

Farms following IPM/IRM practices also report significant savings in the number of sprays as well as cost of plant protection with the exception of farmers in Punjab, where, the cost of plant protection works out to be highest for IPM/IRM farmers due to the changes in pest infestation scenario across agro-climatic regions. Hence, IPM/IRM farmers in Punjab do not report any income gains due to the adoption of the prescribed practices. In Andhra Pradesh, impact of adoption of IRM strategies resulted in the reduction in the number and cost of insecticide sprays (30 per cent each) in IRM villages over non-IRM villages. Cotton yield was higher in IRM village (26.67 qt/ha) compared to non-IRM villages (22.46 qt/ha). Net profit per/ha was 19% more in IRM villages than non IRM villages (Singh *et al.* 2011).

Thus, it emerges from the above analysis that IPM/IRM practices have resulted in some achievements in terms of reduction in the number of sprays, savings in cost of plant protection and the resultant yield and income gains. It may also be argued that the reduction in the use of insecticides is a clear case of environmental benefit as it helps improving the soil conditions and the micro environments in which the farms are located. However, to what extent these economic benefits are translated into social and environmental benefits is yet to be properly investigated in the context of Indian states that are widely different in terms of agro-climatic features of the cotton growing regions as already described in Chapter 2 (Kumar, Chapter 2 in this volume). Based on the above analysis of the immediate outcomes of the technology in terms of reduction in the number of sprays as well as amount of pesticides being sprayed, one may argue that such immediate economic benefits have immense potential for creating sustainable environmental outcomes provided the Bt technology is judiciously used by synchronizing the IPM/IRM principles and practices.

## 3. Dynamics of Farmer Adoption of Bt technology and Farm Management Practices

Given the importance of IPM/IRM practices as important management strategies that would help achieve better environmental management outcomes in the context of Bt adoption as discussed above, the present section tries to juxtapose it with the empirical scenario of farm management practices being adopted by the farmers in Gujarat. We try to explore the following here 'whether the resource use practices being followed by the farmers reflect a judicious and an environmentally benign farm management regime in the post-Bt context?' In doing this, we examine the important trends that emerge in respect of input intensity of the farming practices.

### *3.1 Input Intensity of Cotton Production*

One of the important aspects of farm management practices in the post-Bt scenario is the resource use intensity of cotton farming as evident from the increase in the cost of critical inputs, especially, pesticides, fertilisers and irrigation. While part of the cost increase may be due to inflationary pressures, much of it may be attributed to the increasing use of these inputs, especially, insecticides. While insecticide sprays for control of bollworm have gone down as reported by several studies, those for control of sucking pests and other pests, such as mealybug etc. have increased in the post Bt period which causes a spiralling effect on cost of farming operations. In fact, it emerges that many studies on the performance of Bt cotton undertaken in the Indian context do not point to this empirical reality. This is further evident from an earlier paper (Lalitha *et al.* 2009) which confirms that the increased yield of Bt cotton was also contingent upon increased use of insecticides for control of sucking pests and other pests and unknown plant diseases as well as use of fertilisers (Figure 9.2).

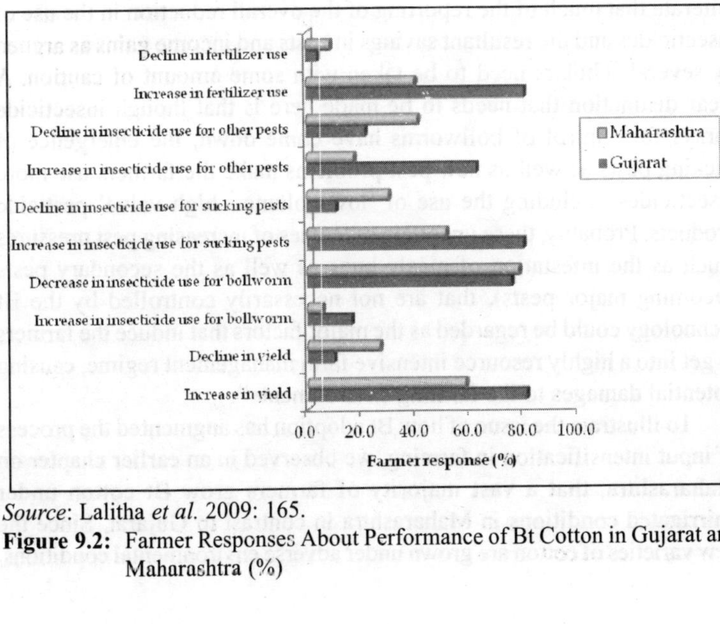


*Source*: Lalitha *et al.* 2009: 165.

**Figure 9.2:** Farmer Responses About Performance of Bt Cotton in Gujarat and Maharashtra (%)

Figure 9.2 shows that vast majority of the sample farmers reported a notable increase in the yield of Bt cotton in Gujarat (about 81%) and Maharashtra (51%). However, it is also important to note that the increase in yield as realised by the farmers has been possible at increased insecticide costs to control sucking pests and other pests. The magnitude of cost increase reported by the farmers has been much higher in case of Gujarat than Maharashtra. On the other hand, larger proportion of farmers from Maharashtra have reported a decline in yield in the post Bt period, which could be mainly attributed to the relatively poorer agro-ecological conditions also, in which farmers grow cotton in Maharashtra compared to Gujarat[10]. It has also been observed especially in respect of Maharashtra that as the sub-optimal agro-ecological conditions that prevail, there has been significant difference between the expected and realised yield of Bt cotton varieties, as reported by the farmers (Viswanathan and Lalitha 2010). Further, a study by Gill *et al.* (2010) examining the economic impacts and sustainability issues of Bt cotton cultivation in Punjab reveals that cotton farmers were still spending large amounts on growing Bt cotton *vis-à-vis* other cotton varieties. It showed that the largest item of expenditure was insecticides/pesticides, which constituted about 28 per cent of the total cash expenditure, followed by labour charges (25%) and irrigation charges (14%).

The above responses of the farmers about the impact of Bt technology reiterate that much of the reporting of the overall reduction in the use of insecticides and the resultant savings in costs and income gains as argued by several scholars need to be taken with some amount of caution. A clear distinction that needs to be made here is that though insecticide sprays for control of bollworms have come down, the emergence of sucking pests as well as new pest problems make the farmers use more insecticides, including the use of 'low volume—high value' pesticide products. Probably, these uncertain outcomes of increasing pest pressures (such as the infestation of mealy bugs as well as the secondary pests becoming major pests), that are not necessarily controlled by the Bt technology could be regarded as the major factors that induce the farmers to get into a highly resource intensive farm management regime, causing potential damages to the farming environments[11].

To illustrate the issue of how Bt adoption has augmented the process of input intensification in farming, we observed in an earlier chapter on Maharashtra, that a vast majority of farmers grow Bt cotton under unirrigated conditions in Maharashtra in contrast to Gujarat. Since the new varieties of cotton are grown under adverse environmental conditions, the farmers reported moderate to high levels of input use along with problems of lack of realisation of the anticipated or potential yield levels from the use of the Bt varieties[12] (Lalitha *et al.* 2009; Viswanathan and Lalitha 2010). In contrast, the high input intensity of cotton production in case of Gujarat is a matter of serious concern, more from the perspective of sustainability of natural resources and environment. Since irrigation water supplies through public irrigation systems has been deficient in terms of coverage of area, large number of cotton farmers pump out water from the already over-exploited and depleted groundwater sources (especially, tube wells) using high-powered submersible pump sets (Viswanathan and Pathak, 2014). The three rounds of panel surveys covering 66 farmers in Gujarat undertaken by the authors during 2003-04, 2007-08 and 2010-11 also revealed that the share of farmers using groundwater sources (including tube wells) had increased from 80 per cent during 2003-04 to 90-92 per cent during the later years. The correspondence between intensity of chemical input use as well as energy use for drawing groundwater among the cotton farmers has also been clearly brought out by another study (Viswanathan and Pathak, 2014), which reported that farms irrigated by tube wells were using more chemical inputs than those using canal irrigation or having no irrigation sources.

To substantiate this point further, an analysis of the resource use pattern of the predominant and alternative crops as a proxy measure of ecological sustainability of agriculture done in Punjab reveals that, cotton consumes the second highest amount of water after basmati rice. In case of fertilizer use, cotton was placed in the third position after chilly and paddy. The monetary cost of pesticide for cotton was reported to be much higher than other kharif crops. It has been indicated that all the alternative kharif crops use less resources (water, fertiliser, pesticides) and are more eco-friendly than cotton and paddy (Ghuman and Romana 2010).

### *3.2 Neglect of Recommended Crop Protection Measures*

The IPM/IRM principles underscore the importance of the judicious use of chemical insecticides during the entire crop cycle spanning about'180 days. Such responsive applications of insecticides are considered to be better alternatives to preventive applications and entail farmers to monitor pest populations and apply chemicals according to the observed pest levels. In this regard, one of the important precautionary measures being proposed under the IPM/IRM programme pertains to the schedule of insecticide sprays. Accordingly, it recommends that no sprays are to be undertaken until 90 days after sowing (DAS) to conserve natural enemies. However, in case of emergency, the use of endosulfan is recommended during 60-90 DAS based on economic threshold level (ETL), as endosulfan is moderately toxic to the natural enemies of insect pests. It also recommends avoiding the use of endosulfan beyond 90 DAS. The recommendations also include no use of organophosphates (OP)/ carbamates/synthetic pyrethroids until 90 DAS.

However, as evident from repeated surveys, the farmer practices are quite contradictory to the above recommendations of judicious use of insecticides. For instance, the panel survey of 66 farmers in Gujarat reveals that almost 70 per cent of the insecticide sprays are undertaken during the early periods of crop growth, i.e. 1-90 days after sowing (Table 9.2). The table also indicates that larger part of the insecticide sprays (more than 60%) take place between second and third months after sowing.

**Table 9.2:** Pattern of Insecticide Sprays Reported by Bt Cotton Farmers in Gujarat

| *Days after sowing* | *Percentage of insecticide sprays reported by farmers* | | |
|---|---|---|---|
| | *2003–04* | *2007–08* | *2010–11* |
| 1 to 30 days | 6.50 | 4.52 | 7.48 |
| 31 to 60 days | 31.26 | 26.24 | 32.24 |
| 61 to 90 days | 35.53 | 33.88 | 33.08 |
| 91 to 120 days | 15.53 | 23.72 | 18.50 |
| 121 to 180 days | 11.17 | 11.64 | 8.69 |
| Total (No. of sprays) | 1030 | 1151 | 1070 |
| Share of endosulfan sprays after 90 days of sowing | 21% | 53% | 37% |
| Share of sprays of organophosphate products until 90 days of sowing | 76% | 63% | 71% |

*Source*: Panel Survey of 66 Farmers in Gujarat.

More interestingly, the table also sheds light on an important violation of the IPM/IRM recommendation of 'avoiding the sprays of endosulfan after 90 days'. It shows that during 2003-04, 21 per cent of the endosulfan sprays took place after 90 DAS, which became 53 per cent during 2007-08 and 37 per cent during 2010-11. Similarly, as against the recommendation that 'no use of organophosphates until 90 days', it has been found that larger share of the insecticide sprays (76% during 2003-04, 63% during 2007-08 and 71% during 2010-11) use organophosphate products, such as monocrotophos, acephate, profenofos, ethiyon, chlorpyrifos, quinalphos, etc. during the initial periods (1-90 days) of the crop cycle. Further, even though the use of monocrotophos and acephate are recommended for sprays after 60 days of crop growth, it has been found that more than half of the sprays involving these products take place in the first two months of crop season.

Based on the mere reporting of the names of various agro-chemical products including the hazardous ones described above being used by the sample farmers, it has been found that the number of products increased by two times from 51 during 2003-04 to 112 during 2007-08 and 115 during 2010-11. Notably, this list of products range from imported insecticide products to local ones, which brings out important concerns about the knowledge of the farmers about the effectiveness as well as genuineness of the various products available in the market.

Though the adverse effects of the continued use of the major pesticide products on human health, farm environments as well as the ecosystems are yet to be properly understood, it may be seen that a major chunk of the insecticides being used singly or in combination by the farmers are highly or moderately hazardous as per the WHO classification of pesticides by hazards (WHO 2009) as also evident from Table 9.3.

As evident from Table 9.3, about 65-71 per cent of the pesticide products being used by the farmers have been classified as extremely, highly or moderately hazardous. Rest of the pesticide products reportedly being used by the farmers could not be classified due to wrong identification of the products by the farmers or even due to the sheer lack of knowledge of the farmers in recognising the product names by the exact names. The issue becomes more serious when we also realise the fact that in most cases, the farmers use these products in cocktails of two or three products together while applying on farms which either intensifies the pest related issue or nullifies the effects *per se*. In either of the cases, farmers tend to adopt more number of sprays to protect the crop and this often clashes with the IPM/IRM recommended schedules or products for the sprays.

The IRM recommendations for following a scientific spray schedule also correspond with scouting for proper identification of the insect pests and beneficial insects before venturing into chemical sprays. Through strictly adhering to the scouting practice, farmers are able to maintain the integrity of the ecosystem subsist in the farms by not adopting pesticide sprays that kills the beneficial insects. As demonstrated by the IRM field data from Wardha and Yavatmal districts of Maharashtra, the early phases (30-90 days) of plant life is quite rich in terms of presence of beneficial insects which goes off as the crop matures (Figure 9.3 and Figure 9.4).

**Table 9.3:** Number of Use of Major Insecticides Reported by Bt Cotton Farmers in Gujarat

| *Name of pesticide* | *Product Group* | *WHO Class* | *Number of use reported* | | |
|---|---|---|---|---|---|
| | | | *2003–04* | *2007–08* | *2010–11* |
| Monocrotophus | Organophosphate | Class Ib | 265 | 421 | 576 |
| Acephate | Organophosphate | Class II | 52 | 333 | 386 |
| Confidor | Neonicotinoids | Class II | 182 | 245 | 171 |
| Acetamiprid | Neonicotinoids | NA | 23 | 142 | 104 |
| Tatamida | Neonicotinoids | NA | 18 | 44 | 24 |
| Imidacloprid | Neonicotinoids | Class II | 15 | 76 | 10 |
| Computor | Neonicotinoids | Class II | — | 99 | 27 |
| Endosulfan | Organochlorine | Class II | 81 | 69 | 94 |
| Starthion | Organophosphate | Class III | — | 86 | 42 |
| Profenofos | Organophosphate | Class II | — | 32 | 134 |
| Hostathion | Organophosphate | NA | 8 | 21 | 6 |
| Quinalphos | Organophosphate | Class II | 24 | 4 | — |
| Dichlorvos (DDVP) | Organophosphate | Class Ib | 2 | 7 | 13 |
| Methyl parathion | Organophosphate | Class Ia | — | 35 | 21 |
| | | Total | 670 (65) | 1614 (65) | 1608 (71) |

*Note:* Figures in parentheses indicate the percentage share of the use of the products to the total number of product use reported by the sample farmers. Class Ia – Extremely hazardous; Class Ib – Highly hazardous; Class II – Moderately Hazardous.

*Source*: Panel survey of 66 farmers in Gujarat.

Figure 9.3 and Figure 9.4 show that the beneficial insects, like Ladybird beetle, Chrysopa eggs, Spiders and Apanteles remain on the plant during the first 45-60 days of the cotton plants and they gradually vanishes as the plants attains almost three months of life. In fact, the IRM spray recommendations fall in line with the presence of beneficial insects as described. Whereas, the farmer spray practices as elaborated above do not appreciate this trajectory of the presence of beneficial insects, thus the insecticide sprays often killing the beneficial insects and in turn promoting the growth and spread of non-beneficial insects.

On the other hand, as IPM data from Gujarat as presented in Table 9.4 show, the incidence of non-beneficial insects and pests, including the varieties of sucking pests and mealy bug occur towards the latter part of the crop growth. For instance, while aphids occur and remain present during the critical period of crop maturity, the intensity of Jassids, Thrips and Whitefly remains much higher between September and November

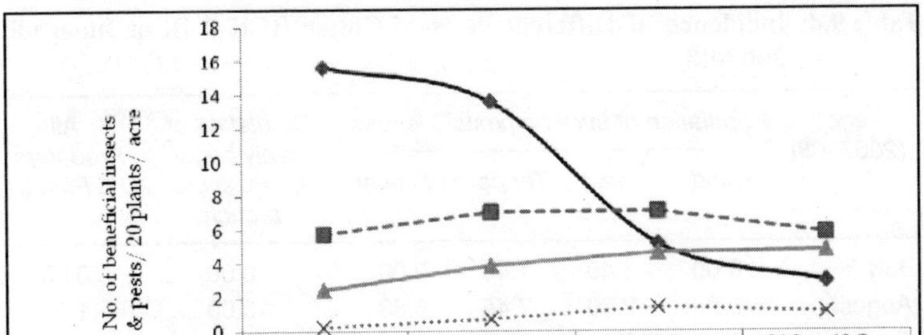


| | 19 Aug-1 Sep | 2-15 Sep | 16-29 Sep | 30 Sep-13 Oct |
|---|---|---|---|---|
| Lady bird beetle | 15.63 | 13.58 | 5.17 | 2.96 |
| Chrysopa eggs | 5.74 | 7.04 | 7.07 | 5.78 |
| Spiders | 2.49 | 3.89 | 4.63 | 4.73 |
| Apanteles | 0.24 | 0.69 | 1.43 | 1.11 |

*Source*: IRM Data, 2006–07 (CICR, Nagpur).

**Figure 9.3:** Number of Beneficial Insects and Pests/20 Plants/Acre, Wardha district, Maharashtra, 2006-07

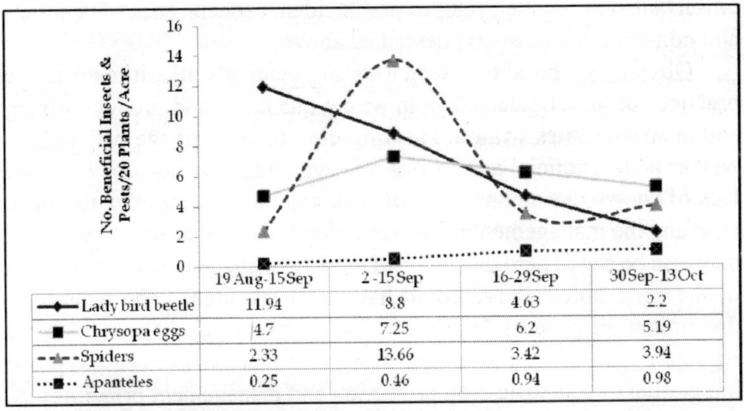


| | 19 Aug-15Sep | 2-15 Sep | 16-29 Sep | 30 Sep-13 Oct |
|---|---|---|---|---|
| Lady bird beetle | 11.94 | 8.8 | 4.63 | 2.2 |
| Chrysopa eggs | 4.7 | 7.25 | 6.2 | 5.19 |
| Spiders | 2.33 | 13.66 | 3.42 | 3.94 |
| Apanteles | 0.25 | 0.46 | 0.94 | 0.98 |

*Source*: IRM Data, 2006-07 (CICR, Nagpur).

**Figure 9.4:** Number of Beneficial Insects and Pests/20 Plants/Acre, Yavatmal district, Maharashtra, 2006-07

months. Similarly, the infestation of mealybugs occurs during the last part of the crop cycle (November-January). This suggests that cotton farmers should be quite vigilant about the pest complexities that occur during the crop cycle and devise a careful strategy of pest management in tandem with the cycle of the beneficial and non-beneficial insects.

But, as emerge from Table 9.2, a larger part of insecticide sprays take place during the first three months of the crop cycle, which are highly

**Table 9.4:** Incidence of Different Pests of Cotton RCH-2 Bt at Junagadh, 2007-08

| *Period (2007–08)* | *Population of sucking pests/3 leaves* | | | | *Population of mealy bugs/ 30 cm. symp. branch* | *All Predators/ 5 Plants* |
|---|---|---|---|---|---|---|
| | *Aphid* | *Jassid* | *Thrips* | *Whitefly* | | |
| July | 0.00 | 3.40 | 1.40 | 0.00 | 0.00 | 0.00 |
| August | 0.00 | 2.80 | 7.45 | 3.80 | 0.00 | 1.94 |
| September | 3.94 | 8.84 | 16.56 | 27.08 | 0.00 | 6.80 |
| October | 5.50 | 17.05 | 17.40 | 58.23 | 0.63 | 9.75 |
| November | 38.95 | 10.60 | 12.55 | 23.55 | 10.03 | 14.44 |
| December | 92.80 | 4.88 | 7.64 | 5.88 | 62.36 | 14.15 |
| January | 46.65 | 1.90 | 5.70 | 4.00 | 50.88 | 14.25 |

*Source*: Personal communication, Department of Agriculture, Government of Gujarat (IPM/IRM data, Junagadh Agricultural University).

unwarranted as per the cycles of pest incidence (occurrence of beneficial and non-beneficial insects) described above.

Obviously, the above scenarios of seemingly illegitimate farmer practices of insecticide sprays in wrong quantities and product mixings and at wrong times without knowing the life cycle of the beneficial as well as non-beneficial insects may be regarded as the outcomes of sheer lack of knowledge among the farmers about the pest complexities of the crop and the management measures needed to adapt or mitigate them. It may also be stated here that much of the farm management contradictions of the sorts noted above could have been avoided if the technology dissemination process had been efficiently managed through proper regulatory as well as institutional systems in place so as to ensure proper integration of scientific crop protection and management principles laid out under the IPM/IRM regimes. The knowledge gap that currently exists in the technology dissemination process is enormous and farmers do not get real time solutions to much of their problems, be it in terms of scientific crop management practices or resolving the pest related complexities. That the significant gap in the expected and realized yield of Bt cotton as already reported by the farmers from states such as Maharashtra (Viswanathan and Lalitha 2010) underlines the fact that the farmers are unable to make optimum use of the frontier technology that is available to them in the form of Bt, in view of the lack of irrigation facilities or congenial environmental conditions or also due to the absence of much required extension and other institutional support services.

### *3.3 Issues of Improper Refugia Practices and Cotton Monoculture*

A third important issue having significance in realising sustainable environmental outcomes in the post-Bt scenario is the concerns arising from the lack of or improper refugia practices followed as well as the potential threat of cotton monoculture to the agro-ecosystems and environment. While the problem of lack of adoption of scientific refugia practices has been a major source of concern and pursuits in academic and policy circles, the potential challenges posed by cotton monoculture are yet to receive adequate attention the world over. However, it may be perhaps difficult at this stage to make any conclusive statements that would help design policies and strategies, such that cotton cultivation in the post-Bt context is seen from the perspective of an agriculture system that enhances ecosystem integrity.

As a large number of studies repeatedly show, the adoption of refuge practices has not only been very low but also much controversial one as regards the percentage area to be used for growing refuge as well as the crops suited for various soil conditions. For instance, while Showalter *et al.* (2009) argues that refuge preferences are highly conditioned by the geographical specificities and growing conditions across regions; Qaim and Janvry (2005) brings out the case of mandatory compliance as exist in countries, such as Argentina. While different opinions[13] prevail over the percentage of area and the crops to be adopted as refugia, the extent of adoption levels as reported from various studies in India have never crossed 20-30 per cent. The panel survey data from Gujarat also conforms to this with a percentage of refuge adoption reported at 32 per cent during 2007-08 and 39 per cent during 2010-11. More importantly, the survey also revealed that smaller size class of farmers always tend to dis-adopt refuge (reported adoption rate being hardly 5%), as they fear that growing refuge would disrupt the entire cotton production practices. Although the farmers are provided with the recommended quantity of non-*Bt* seeds in a separate packet along with each packet of Bt-cotton seeds, the non-Bt seeds are not planted as refuge. Instead, the farmers sow only the Bt seeds in the entire field as their contention is that there would be yield loss from the 'refuge' crop (Manjunath 2011). Thus in a vast majority of the cases, the farmers merely throw off the refuge packets or use the same for gap filling. The major concern in the context of the non-adoption of refugia practices on a wide-scale is of resistance development as almost 90 per cent of the total cotton area in India has already been grown with Bt varieties with more area brought under the new technology year after year.

This leads us to the related issue of cotton monocropping as the repeated cultivation of the same crop without rotation can cause a variety of soil as well as plant infestations and ecosystem alterations. Though this issue is not very specific to cotton alone (as states like Punjab also experience problems of wheat-rice monoculture in the intensive irrigation areas), the wide-scale adoption of the Bt crop and the increasing adoption of the same as a monoculture raises larger questions as regards the sustenance of the technology on the one hand and the future challenges for the cotton growing regions on the other. It may also be important to note here that such a system of cotton monoculture has also been posing issues of increased use of scarcer inputs, including land and water and also causing problems of soil degradation at an alarming pace. However, apparently, these issues are yet to receive adequate attention of researchers and policy makers, who by and large, are baffled by the 'economic performance of the technology'.

## 4. Conclusions and Policy Imperatives

Realizing the fact that a large section of the poor cotton farmers would have already made an escape from the poverty trap by now (as emerge from a large number of studies), the emerging challenges on the agrarian front along with the climate change induced threats should provide strong signals for Indian agriculture to ponder over the next trajectory of agricultural growth under the Bt technology paradigm. As argued in the paper, the Bt technology along with proper integration of IPM/IRM practices would have made significant environmental management outcomes and welfare gains in the Indian context. However, the empirical reality appears to be far from the technology induced desirable outcomes as evident from three major arguments made in this Chapter.

The analysis also highlights the imperatives of broad-basing IPM/IRM interventions as a mandatory farmer practice in the process of legitimization of the use of Bt technology as observed from other cotton growing countries, especially, the US. In this regard, it is important to streamline the policies and programmes as well as evolving/revamping regulatory/institutional interventions for effecting sustainable environmental management outcomes in the post-Bt context. A closer examination of the empirical literature reveals that even in the states of Maharashtra, Punjab and Andhra Pradesh, where IPM/IRM practices are being adopted by Bt cotton farmers, it was observed that participation in the programmes is quite voluntary with negligible adoption by the small and marginal farmers. The lack of extension support systems along with knowledge intensity of the IPM/IRM interventions make them less popular among the farmers, who are constrained by lack of education and basic skills to adopt the scientific ways of farm management prescriptions as laid out in the programme.

At the same time, the knowledge and practice intensities involved in IPM/IRM practices have also been identified as the major obstacles in their adoption among the farmers (Peshin *et al.* 2009; Raven *et al.* 2008; Atreya *et al.* 2011). This is an area requiring major interventions in terms of simplification of the practices through innovative methods and technological solutions. There is also a need for making the IPM/IRM practices simpler from the point of easy adoption in a context when even the routine farm management operations are seriously affected by the growing shortage of farm labour and the increasing cost of farming operations side by side. At the same time, the real challenge facing Indian agriculture in the post-Bt scenario in strengthening the IPM/IRM interventions is the almost complete absence of the state-run institutional support systems and extension agencies that creates a huge vacuum. The providers of the Bt technology on the other hand, neither have a foolproof system of technology dissemination targeted towards the remotest areas nor have the CSR systems (ethos) for building knowledge and skill capacities among the farmers.

Turning to a more politically loaded question, the cotton crop in India has also been facing immense criticisms for the indiscriminate use of hazardous chemicals in the form of insecticides. While some states have already announced a ban on pesticides, such as endosulfan, it has been found that still a large segment of cotton farmers are attuned to use of the pesticide even refuting the scientific recommendations under the IPM/IRM practices. Notwithstanding the larger evidences that Bt technology had resulted in significant savings in the use of insecticides, the analysis clearly demonstrates that majority of the Bt farmers still continue to use more number of hazardous pesticides in single or in combinations. While we do not refute the mainstream literature suggesting the significant decline in pesticide use for control of bollworm, the farm level dynamics also demonstrates that large number of farmers still continue to use more number of insecticides to control the sucking pests and other plant diseases that were little evident before the large scale adoption of the technology. Given these trends, it is all the more important to strongly make a case for streamlining the use of insecticides in cotton in the post-Bt context and evolve regulatory systems to enforce the adoption of the technology with mandatory compliance to the IPM/IRM practices at least in the case of pesticide use.

This also raises an important political economy issue that how far the national and the state governments would make an ethical stand on aspects of continued promotion and use of hazardous insecticides that are domestically sourced or even imported. It is also important to consider whether the governments or regulatory systems have alternative thoughts on: (a) encouraging or incentivising the wider use of alternative crop protection strategies in the form of use of bio-pesticides or no-pesticide use; and (b) reinventing the wheels in terms of popularising the adoption of the traditional (*desi*) cotton varieties with improved economic performance at par with Bt cotton, etc.? As some argue, with the introduction of Bt cotton, novel eco-friendly pesticides and IRM strategies coupled with the trends in technology dissemination through area-wide farmer participatory approaches and Farmer field schools, IPM programmes can become successful and sustainable (Kranthi 2007).

To conclude, we make a plea for a paradigm shift in Indian cotton farming towards an ecosystem management perspective, wherein, the Bt technology is being adopted with due consideration to the integrity of the ecosystems and environment subsisting in the cotton growing regions of the country. This calls for streamlining Bt adoption with proper integration of IPM/IRM practices so as to make sustainable environmental management outcomes in the future. As argued by ecologists, growing Bt cotton in an ecosystem management perspective incorporating off-cum-growing season habitat management through synchronisation of soil, water, nutrient oriented production practices, at farm and community levels, could help yield sustainable environmental outcomes.

## Notes

1. As per the latest figures of the Union Ministry of Agriculture, the extent of Bt adoption has been almost 88.5 per cent at the national level in the total area of 11.14 million ha with states such as Maharashtra, Gujarat, Andhra Pradesh and Punjab reporting highest rates of Bt adoption (Shah 2012; Lalitha and Viswanathan 2013).
2. Earlier assessments of crop losses in cotton due to insect pests and diseases in India were estimated to be 18 per cent of the agricultural output per annum during the pre-Bt period. Raheja and Tewari (1997) observe that American bollworm (*H. armigera*) alone caused an annual loss of about Rs. 1000 crore. The production losses have shown to be much more, as

some earlier studies had reported losses worth Rs. 6,000 crore due to insect pests (Krishnamurthy Rao and Murthy 1983 as cited in Birthal and Sharma 2004), which had increased to Rs. 20,000 crore in 1993 (Jayaraj 1993, as cited in Birthal and Sharma 2004) and to Rs. 29,000 crore in 1996 (Dhaliwal and Arora 1996).

3. By and large, most studies on the environmental effects of the Bt technology have focused on the critical issue of use of pesticides following the damage control model proposed by (Lichtenberg and Zilberman 1986). While appreciating that pesticides act as a damage control input has an indirect effect on output rather than a yield-increasing input, many of these studies highlighted the contribution by the Bt technology in increasing the share of potential output farmers could realise by reducing damage from human and natural causes. Following this model, an earlier study (Qaim and Zilberman 2003) indicated that in India, farms that planted Bt cotton could reduce their pesticide applications by 70 per cent with an incremental yield effect of 80 per cent higher than those of conventional cotton. Similarly, in China, Huang *et al.* (2002) reported a return of $5.96 per $1.00 expenditure on Bt cotton with a reduction in pesticide use by 70 per cent. Similar effects of Bt cotton have been reported from Mexico (Traxler *et al.* 2001) and South Africa (Thirtle 2003).
4. It has been estimated that cotton consumed at least 10 per cent of all insecticides used globally. Pesticides worth US $600 million were used annually in India for pest management, of which, nearly 50 per cent of the total insecticides used were applied on cotton crop alone, which occupies only 5 per cent of the total cropping area in the country (Ghosh 2001). India is one of the few countries still engaged in the large scale manufacture, use and export of some of the toxic chlorinated pesticides, such as p,p'-dichlorodiphenyltrichloroethane (DDT), hexachlorocyclohexane (HCH) and pentachlorophenol (PCP). Even in the 1990s more than 70 per cent of the gross tonnage of pesticides used in agricultural applications in India consisted of formulations which are banned or severely restricted in the east and west (Abhilash and Singh 2009).
5. In a review of farmer surveys that report changes in yields and production practices, 45 results show decreases in the amount of insecticide and/or number of insecticide applications used on Bt crops compared to conventional crops in Argentina, Australia, China, India and the US. The reductions range from 14 to 75 per cent in terms of amount of active ingredient and 14 to 76 per cent for number of applications (Carpenter 2011).
6. The best way for growers to preserve the benefits and insect protection of *Bacillus thuringiensis* (Bt) technology is to incorporate Integrated Pest Management (IPM) practices and to develop and implement Insect Resistance Management (IRM). The IRM plan is used to decrease the likelihood of an insect developing the ability to survive pesticide application or the ingestion of transgenic crops containing the Bt trait.

Every grower of a Monsanto Bt corn or cotton product must comply with licensing terms and must have an Insect Resistance Management (IRM) plan (Monsanto 2013).

7. There is little consensus on how large the refuges should be to provide ample protection against pest resistance. This is due to the fact that optimal refuge size is a function of the characteristics of the pest population, the region, and the crop. The US Environmental Protection Agency (EPA), which regulates Bt crops since their inception, has mandated refuge equal to 20 per cent of the land devoted to Bt crops. Recommendations from economists and scientists range from 5 to 80 per cent, depending on whether supplemental pest control is allowed on the refuge (Hurley *et al.* 2000).
8. In the US, based on data presented at http://www.msstate.edu/Entomology/Cotton.html, the number of insecticide applications has decreased by 50 per cent in the last 10 years. The global reduction of the environmental impact of cotton due to insect and herbicide resistant cotton is documented by Brooks and Barfoot (2005).
9. Since insect resistance to insecticides had emerged as a major threat to pest control programmes, thereby rendering insect pest control ineffective, inefficient and unsustainable, IPM packages were refined to include IRM (Insecticide Resistance Management) as a major component. IRM is more relevant to the management of crisis caused by insect resistance to insecticides. Therefore, specific IRM programmes are designed for regions affected by severe resistance problem. While IPM is seen as a proactive method with emphasis on biopesticide and biological control interventions, IRM is meant to overcome the existing 'resistance' crisis through specific strategies to ensure efficient pest control and mitigate the problem of resistance. Governments, especially in India, wholeheartedly support the IPM/IRM initiatives (Russell 2004).
10. It is widely known that Maharashtra grows much of its cotton under rainfed conditions with a state average of hardly 10 per cent of cotton grown under irrigated conditions. Yet, the study by Lalitha *et al.* (2009) show relatively better irrigation ratios for cotton in the districts of Wardha (56%), Akola (51%), Amaravati (32%) and Yavatmal (19.5%) in Maharashtra. However, these ratios are much lower if we compare the ratio of Bt cotton grown with irrigation in the districts of Gujarat (60-91%) as reported by the sample farmers (Lalitha *et al.* 2009: 138).
11. Entomologists, like Kranthi, endorses to this point by stating that 'the reduced use of pesticides on GM cotton and the proliferation of GM cotton hybrids that are susceptible to these insects may have contributed to the emergence of these pests. The inappropriate choice of hybrids and the arbitrary and prolific spread of GM cotton hybrids have created conditions congenial for the rapid multiplication of these new insects', especially, mealybugs and other pests and diseases that were least reported in the pre-Bt period (as cited in Mudur 2010).

12. In fact, the farmer expectations about the yield performance of the Bt varieties have been built upon the information (if not scientifically proven) they acquired from the seed sellers who form the major source of information in the absence of extension agents.
13. In India an area of sprayed 20 per cent refuge of non-Bt with Bt-cotton has been recommended by the Genetic Engineering Approval Committee (GEAC). However, a stochastic model 'Bt-Adapt' developed at CICR, Nagpur (Kranthi and Kranthi 2004) showed that the 20 per cent refuge crop would not offer a significant advantage in delaying resistance development.

## References


Abhilash, P.C. and Nandita Singh. 2009. 'Pesticide use and application: An Indian scenario'. *Journal of Hazardous Materials,* 165 (2009) 1-12.

Anonymous. 2007. 'Insecticide resistance management in cotton pests: success story'. *DARE/ICAR Annual Report*, Directorate of Information and Publications of Agriculture (DIPA), Indian Council of Agricultural Research, New Delhi, India, p. 24.

Ashok, K.R., K. Uma, M. Prahadeeswaran and H. Jeyanthi. 2012. 'Economic and Environmental Impact of Bt Cotton in India'. *Indian Journal of Agricultural Economics*, 67 (3): 405-428 (July-September).

Atreya, Kishor, Bishal K. Sitaula, Fred H. Johnsen and Roshan M. Bajracharya. 2011. 'Continuing Issues in the Limitations of Pesticide Use in Developing Countries'. *Journal of Agricultural and Environmental Ethics*, 24 (1): 24:49-62.

Baffes, John. 2011. 'Cotton, Biotechnology and Economic Development'. *Policy Research Working Paper 5896*, The World Bank, December.

Belis-Bergouignan Marie-Claude, Vanessa Oltra and Maider Saint Jean. 2004. 'Trajectories towards clean technology: example of volatile organic compound emission reductions'. *Ecological Economics*, 48 (2004): 201-220.

Bernstein, Henry and Philip Woodhouse. 2007. 'Africa: Eco-populist Utopias and (Micro-) Capitalist Realities', *Social Register 2007*, pp. 147-69.

Brooks, G, and P. Barfoot. 2005. 'GM Crops: The Global Economic and Environmental Impact—The First Nine Years 1996–2004'. *AgBioForum*, 8(2&3): 187–196.

Birthal, Pratap S. and O. P. Sharma. 2004. *Integrated Pest Management In Indian Agriculture*. Proceedings 11, National Centre for Agricultural Economics and Policy Research (NCAP), New Delhi, India National Centre for Integrated Pest Management (NCIPM), New Delhi, p. 279.

Carriere, Y., Christa Ellers-Kirk, Kamlesh Kumar, Shannon Heuberger, Mike Whitlow, Larry Antllla, Timothy J. Dennehy and Bruce E. Tabashnik. 2005.

'Long-term evaluation of compliance with refuge requirements for Bt cotton'. *Pest Management Science*, 61 (4): 327-330.

Dhaliwal, G.S. and Ramesh Arora. 1996. *Principles of insect pest management*. National Agricultural Technology Information Centre, Ludhiana.

Ghuman, R.S. and G.S. Romana. 2010. 'Sustainability of the Existing and Alternative Cropping Systems in South-west Punjab', in Sucha Singh Gill, Lakhwinder Singh and Reena Marwah (eds), *Economic and Environmental Sustainability of the Asian Region*, London and New York: Routledge, pp. 317-336.

Gill, Sucha Singh, Sukhwinder Singh and Jaswinder Singh Brar. 2010. 'Cotton Economy, Cotton Producers and Bt Cotton Cultivation in Indian Punjab: An Assessment of Economic Impact and Sustainability Issues', in Sucha Singh Gill, Lakhwinder Singh and Reena Marwah (eds.), *Economic and Environmental Sustainability of the Asian Region*, London and New York: Routledge, pp. 419-442.

Huang, J., R. Hu, C. Fan, C. Pray and S. Rozelle. 2002. 'Bt Cotton Benefits, Costs, and Impacts in China'. *AgBioForum*, 4: 153-166.

Hurley, T. M., B.A. Babcock, and R.L. Hellmich. 2000. 'Bt Corn and Insect Resistance: An Economic Assessment of Refuges', *Journal of Agricultural and Resource Economics,* 26: 176-194.

ICAC. 2004. *ICAC Report of the Second Expert Panel on the Biotechnology of Cotton*. International Cotton Advisory Committee. Available at https://www.icac.org/cotton_info/tis/biotech/documents/expert_panel_2/english_summary.pdf (accessed on 4 August, 2013).

Kouser, Shahzad and Matin Qaim. 2012. 'Valuing Financial, Health and Environmental Benefits of Bt Cotton in Pakistan'. Paper prepared for presentation at the International Association of Agricultural Economists (IAAE) Triennial Conference, 18–24 August, 2012, Foz do Iguacu, Brazil. Available at http://ageconresearch.umn.edu/bitstream/126544/2/Kouser_manuscript.pdf (accessed on June, 24 2013).

Kranthi, K.R. 2007. 'Insecticide Resistance Management in Cotton to Enhance Productivity'. Paper presented at the Model Training Course on 'Cultivation of Long Staple Cotton (ELS)', December 15-27, Central Institute for Cotton Research, Regional Station, Coimbatore.

Kranthi, K.R., S.K. Banerjee and D. Russell. 2000. 'IRM Strategies for Sustainable Cotton Pest Management in India'. *Pestology* 24: 58-67.

Kranthi, K.R. and N.R. Kranthi. 2004. 'Modelling Adaptability of Cotton Bollworm *Helicoverpa armigera* (Hubner) to Bt-cotton in India'. *Current Science* 87 (8): 1096-1107.

Kranthi, K.R. and D. Russell. 2009. 'Changing Trends in Cotton Pest Management'. In R. Peshin and A.K. Dhawan (eds), *Integrated Pest Management: Innovation-Development Process,* Chapter 17, pp. 499-541.

Lalitha, N and P.K. Viswanathan. 2013. *Economic Evaluation of Bt Cotton Adoption in Gujarat*. Unpublished report submitted to the Department of Agriculture, Government of Gujarat, January.

Leontief, W.W. 1963. *Studies in the Structure of American Economy*. New York: Oxford University Press.

Lichtenberg, E. and D. Zilberman. 1986. 'The Econometrics of Damage Control: Why Specification Matters'. *American Journal of Agricultural Economics*, 68: 261-273.

Manjunath, T.M. 2012. 'Bt-cotton: Refuge in Seed-mix—Some Clarifications on itş Need and Viability'. *Current Science*, 102 (4): 541–542 (February).

MoA. 2011. *IPM at a Glance*. Directorate of Plant Protection, Quarantine and Storage, Department of Agriculture & Cooperation (DAC), Ministry of Agriculture, Government of India. Available at http://ppqs.gov.in/Ipmpest-main.htm (accessed on 28 June, 2013).

Mudur, G.S. 2010. 'Cotton Lessons for Bt Brinjal'. *The Telegraph*, Tuesday, 16 February, 2010. Available at http://www.telegraphindia.com/1100216/jsp/nation/story_12110833.jsp (accessed 4 August, 2013).

Peshin, R. and R. Kalra. 2000. *Integrated Pest Management: Adoption and its Impact on Agriculture.* New Delhi: Classical Publishing Company.

Peshin, Rajinder, A.K. Dhawan, K.R. Kranthi and Kuldeep Singh. 2009. 'Evaluation of the Benefits of an Insecticide Resistance Management Programme in Punjab in India'. *International Journal of Pest Management*, 55:3, 207–220, DOI: 10.1080/09670870902738786.

Peshin, Rajinder. 2012. 'Farmers' Adoptability of Integrated Pest Management of Cotton Revealed by a New Methodology'. *Agronomy for Sustainable Development*, DOI 10.1007/s13593-012-0127-4.

Peshin, Rajinder. 2005. *Evaluation of Dissemination of Insecticides Resistance Management Technology in Cotton Crop in Punjab*. Ph.D. Dissertation, Punjab Agricultural University, Ludhiana.

Peshin, R., A.K. Dhawan, K. Vatta, and K. Singh. 2007. 'Attributes and Socio-economic Dynamics of Adopting Bt-cotton'. *Economic and Political Weekly*, 42(52):73-80.

Pretty, J. 2005. 'Sustainability in Agriculture: Recent Progress and Emergent Technologies'. *Issues in Environmental Science and Technology*, 21: 1-15.

Qaim, Matin, and D. Zilberman. 2003. 'Yield Effects of Genetically Modified Crops in Developing Countries'. *Science*, 299: 900-902.

Qaim, Matin and Alain De Janvry. 2005. 'Bt Cotton and Pesticide Use in Argentina: Economic and Environmental Effects'. *Environment and Development Economics*, 10: 179-200.

Raheja, A.K. and G.C. Tewari. 1997. *Awareness Programme on Pesticides and Sustainable Agriculture*. Indian Council of Agricultural Research, New Delhi, p. 69.

Raven, P.H., Berg, L.R., & Hassenzahl, D.M. 2008. *Environment*, USA: Wiley.

Russell, D.A. 2004. 'Integrated Pest Management for Insect Pests of Cotton in Less Developed Countries'. In A.R. Horowitz and I. Ishaaya (eds), *Insect Pest Management—Field and Protected Crops*, Springer Verlag, Berlin, Heidleberg, New York: 141-180.

Scoones, Ian and Camilla Toulmin. 1999. *Policies for Soil Fertility Management in Africa*. DFID, London.

Shah, Deepak. 2012. 'Bt Cotton in India: A Review of Adoption, Government Interventions and Investment Initiatives'. *Indian Journal of Agricultural Economics*, 67 (3): 365-375 (July-September).

Showalter, Ann M., Shannon Henberger, Bruce E. Thabushnik and Yves Carriere. 2009. 'Agronomic Performance and Sustainability of Transgenic Cotton for Insect Control'. In Robert Tripp (ed.), *Biotechnology and Agricultural Development: Transgenic Cotton, Rural Institutions and Resource Poor Farmers*, Routledge Explorations in Environmental Economics. London: Routledge, pp. 49-71.

Singh, T.V.K., N.V.V.S.D. Prasad, S. Sharma and S. Dayakar. 2011. 'Impact of IRM Strategies on Bt Cotton in Andhra Pradesh'. In K.R. Kranthi, M.V. Venugopalan, R.H. Balasubramanya, Sandhya Kranthi, Sumanbala Singh and Blaise Desouza (eds), *Proceedings of the Fifth World Cotton Research Conference on Technologies for Prosperity*, Mumbai, November 7-11, Indian Society for Cotton Improvement, Mumbai, Excel India Publishers, New Delhi, pp. 275-278.

Subramanian, A. and M. Qaim. 2010. 'The Impact of Bt Cotton on Poor Households in Rural India'. *Journal of Development Studies*, 46 (2): 295-311.

Thirtle, C., L. Beyers, Y. Ismael, and J. Piesse. 2003. 'Can GM Technologies Help the Poor?: The Impact of Bt Cotton in Makhathini Flats, KwZulu-Natal'. *World Development* 31: 717-734.

Traxler, G., S. Godoy-Avila, J. Falck-Zepeda and J. de J. Espinoza-Arellano. 2001. 'Transgenic Cotton in Mexico: Economic and Environmental Impacts'. Paper presented at the 5th ICABR International Conference on Biotechnology, Ravello, Italy.

USEPA. 2007. Insect Resistance Management Fact Sheet for Bacillus thuringiensis (Bt) Corn Products, US Environmental Protection Agency, Washington DC. Available at http://www.epa.gov/oppbppd1/biopesticides/pips/bt_corn_refuge_2006.htm (accessed 4 August, 2013).

van den Berg, H. and J. Jiggins. 2007. 'Investing in Farmers: The Impacts of Farmer Field Schools in Relation to Integrated Pest Management'. *World Development*, 35: 663-686.

Vatta, Kamal, A.K. Dhawan and Rajinder Peshin. 2009. 'Economic Evaluation of Integrated Pest Management Programs'. In R. Peshin and A.K. Dhawan (eds.), *Integrated Pest Management: Dissemination, and Impact,* DOI 10.1007/978-1-4020-8990-9 4, C_ Springer Science+ Business Media B.V., pp. 103-117.

Vennila, S. 2008. 'Pest Management for Cotton Ecosystems or Ecosystem Management for Cotton Protection?'. *Current Science*, 94 (11): 1351-52. June.

Viswanathan, P.K. and N. Lalitha. 2010. 'Does GM Technology Pay Rich

Dividends? Reflections from Bt Cotton Farmers in Maharashtra'. In Sucha Singh Gill, Lakhwinder Singh and Reena Marwah, *Economic and Environmental Sustainability of the Asian Region*, London/New Delhi: Routledge, pp. 391-418.

Viswanathan, P.K. and Jharna Pathak. 2014. 'Economic Growth and the State of Natural Resources and the Environment in Gujarat: A Critical Assessment', (Chapter 10) in: Indira Hirway, Amita Shah, and Ghanshyam Shah (eds.), Growth or Development: Which Way is Gujarat Going, New Delhi: Oxford University Press, pp. 380-433.

WHO. 2010. *The WHO Recommended Classification of Pesticides by Hazard and Guidelines to Classification 2009*. World Health Organisation, Geneva, Switzerland. Available at http://www.who.int/ipcs/publications/pesticides_hazard_2009.pdf (accessed on 29 July, 2013).

# APPENDIX

**Box 9.1:** Components of IPM Technologies Developed for Cotton in India

| S. No. | Components of IPM |
|---|---|
| 1 | Deep summer ploughing to expose soil borne pests, pathogens and their destruction. |
| 2 | Avoiding monocropping and proper crop rotation, selection of pest resistant cultivars. |
| 3 | Using certified acid delinted seeds for good plant stand and preventing seed-borne disease-causing organisms and pink bollworm. |
| 4 | Seed treatment with Imidacloprid to prevent the early season sucking pests and Trichoderma to prevent seedling rot. |
| 5 | Monitoring with pheromone traps and assessment of pest population through scouting and ETL based application of appropriate technology to reduce the pest build up. |
| 6 | Application of biopesticides (Trichogramma, neem, nuclear polyhedrosis viruses [NPV] )in the early phase of the crop growth. Phased application of proper pesticides ensuring correct dosage and coverage depending upon the pest and age of the crop. |
| 7 | Adoption of recommended agronomic practices; Avoiding closer spacing, high dose of nitrogenous fertilisers and excess irrigation. |
| 8 | Inter-cropping with cowpea/sorghum/soybean to enhance the natural enemy population of cotton pests and to diversify the cropping system. |
| 9 | Trap cropping of castor for Spodoptera and Bhendi/pigeon pea for Helicoverpa. |
| 10 | Hand collection and destruction of grown up larvae and damaged plant parts. |
| 11 | IPM strategy consisted of sex pheromone, release of Trichogramma and spraying of neem based insecticides and need based synthetic insecticides. |
| 12 | This proved to be effective in managing cotton pest complex |
| 13 | Basal application of neem cake at 150 kg/ha coupled with drenching 1per cent neem oil helped in minimising stem weevil damage. |
| 14 | Major emphasis is given for developing varieties/hybrids resistant to key pests. |
| 15 | Role of egg larval parasite—Chelonus blackburni—in control of bollworms was established. |
| 16 | Growing of intercrops like cluster bean and Maize reduced pest infestation. |
| 17 | Cotton intercropped with cowpea, blackgram and greengram to lessen incidence of jassids and bollworms. |
| 18 | New Pesticide molecules. viz. Imidacloprid (10g/kg of seed) and Thiomethoxam (4.3 g/kg of seed) were effective in controlling jassid and whitefly population. |
| 19 | Topsin M (0.05%) and Carbendazim (0.05%) were effective in control of Myrothecium and Alternaria leaf spot. |
| 20 | Estimation of yield loss due to Grey mildew was found to be of the order of 20 to 40 per cent in Central Zone. Prochloraz and Propiconazole gave significant control of Grey mildew. |
| 21 | Flavobacterium sp. was found effective in control of bacterial blight. |

*Source:* All India Coordinated Cotton Improvement Project, *Technical Bulletin No. 2/2010*, CICR, Nagpur, Maharashtra.

**Box 9.2:** Features and Implementation Aspects of IRM in Maharashtra

**I. Features of IRM Programme**

a) Strategies were demonstrated in the farmer's own field.
b) Technical back-up support was provided throughout the cropping season.
c) No free material input was provided.
d) Whosoever was willing could participate in the programme.
e) Imparting confidence amongst the farmers, through knowledge based strategies (identification, selection of chemicals, resistance).
f) Simplicity of scouting for ET.
g) Drastic reduction in the spray expenditure of the participating farmers.
h) Make farmers realise that healthy and environment issues are also related with cotton production.

**1. Preparatory work:**

a) Initial meetings to acquaint the farmers with IRM programme.
b) Selection of willing farmers and enrolment of farmers.
c) Selection of local field workers and their training.

**2. Rapport building measures:**

a) Availability of plants (fruits and other) at cost in IRM villages.
b) Availability of bio-fertilizers at cost.
c) Demonstration of improved agricultural tools.
d) Demonstration of improved cook stoves.

**3. Demonstration and training for farmers**

a) Identification of insect pests and beneficial insects.
b) Scouting methods for sucking pests and bollworms.
c) Information about insecticides groups and their merits and demerits.
d) Dose of insecticides and type of nozzles to be used.
e) Demonstration of different types of pheromone traps.

**4. Publicity and awareness campaign:**

a) Hand-bills for different window strategies.
b) Posters to stop indiscriminate spraying.
c) Wall paintings (depicting 08 strategies per village).
d) Banners at market places.
e) Labels of beneficial insects for school children/school-activity book.
g) Slide-show regarding insect-pests/poster exhibition, Street plays/C.D. shows on IRM strategies.

**5. Efforts made to reach out to more farmers:**

a) Treated IRM as an educational programme.
b) Educated farmers not only about cotton crop-pest but also about insects, their distribution and dominance, ability to develop resistance, role of beneficial insects and chemicals hazards to ecosystem.
c) Involvement of school children particularly for identification to reach out to more farmers, i.e. from child to parent.
d) Participation of women were ensured through Self-Help groups
e) Dealers of pesticides in the area were informed about IRM strategies and need of the programme.

*Source*: IRM Programme in Maharashtra (Atul Sharma, CICR personal communication).

# 10

# In Lieu of Conclusion

*N. Lalitha* and *P.K. Viswanathan*

Agri-biotechnology has a significant role to play in the days to come in improving land and crop productivity particularly in view of the growing population and shrinking land that is available for agriculture. Specifically, advances in genetic engineering has opened new vistas in agriculture, viz. sequencing of genomes of microbes that has made introduction of genetic variation in the plants possible. This explains the flood of research in a number of insect and disease resistant crops the world over and the few genetically modified (GM) crops with the insect or disease resistance traits adopted in different countries. Compared to this, breeding for orphan crops is lagging behind in research even though these are important staple crops for farmers in small holder environment (Varshney *et al.* 2012).

Despite the genomics revolution that is taking place in the world, several countries, especially, the developed ones, have chosen not to be swayed by the genomics revolution in agriculture. While this non-acceptance of the technology raises certain pertinent issues about its legitimacy, it is by now widely felt that though crop biotechnology promises several possibilities, it also requires a comprehensive approach towards institutional choices, regulation and monitoring. The fact that though agriculture is the mainstay of a number of countries, so far GM crops have been cultivated in 28 countries only (Chapter 1) and in as many as 16 of them with less than 1 million hectares of land, partially explains the apprehensions in accepting the GM technology. Importantly, it implies that governmental and institutional commitments are important for adoption of biotechnologies in developing countries (Varshney *et al.* 2012) dominantly characterised by large number of small land holders. Since small land holders form the mainstay of the agriculture sector in the developing countries, agricultural biotechnology adoption can play a key role in poverty reduction in rural communities and in achieving food security. Biotechnology in agriculture could address the issue of food security from three angles, viz. (1) by providing GM food technologies; (2) by increasing the economic gains of farmers who have adopted Bt technology and raising their income levels thereby ensuring food security; and (3) GM crops could affect food safety and food quality (Qaim and Kouser 2013). However, legitimisation and widescale adoption of the GM technologies in the developing countries in particular, are beset with several constraints, like: (a) Inadequate policies for agricultural R&D at government and institutional level; (b) Poor scientific, political and public awareness of the opportunities and risks of different crop biotechnologies; (c) Inconsistent policy and regulatory regimes regarding IPR enforcement, protection of plant and animal health, biosafety, food safety and bioethics; (d) Deficiencies in economic and physical infrastructure that impede farmer ability to adopt biotechnologies; (e) Weakness of research institutions that do not allow research projects to be implemented; and (f) Insufficiently educated/trained human resources and lack of schemes for capacity building (FAO 2011).

In India, despite the prevalence of inadequacies of the above mentioned nature, GM cotton has spread to more than 10 million hectares in a short span of less than a decade. Bt cotton technology has been introduced at a time when the cotton sector was reeling under a major crisis of persistent low productivity levels despite massive allocation of productive farm lands for growing the crop coupled with heavy usage of agro-chemicals, especially, fertilisers and pesticides with exceptions of a few states. On the other hand, there was a drastic reduction in the share of agricultural GDP contribution from 30 per cent in 1990-91 to 14.5 per cent in 2010-11 (GOI 2012: 1). R&D investment in the agricultural sector has been stagnating both in the public and private sectors. However, it may be argued that much of the dynamism that is seen in the wide-scale promotion of the Bt technology in India has come from the private sector, with the state only intervening in terms of the official price control as the cost of the technology (Bt seed prices) had gone beyond the affordable limits of the resource poor farmers in the initial years.

Interestingly, though the official price control had been levied only by a few states, almost all other cotton growing states have immediately adopted the price control strategy making the technology more accessible even to the small and marginal farmers. At the same time, the fact that more number of companies continued to offer more varieties indicate that price control has not prevented the companies from entering the Bt cotton seed market. Thus, it may be observed that even in the absence of a highly proactive state machinery to promote the technology, both the performance of the seeds, and the seed price control in the early years in their own way have contributed to the wider adoption of Bt cotton which tends to be scale neutral.

Ever since the commercialisation of Bt cotton in India, many researchers have commented on the 'success and failure' of the technology based on cross-sectional studies. Many of the studies done in the Indian context have examined the economic impact of Bt cotton adoption. Despite the differences in the methodologies adopted and the size of the samples, and the serious non-comparability issue (because, the comparison of Bt and non-Bt cotton adopters has strictly not been with the same non-Bt hybrids of the Bt hybrids in most studies), independent researches done by scholars within and outside India have pointed out the economic gains by the Bt adopters. Based on the observed trends, it can be said that Bt cotton acted as a vehicle in improving the economic well-being of the farmers and in improving the food security among the cotton growers. A study cited in Baffes (2011) that reviewed 51 estimates based on 23 studies that focused exclusively on India, concluded that 'on average, use of biotech cotton reduces the number of chemical applications and pesticide costs by 36 per cent each, increases yields by 34 per cent, raises net returns by 84 per cent, while it increases the costs of production by 15 per cent'. Introduction of Bt cotton has resulted in India and China dominating the world cotton market accounting for half of world's cotton output (Baffes 2011).

This volume assumes relevance in the backdrop of the growing apprehensions about the long-term gains and socio-economic as well as environmental outcomes of the technology. Notwithstanding the voluminous literature on the technology adoption and its farm level performance across countries and regions, including India, this volume is quite unique and timely in terms of consolidating some of the most recent theoretical, R&D as well as empirical analyses of the performance outcomes and farm level responses following the wider diffusion of Bt cotton across the Indian states as examined by several scholars in a multi-disciplinary perspectives. In lieu of conclusion, it may be relevant to provide a synoptic view of all the chapters covered in this volume to streamline the issues for further investigations as well as for facilitating policy formulations at appropriate levels.

While setting the stage for discussing the development and diffusion of the Bt technology in case of cotton, Kumar (Chapter 2) highlighted the historical role played by the public sector research institutions in seed provision, innovations and strategies in evolving cotton varieties suiting the specific ecological conditions and of late in researching on GM cotton varieties. The chapter ends with a word of caution that sounds logical to think that the burgeoning population in the world would need a judicious agricultural production strategy in the future such that expansion of area under cotton will be at a huge cost of food security of countries producing cotton. This necessitates increasing the productivity of cotton further, where modern technology with a blend of traditional wisdom will have to be carefully used if we have to meet the goals without risking our natural resources.

The importance of regulatory mechanisms in legitimising the promotion and adoption of Bt technology has been highlighted by Indira in Chapter 3. As indicated, the regulatory mechanisms have been improvised over the 10-year period to provide greater role to the state government and its institutions (like the state agricultural universities approving suitable Bt varieties for different regions). At the same time, the chapter also points to the flaws in the enforcement of regulatory measures at various levels, which mainly include: (a) planting of refuge; (b) vigilance with regard to illegal seeds; (c) building laboratories for independent testing facilities; and (e) working cogently to build institutional relations between central and state regulatory authorities. The fact that the Government of India still remain undecided on the commercial release of Bt Brinjal (on the ground that adequate tests were not carried out) further intensifies the apprehensions on the sanctity and the settlement of the long-drawn controversies around GM technology.

To a large extent, the concerns about the economic performance of Bt technology in the Indian states have been almost settled with a large number of empirical studies (though constrained by several methodological issues) providing evidences of reduction in the use of pesticides and the resultant yield and income gains. However, the short term as well as long term environmental issues arising from Bt cultivation is a potential research area needing more and more micro-level studies. In this regard, Chapter 4 by Ashok *et al.*, provides some useful insights about the environmental performance of the Bt technology. Using empirical data from four states, viz. Gujarat, Maharashtra, Tamil Nadu and Andhra Pradesh, the chapter derives the environmental impact quotient (EIQ) for Bt and non-Bt farms and suggests that EIQ was significantly lower for Bt cotton because of reduced pesticide consumption. The findings with respect to the EIQ of Bt farms as discussed in this Chapter might provide useful insights for researchers and scientists to develop protocols for examining the potential long term environmental impacts arising from the continued use of Bt technology under diverse agro-ecological conditions. As the environmental impacts might differ across farms, regions with diverse ecology and environments, the crop protection measures also could vary, which makes it imperative to have more localised understanding of the environmental performance of Bt varieties based on long term surveillance of farms and farm management practices.

Chapters 5 to 9 provide a holistic perspective about the dynamics as well as techno-economic performance of Bt cotton in India over the past 10 years based on a mix of cross-sectional and longitudinal data gathered from prominent cotton growing states. These chapters prominently underscore: (a) the need for a careful scrutiny of the farm level performance of Bt technology in view of the its dynamic interaction with a host of other factors like changing pest ecology, price risks, multi-crop systems and market dynamics (Gaurav and Mishra, Chapter 5); (b) the imperative of strengthening R&D and capabilities of local seed providers as well as supporting the indigenous research by farmer scientists for developing and use of own hybrids in order to overcome the existing imperfections in the cotton seed markets (Arora and Bansal, Chapter 6); (c) the concerns arising from growing pesticide use by Bt farmers for control of sucking pests and other unknown pests and diseases, the persistent knowledge gap in matters of refugia practices, appropriate quantities, time schedules and type of products to be used for sprays, etc. (Lalitha and Viswanathan, Chapter 7); (d) the emerging issues of labour shortage for various farm management operations, especially harvesting of cotton and the imperative of innovations through mechanization in weeding and harvesting operations to resolve the labour market impasse in the event of increased adoption of the Bt technology (Vatta and Sidhu, Chapter 8); and (e) the paradigm shift needed in approaching farming with Bt technology synchronising the IPM/IRM practices to achieve sustainable environmental management outcomes in the long run (Viswanathan and Lalitha, Chapter 9).

A critical assessment of Bt technology in retrospect may be in order here so that it provides the right signals for the future course of technology development and dissemination. A larger concern that is yet to receive adequate attention in academic as well as policy circles is that how far the Bt (GM) technology could offer sustainable solutions for the environmental issues caused by the technology paradigm evolved under the green revolution (GR) era. As much of the environmental problems posed in the GR era were mostly pertaining to the issues of intensification of agro-chemicals in farming, the new technology paradigm under the GM revolution should have the onus of resolving this issue at least in the context of the GM crops, especially, cotton in countries, including India. In this regard, one of the shortcomings with dissemination of Bt cotton technology in India from our point of view appears to be that Bt cotton has not been provided as a pack of technology—to be used along with fixed number of irrigation, fertiliser, pesticide and the integrated pest management programmes. Instead of promoting the technology in a holistic farm management perspective, it emerges that only the seed technology was overtly promoted and discussed the most, relegating the importance of specific socio-economic as well as ecological characteristics of the farming environments. As a result, the biosafety concerns and the institutional mechanisms that should have been in place seem to be totally neglected.

Secondly, while the Central regulatory regime has elaborate checks till the commercial approval of seeds, at the State level, the required institutions that needed to have been fiercely monitoring the adoption were either lacking or not functioning optimally. With the result, the 'bottom up' approach, where the feedback from the taluka and district level regulatory authorities to the Central regulatory authorities, that could have resulted in mid-course correction measures, have been either inadequate or totally absent.

Third, due to this neglect, the important aspect in the diffusion of technology, namely the extension service has not been paid the desired attention either by the providers of the technology or the state level agencies working at the grassroots level. An important drawback of the lack of extension is that farmers learn about the technology by the costly process of experimenting it in their own fields. It is important to highlight here that the technology, R&D and extension support systems that have been evolved in India in the heydays of GR have almost seem to be dormant in the present context when the farmers need them urgently to take advantage of the Bt technology. On the other hand, the providers of the Bt technology (mostly being in the private domain) have also not been quite proactive to fill in this vacuum through providing alternative channels of agricultural extension and support with a wider coverage of farming population and areas. This brings out the dire need for revamping and deepening the existing extension services beyond the narrow confines of mere provision of seeds and agro-chemicals. Private extension services are present on a very limited scale, where farmers under contract farming use such services and they also pay for such services. These could be for very specific crops. Producer co-operatives also provide extension services with limited reachability. Given the complexities in crop management especially pest control in the post Bt context, it is most likely that the farmers would be willing to pay for the extension service, if the service is of economic importance to them and not something of the nature of public good and would be associated with large scale commercial agriculture (Chapman and Tripp 2003).

Looking at the sheer scale of adoption trends in India, it can be said that Bt cotton adoption has been scale neutral, as even the small and marginal farmers could access the technology. However, differences in the resource endowments and other features underscore that there are wide variations in the performance of the technology across the country. This is also due to the poor linkages between agricultural researchers, extension service providers, producers and farmers resulting in poor adoption and absorption of technologies. FAO (2011) observes that since most transgenic technologies have been developed outside developing countries, North–South linkages in building capacities and technology flow are essential. However, efforts to build links between public sector crop research institutes in the developed and developing countries have been only partially effective at least in the Indian context.

**Taking Forward the Positive Spillovers**

Now it is logical to think in terms of taking the Bt technology forward with its positive spillovers by overcoming the innumerable issues stated above. This becomes important as cotton continues to be a major source of economic dynamism in many parts of India. One positive spill over of introduction of Bt trait in cotton varieties has been that it has increased the number of varieties seeking registration with the Indian plant varieties registry. Though private sector R&D is slowly improving, much of the present and proposed research is in different types of insect resistant crops. In cotton particularly, more research is undertaken around the different types of hybrids. Similar to many other sectors, here again, there are a handful of companies which are responsible for the gradual increase in R&D and more registration of varieties.

With a decline in area under Bt cotton and the consequent dip in seed sales in the last couple of years as reported from a few places, top bio-agri companies seem to adopt various types of public-private partnerships. For example Nuziveedu, has strategic partnerships with state governments of Uttar Pradesh and Maharashtra. Under the public-private partnership with the Maharashtra government, Nuziveedu was given a project to improve cotton productivity by 30 per cent in the drought prone Vidarbha region covering 10,000 acres in 2012. Following the success of the project where the average increase in yield was 50 per cent, in 2013, the company were to cover a total of 20,000 acres (Rao, M. Prabhakar 2013). Rasi Seeds and Ankur Seeds have a number of collaborations with both national and international institutes to develop hybrid varieties, some of which are supported by the Department of Biotechnology.

Public sector research in cotton is also appearing to overlap with the private sector research, though the public sector also focuses on the varieties instead of hybrids. Particularly, if the Bt trait is introduced in the varieties, it would enable the farmers to use the farm saved seeds more. The GM trait in traditional varieties could further reduce the use of pesticides. But the recent release by the public sector (Surat research station) has been in the hybrid variety only. The earlier public sector Bt, Bikaneri Nerma, supposedly developed by the University of Agricultural Sciences, Dharwad and was approved for commercial sale, ran into controversies because of the presence of the Monsanto event in it and was eventually withdrawn. However, the most promising innovation in the public sector has been the lectin gene in the recent years. The Centre for Plant Molecular Biology of the Tamil Nadu Agricultural University has developed the lectin gene for the control of sucking pests in rice and cotton (Naranjo 2008; Tripp 2009). If found successful, it could have a significant impact in further reducing the insecticide use in cotton in the future.

The lack of implementation of biosafety regimes aggravate several fears of growing GM crops that stem from the ethical, social, economical and political concerns. One is that the possibility of genetic interchange as the GM crops are capable of crossing with other related species and could even replace those related species. With widespread adoption and with the lack of biosafety adoption, these insects could develop resistance. However, issues with conventional plant breeding technologies have also been noted. The European Commission Directorate-General for Research is cited as pointing out that 'the main conclusion to be drawn from the efforts of more than 130 research projects, covering a period of more than 25 years of research and involving more than 500 independent research groups is that biotechnology, and in particular GMOs are not *per se* more risky than, e.g. conventional plant breeding technologies' (Herring 2011).

Most of the apprehensions regarding the environmental impact of Bt cotton could have been reconciled if the regulatory systems had been put in place appropriately. For instance, the refuge adoption even after 10 years of Bt adoption has remained low, which could have been reversed if appropriate extension services had been provided, thus yielding an enhanced shelf life of Bt technology along with the reduced use of pesticides. As discussed in Chapter 9 of this volume, GM technology becomes more fruitful if it is adopted with IPM and IRM technologies and in the context of Maharashtra, it has been proved that adoption of the IRM technologies results in the reduced pesticide use. Agro-ecological conditions could have also been better. Lack of IRM and IPM technologies and continued use of harmful combinations of pesticide would reduce the life of existing GM traits that are available. Adopting the refuge would also reduce the possible contamination with organic and other cotton that is grown by farmers.

Bio-safety regimes are important from two perspectives. One is that the plant does not develop resistance to pests early if biosafety regimes are adopted. In the absence of the same, Bt cotton may indirectly cause existing primary pests or the secondary pests to increase. This is because, reduction in the pesticide sprays during flowering and boll formation stage has resulted in the emergence of tobacco caterpillar, mealy bugs, thrips and leaf hoppers which could be controlled by insecticides and pest management practices. However, lepidopteron pests that cause damage within the fruiting structures are difficult to control and Bt cotton is a key tactic for their suppression (Naranjo *et al.* 2008). Secondly, appropriate bio-safety regimes would also ensure that the technology lasts longer. It should be noted that each event before it reaches the regulatory stage takes nearly about 8 to 10 years of research in the laboratories. After this, the regulatory regime takes another 7 to 10 years. So fruits of nearly 20 years of labour would be lost if appropriate precautions are not taken.

Naranjo *et al.* (2008) note that Bt cotton technology has not developed resistance in countries where appropriate refuge practices have been adopted. Even in China where 'unstructured natural refuges' have been used, the technology has proved to be resilient. It has been further indicated that that the decomposition of Bt and non-Bt cotton on residues on soil communities were similar and cry1 AC proteins did not accumulate in biologically active elements after multiple years of Bt cotton cultivation.

While resistance to pests by the Bollgard II could also be a reason for the farmers shifting to Bollgard II variety, companies have also started phasing out the BG I seeds. 'With growing acceptance of BG II, Mahyco has started phasing out BG I seeds' (Barwale 2013). This leads to the situation that only BG II will be cultivated throughout the Bt cotton growing regions of the country in the emerging scenario. And if BG II fails, perhaps in the few years to come, the next technology that is available is the Round-up Ready Flex (RRF) cotton, which has reached the advanced regulatory requirements in India. Though apprehensions are there about the labour displacement effect of this variety, as shown, there are several cotton growing regions where severe shortage of labour is already being felt particularly during the weeding and harvesting operations (Chapter 8), raising the wage rates substantially, affecting the cost of production especially of farmers producing Bt in marginal environments. For instance, in Bhavnagar in Gujarat, where lot of farm workers have taken up jobs in diamond polishing/cutting industry, many farm households have either given up farming or started leasing out the cotton lands. Perhaps, RRF could offer an alternative in such circumstances.

While uncertainties about the next seed in the Bt pipeline prevail, the year 2012 had been a double whammy for the agri-biotech sector when the two independent committees of Government of India declared a moratorium on GM and GM food crops till the regulatory loopholes are addressed appropriately. Particularly, the Technical Expert Committee appointed by the Supreme Court of India has also recommended that no field trials should be allowed in crops where India is the centre of origin. It is not clear whether these recommendations would be accepted by the government without any discussion. But already these reports have had an impact as several state governments have proactively stopped providing clearances for the GM trials in their respective states. 'It has been noted that government scientists in India are faced with increasing difficulty in finding locations for the trials of GM rice. Because of increasing costs due to the need to protect the fields from anti-biotech organisations, many research organisations are pulling back from trying to develop GM varieties and simply publishing their research results or working on industrial crops such as cotton where GM varieties can be commercialised. The private sector also is cutting back because of consumer resistance to GM products and the rising cost of commercialising new products' (Huang *et al.* 2008: 241-263).

In the case of Bt brinjal, even before the Central Government's decision to put the same on hold, state governments of Odisha, Bihar, West Bengal,

Madhya Pradesh, Chhattisgarh, Karnataka and Kerala have decided not to allow GM crops in their states. While from the experience of Bt cotton it could be said that spreading of GM varieties may not be preventable, it is essential at least from those states that they agree to participate in the GM crop trial. Further institutional monitoring need to be in place to ensure compliance with safety guidelines and periodic reporting to the taluka/ district level biosafety committees and state authorities.

One silver lining in the biotech policy scenario is that the Biotechnology Regulatory Authority of India (BRAI) bill was tabled in Parliament on 22 April, 2013[1]. The preparatory work on such a bill dates back to the year 2003, when the taskforce headed by Dr. M.S. Swaminathan, recommended setting up of an independent regulatory authority for biotechnology. The Ministry of Science and Technology drafted the bill, but it was soon withdrawn due to severe criticism. In 2009, the Ministry drafted another bill, but it was not tabled in the parliament.

The BRAI bill seeks to create an independent regulatory authority called Biotechnology Regulatory Authority of India (BRAI) besides a 17-member inter-ministerial governing board to oversee the authority's performance and a Biotechnology Regulatory Appellate Tribunal, where BRAI decisions could be challenged. The bill leaves the commercialisation of biotechnology products in agriculture and health to be decided by the central and state governments.

An important aspect of the bill is that it has the provision to ensure that the risk assessment and risk management are easily accessible so that product developers, stakeholders and the public can be confident that the regulatory system is sound and appropriate. While the biotech industry hails the bill that it will bring single window clearance on approvals, critics point out that certain confidential information which are left to the discretion of the BRAI officials to share would make the biotech decisions to be non-transparent. Critics also point out that the 'bill seeks to violate the spirit of federal polity' and includes no mechanism for biosafety management.

Now, from a farmer's perspective, one of the outcomes of introduction of Bt cotton is that farmers with all types of holdings are now dependent on the formal seed markets for their seed supply. 'If resource poor farmers are to participate in commercial input markets, they need to know how to voice their opinions and to pursue complaints. Better consumer protection and education programmes are required to support farmers who will become increasingly dependent on private input markets. In addition, if resource poor farmers are a target for public biotechnology research, these farmers require increased representation in the decision making of public research institutes' (Tripp 2001: 262). Here, it should to be noted that most of the Department of Biotechnology's outreach programmes are aimed at the industry, university and researchers. This signifies that the farmer-oriented programmes need immediate attention and this would be possible only if the biotechnology research programmes also have the component of socio-economic assessment of the technology incorporated right from the beginning of the project. This would enable efficient diffusion and adoption of the technology with the much desired social outcomes.

Further, it is also essential to learn from the experience of other countries that have adopted a variety of GM crops. The biosafety clearing house (BCH), recommended under the Cartagena Protocol should act as a signal board for the developing countries. Under the Cartagena Protocol, each country has to set up the BCH, which facilitates the exchange of scientific, technical, environmental and legal information on, and experience with, living modified organisms (LMOs); and to assist parties to implement the Cartagena Protocol. Under this, India has filed 27 records, of which eight pertain to law, regulation or guideline and five records each pertain to the country's decision on open release of GMOs and risk assessment done regarding the GMOs that were released.

Some countries like Argentina, Canada and the US which have not ratified the Cartagena Protocol have also provided records under relevant aspects. India has provided risk assessment information of all the five events that have been released in the open environment. Interestingly, India does not have any entries under the heading biosafety expert mentioned in the country profile even after a decade of experience with GM cotton and also when a number of food crops are undergoing trials[2]. As all the countries are supposed to use the uniform codes and standards of providing information on GMOs, the BCH, if effectively organised, would be a valuable source of information for the developing countries in regulation and monitoring.

From a futuristic perspective of the sustainability of the technology and its adoption, India needs more independent research, uniformity in methodology carried out in different parts of the country to assess the performance of Bt cotton from a trans-disciplinary framework. A critical review of the empirical studies on the dynamics of adoption of Bt technology especially in the Indian context done by the social scientists, would reveal that much of the discussions have been centred around proving or disproving the benefits of the technology measured in terms of the extent of reduction in the use of insecticides and the resultant incremental yield outcomes. Some of the most recent papers in this context (Stone 2012 and 2013; Herring and Rao 2012; Herring 2013; Rao, N. Chandrasekhara 2013:66-69) bear testimony to this point.

Surprisingly, there has been hardly very little attention given by researchers or policy makers to take the discourse to the next level with more long term impacts by trying to address some of the critical issues emerging from the universal adoption of the Bt technology in the present context. These issues mainly engross the domains of policies, institutions and markets. The discussions and deliberations on the policies should cover the broad spectrum of legitimisation of the technology as well as evolving more sustainable crop protection and management along with synchronisation of IPM/IRM practices. The need for R&D, institutional, regulatory interventions as well as extension systems are to be much more robust in the emerging context when institutional or regulatory systems are either non-existent or dormant. Similarly, the need for reforming the marketing, trade and value chains has never been the point of discussion in the post-Bt scenario. While most of the research confined to understanding the behaviour of the seed and other input markets, the potentials of improving the functioning of the output markets, changing quality of cotton output, issues of price formations in the output markets, the farm gate price realisations, etc. have never been the sources of scholarly attention. These are some of the research gaps that exist in India's tryst with Bt cotton.

Yet another issue needing a mention in this regard is the level of saturation that the technology adoption might encompass in the near future. A recent survey of the top ranking Biotechnology companies by *Biospectrum*, points out that the agri biotech sector had registered a lower growth rate of 5 per cent in 2012-13 and one of the reasons cited is the policy logjam and the saturation in the area under Bt cotton (Narayanan 2013). In this regard, strengthening the breeding and delivery capacity of the national programmes will help remove the bottlenecks that have hampered the adoption of biotechnologies (Varshney *et al.* 2012). Particularly, the outcome of Bt research in desi cotton varieties which is undergoing experimentation in the premier cotton research institute of the country is anticipated with lot of interest by both farmers and researchers. In view of the shrinking land, increasing population and achieving the food security, it is essential that India adopts the technological advances. At the same time, the technology so adopted should be economically viable, socially acceptable and environmentally sustainable.

To sum up, it can be said that 'science cannot declare any technology completely risk free. Genetically engineered crops can reduce some environmental risks associated with conventional agriculture, but will also introduce new challenges that must be addressed. Society will have to decide when and where genetic engineering is safe enough' (cited in GOI, 2009). While it is essential to adopt a precautionary principle, it should also be realised that no technology would be 100 per cent safe and therefore, safety guidelines need to be drawn and strictly adhered to. Perhaps, this could be the only option for India to take advantage of the technological developments that are taking place, otherwise, the country could develop 'scientific apartheid' in adopting the relevant technologies in agriculture.

## Notes

1. Compiled from Sood (2013: 10) and Koul (2013: 20).
2. Available at http://bch.cbd.int/about/countryprofile.shtml?country=in, accessed on 10 July 2013.

## References

Baffes, John. 2011. 'Cotton, Biotechnology and Economic Development'. Policy Research Working Paper No. 5896, The World Bank, Washington D C.

Barwale, R. 2013. 'Controversies Hit Market Share'. *Biospectrum*, Vol. 11, Issue 6, June, p. 55.

Chapman, R. and R. Tripp. 2003. 'Changing Incentives for Agricultural Extension: A Review of Privatised Extension in Practice, Agricultural Research and Extension Network.' Network Paper No. 132, Overseas Development Institute, London.

FAO. 2011. *Biotechnologies for Agricultural Development*. Proceedings of the Food and Agriculture Organisation (FAO) International Technical Conference on 'Agricultural Biotechnologies in Developing Countries: Options and Opportunities in Crops, Forestry, Livestock, Fisheries, and Agro Industry to face the Challenges of Food Insecurity and Climate Change, ABDC 10' Rome.

GOI. 2009. 'Genetically Modified Crops-Issues and Challenges in the Context of India', Research Unit, Rajya Sabha Secretariat, New Delhi.

GOI. 2012. *State of Indian Agriculture 2011-12*. Government of India, Ministry of Agriculture and Cooperation, New Delhi, p. 273. Available at http://agricoop.nic.in/sia111213312.pdf (accessed 21 August 2013).

Herring, R. 2011. 'Politics of Science and the Science of Politics: Why is Agricultural Biotechnology So Vulnerable?' Keynote address in the 'Conference on Biotechnology in Indian Agriculture: Performance, Potential and Concerns', organised by Centre for Economic and Social Studies (CESS) Hyderabad, International Food Policy Research Institute, Department of Agriculture, Government of Andhra Pradesh and held at CESS, Hyderabad on 18-19, January, 2012.

Herring, Ronald J. 2013. 'Reconstructing Facts in Bt Cotton: Why Scepticism Fails'. *Economic & Political Weekly*, 48 (33): pp. 63-66.

Herring, Ronald J. and N. Chandrasekhara Rao. 2012. 'On the "Failure of Bt Cotton": Analysing a Decade of Experience'. *Economic & Political Weekly*, 47 (18): pp. 45-53.

Jikun, Huang, Ruifa Hu, Rozelle Scott and Carl Pray. 2008. 'Genetically Modified Rice, Yields and Pesticides: Assessing Farm-level Productivity Effects in China'. *Economic Development and Cultural Change*, Vol. 56, No.2, January 2008, pp. 241-263.

Koul, Rahul. 2013. 'BRAI Bill Makes a Comeback, Industry Hopeful'. *Biospectrum*, Vol. 11, No. 5, May, p. 20.

Naranjo, S.E., J. R. Ruberson, Hari Sharma, Lewis Wilson and Kongming Wu. 2008. 'The Present and Future Role of Insect Resistant Genetically Modified Cotton in IPM'. In J. Romeis. A.M. Shelton and G.G. Kennedy (eds.), *Integration of Insect Resistant Genetically Modified Crops Within IPM Program*, Netherlands: Springer, pp. 159–194.

Narayanan, Suresh. 2013. 'Growth Pangs Hit India's Biotech Industry'. Bioedit, *Biospectrum,* Vol. 11, Issue 6, June, p. 7.

Qaim, M. and S. Kouser. 2013. 'Genetically Modified Crops and Food Security', Plos One 8 (6). Available at http://www.plosone.org/article/info%3Adoi%2F10.1371%2Fjournal.pone.0064879.

Rao, M. Prabhakar. 2013. 'Undisputed Leader in Bt Cotton'. *Biospectrum*, Vol.11, Issue 6, June, p.40

Rao, N. Chandrasekhara. 2013. 'Bt Cotton Yields and Performance Data and Methodological Issues'. *Economic & Political Weekly*, 48 (33): pp. 66-69.

Sood, Jyotica. 2013. 'Stage Set for GM Crops'. *Down to Earth*, 16-31 May 2013, p.10.

Stone, Glenn Davis. 2012. 'Constructing Facts: Bt Cotton Narratives in India'. *Economic & Political Weekly,* 47 (38): pp. 62-70.

Stone, Glenn Davis. 2013. 'A Response to Herring and Rao'. *Economic and Political Weekly*, 48 (33): pp. 70-72.

Tripp, R. 2001. 'Can Biotechnology Reach the Poor? The Adequacy of Information and Seed Delivery'. *Food Policy*, 26 (2001), pp. 249-264.

Varshney, R.K., Jean Marcel Ribaut, Edward S. Buckler, Roberto Tuberosa, J. Antoni Rafalski, and Peter Langridge. 2012. 'Can Genomics Boost Productivity of Orphan Crops?', *Nature Biotechnology*, Vol. 30, No. 12: 1172-1176, December.

# Index